Applications of Fibonacci Numbers

Applications of Fibonacci Numbers

Volume 8

Proceedings of 'The Eighth International Research Conference on Fibonacci Numbers and Their Applications', Rochester Institute of Technology, Rochester, New York, U.S.A., June 22–26, 1998

edited by

Fredric T. Howard

Department of Mathematics and Computer Science,
Wake Forest University,
Winston-Salem, North Carolina, U.S.A.

SPRINGER SCIENCE+BUSINESS MEDIA, B.V.

A C.I.P. Catalogue record for this book is available from the Library of Congress.

ISBN 978-0-7923-6027-8   ISBN 978-94-011-4271-7 (eBook)
DOI 10.1007/978-94-011-4271-7

Cover figure by John C. Turner

Printed on acid-free paper

TABLE OF CONTENTS

A REPORT ON
THE EIGHTH INTERNATIONAL CONFERENCE
ON
FIBONACCI NUMBERS AND THEIR APPLICATIONS

Rochester Institute of Technology, which has been internationally respected as a world leader in career-oriented and professional education since 1829, was the inspired choice of setting for our Eighth Conference. We are indeed grateful to Rochester Institute of Technology, as well as to the Fibonacci Association, for sponsoring our conference. A special word of thanks goes to Dr. Albert Simone, President of Rochester Institute of Technology, Wiley R. McKinzie, Dean of Applied Science and Technology, and Dr. Walter A. Wolf, Chair of The Computer Science Department.

The participants came from eighteen different countries: 28 from the USA, three each from Australia, Canada and Japan, two each from Germany and Italy, and one participant from each of Austria, Belarus, Cyprus, Denmark, France, Greece, Iceland, New Zealand, Poland, Romania, and Scotland. Five of the presenters were women. There were four mathematicians who have attended all of the eight conferences, many who have attended several and, most happily, several who have attended for the first time. The magnetism of Fibonacci-type mathematics drew even some who did not present a paper. The ages ranged from a few who were in their twenties to one who will soon earn the title of nonagenarian.

There have been two major changes since our last conference, both involving Jerry Bergum. After eighteen years as editor of the Fibonacci Quarterly, he has handed over the baton to Curtis Cooper. We wish Curtis all success and fulfillment in his new role. At the same time, Jerry has been succeeded as conference organizer by Fred Howard, who also has our best wishes. Fred is already widely respected for his wisdom and kindliness.

We hope that Jerry will attend many more of our conferences. We deeply appreciate all he has done. He has been in a very real sense the heart and soul of our Association. We would also like to renew our thanks to Calvin Long for his continuing work as our President. Our discussions have been illuminated by his fine mathematical insights.

This was a conference where all of the talks were attended by almost all of the participants, who appreciated the diversity of topics covered and the remarkable level of quality of the papers. All of the presentations displayed the high enthusiasm of the speakers for their studies, and they all showed enjoyment over the opportunity of sharing their ideas with each other.

As well as working hard (41 talks in five days!), the group also enjoyed some delightful social events, the highlight being a "cook-out" at the Andersons' home with Peter and Jane, our gracious hosts. Through his wit and warmth, Peter immediately set the stage for a conference where we not only saw a fellow-mathematician in each other (which would already be enjoyable) but, moreover, a friend. We are deeply grateful to Peter and his helpers for all their hard work in preparing those delightful outings for us, and the extra care they took in looking after us in Rochester.

The friendships created by this sequence of Fibonacci conferences has produced many worthwhile results in this area of mathematics. At this conference we have enjoyed renewing old friendships and beginning new ones. "The Goddess Mathesis" (to use Howard Eves' term) looks favorably on these friendships.

Finally, we had to part. But now we greatly look forward to meeting again in two years: in Luxembourg in 2000.

Herta T. Freitag

LIST OF CONTRIBUTORS TO THIS PROCEEDINGS*

Professor Arnold Adelberg
Grinnell College
Department of Mathematics & Computer Science
P.O. Box 805
Grinnell, Iowa 50112-0806

Professor Octavian Agratini
Babes-Bolyai University
Faculty of Mathematics and Informatics
str. Kogalniceann
Nr. 1, 3400 Cluj-Napoca, Romania

Professor Cecil O. Alford (pp. 121-128)
School of Electrical and Computer Engineering
Georgia Institute of Technology
Atlanta, GA 30332-0250

Professor Peter G. Anderson (pp. 337-352)
Rochester Institute of Technology
Department of Computer Science
102 Lomb Memorial Drive
Rochester, NY 14623

Shiro Ando (pp. 1-10)
5-29-10 Honda
Kokubunji-shi
Tokyo 185-0011
JAPAN

Professor Angel Andreu
Department of Mathematics
Monroe Community College
1000 East Henrietta Rd.
Rochester, NY 14623

---

*This list includes all authors and coauthors of papers presented at the conference even if their paper was rejected, published elsewhere or not submitted to the proceedings. Those who attended but did not present a paper are also in this list.

Professor Demetrios L. Antzoulakos (pp. 27-42)
Department of Statistics and Actuarial Science
University of Piraeus
18534 Piraeus
Greece

Professor Vassia K. Anatassova (pp. 11-26)
Centre of Biomedical Engineering
Bulgarian Academy of Sciences
P.O. Box 12, 113 Sofia
Bulgaria

Professor K. T. Atanassov (pp. 43-46)
Centre of Biomedical Engineering
Bulgarian Academy of Sciences
Sofia-1113
Bulgaria

Professor Gerald E. Bergum
Box 2201
Computer Science Department
South Dakota State University
Brookings, SD 57007-1596

Dr. Marjorie Bicknell-Johnson (pp. 47-52; 53-60; 141-148)
665 Fairlane Avenue
Santa Clara, CA 95051

Professor John A. Biles (pp. 61-74)
Information Technology Department
Rochester Institute of Technology
102 Lomb Memorial Drive
Rochester, NY 14623-5608

Professor Mihai Caragiu (pp. 75-82)
Department of Pure and Applied Mathematics
Washington State University
Pullman, WA 99164-3113

Professor Walter Carlip
18 Garfield Avenue
Athens, OH 45707

Professor Charles Cook
1 Louise Circle
Sumter, SC 29150

Professor Curtis Cooper (pp. 83-94)
Department of Mathematics and Computer Science
Central Missouri State University
Warrensburg, MO 64093

Professor Karl Dilcher
Dept of Mathematics, Statistics & Computer Science
Dalhousie University
Halifax NS
Canada B3H 3J5

Professor Ernest Eckert
Sondervangsvej 43
9000 Aalborg
Denmark

Professor Michele Elia (pp. 95-102)
Dip. di Elettronica
Politecnico di Torino
C.so Duca degli Abruzzi 24
I-10129 Torino, Italy

Mr. Larry Ericksen (pp. 103-120)
P.O. Box 172
Millville, NJ 08332

Professor Daniel C. Fielder (pp. 121-128; 149-154)
School of Electrical and Computer Engineering
Georgia Institute of Technology
Atlanta, GA 30332-0250

Mr. Piero Filipponi (pp. 95-102; 129-140)
Fondazione Ugo Bordoni
Via B. Castiglione 59
I-00142 Roma, Italy

Ms. Herta T. Freitag (pp. 141-148; 149-154; 155-164)
B40 Friendship Manor
320 Hirshberger Road
Roanoke, Virginia 24012

Professor George W. Grossman (pp. 165-178)
Department of Mathematics
Central Michigan University
Mt. Pleasant, MI 48859

Professor Helen Grundman
Dept. of Mathematics
Bryn Mawr College
101 North Merion Avenue
Bryn Mawr, PA 19010-2899

Professor Heiko Harborth
Technishce Universitat Braunschweig
Germany

Professor Deborah L. Harrell
Department of Mathematics
Salem College
507 Wachovia St.
Winston-Salem, NC 27101

Professor Evelyn Hart
Colgate University
Department of Mathematics
13 Oak Drive
Hamilton, NY 13346-1398

Professor A.F. Horadam (pp. 129-140; 179-194; 307-324)
The University of New England
Armidale
Australia 2351

Professor Yasuichi Horibe (pp. 195-200)
Department of Applied Mathematics
Science University of Tokyo
1-3 Kagurazaka, Shinjuku-ku
Tokyo 162-8601 Japan

Professor F. T. Howard (pp. 201-212)
Mathematics and Computer Science
Wake Forest University
Winston-Salem, NC 27109

Kristin Halla Jonsdottir
Sefgardar 28
170 Seltjarnarnes
Iceland

Professor Robert E. Kennedy (pp. 83-94)
Department of Mathematics and Computer Science
Central Missouri State University
Warrensburg, MO 64093

Professor William A. Kimball (pp. 213-218)
University of Maryland
University Park at Schwabisch Gmund
Universitatspark 8, 73525 Schwabisch Gmund
Germany

Professor Clark Kimberling (pp. 219-232)
University of Evansville
1800 Lincoln Avenue
Evansville, IN 47722

Professor Ron Knott
University of Surrey
Department of Computing
Guildford, Surrey GU2 5XH
United Kingdom

Professor Harris Kwong
State University of New York
College at Fredonia
Department of Mathematics and Computer Science
Fredonia, NY 14063

Dr. Jack Lee (pp. 233-240)
280 86th Street
Brooklyn, NY 11209

Professor Calvin Long
Department of Mathematics
Northern Arizona University
Flagstaff, AZ 86011

Professor Florian Luca (pp. 241-250)
Mathematics Department
Bielefeld University
Postfach 10 01 31
33 501 Bielefeld
Germany

Dr. R.S. Melham (pp. 251-258)
School of Mathematical Sciences
University of Technology, Sydney
PO Box 123, Broadway
NSW 2007 Australia

Mr. Mark D. Morgan
8911 Tamar Drive #302
Columbia, MD 21045

Professor Siguna Müller (pp. 259-276)
Department of Mathematics
University of Klagenfurt
Universitatsstasse 65-67
A-9020 Klagenfurt, Austria

Professor Kenji Nagasaka
Hosei University
Koganei-Shi
Tokyo, 184-8584
Japan

Professor Shigeru Nakamura
Tokyo University of Mercantile Marine
Etchujima
Kotoku
Tokyo
Japan

Professor Sivaram K. Narayan (pp. 165-178)
Department of Mathematics
Central Michigan University
Mt. Pleasant, MI 48859

Professor Richard Orr
Rochester Institute of Technology
Department of Mathematics
Lomb Memorial Drive
Rochester, NY 14623

Professor Andréas N. Philippou (pp. 27-42)
Department of Mathematics
University of Patras
26100 Patras
Greece

Professor George M. Phillips (pp. 141-148; 155-164)
Mathematical Institute
University of St. Andrews
St. Andrews
Scotland

Dr. Stanley Rabinowitz (pp. 277-292)
12 Vine Brook Road
Westford, MA 01886-4212

Ms. Margaret Ribble
1007 Court Street
Maryville, TN 37803

Professor J. Adair Robertson
Peace College
15 East Peace St.
Raleigh, NC 27604

Professor Andrzej Rotkiewicz (pp. 293-306)
Institute of Mathematics
Polish Academy of Sciences, ul. Sniadeckich 8
00-950 Warszawa, Poland

Professor Laura Sanchis
Colgate University
13 Oak Drive
Hamilton, NY 13346-1398

Professor Daihachiro Sato (pp. 1-10)
Department of Mathematics & Statistics
University of Regina
Regina, Saskatchewan, S4S 0A2
Canada

Professor A.G. Shannon (pp. 43-46; 307-324)
University of Technology
Sydney, 2007
Australia

Professor Lawrence Somer (pp. 325-336)
Department of Mathematics
The Catholic University of America
Washington, DC 20064

Mr. Colin Paul Spears
3407 Buena Vista Ave.
Glendale, CA 91208

Professor John Szybist (pp. 337-352)
Intrinsix Corporation
160 Allens Creek Rd.
Rochester, NY 14618

Professor Tomakazu Takahashi
Hosei University, Graduate School
Engineering Division, Systems Engineering
3-7-2, Kajino-cho, Koganei-shi
Tokyo

Professor J.C. Turner (pp. 11-26; 353-368)
Department of Mathematics and Statistics
University of Waikato
Private Bag 3105
Hamilton, New Zealand

Professor Theresa Vaughan
Department of Mathematics
University of North Carolina at Greensboro
Greensboro, NC 27410-5608

Professor William A. Webb (pp. 75-82; 213-218)
Department of Pure and Applied Mathematics
Washington State University
Pullman, WA 99164-3113

Professor Diana Wells
Box 8376
University of North Dakota
Grand Forks, ND 58202

Professor Chizhong Zhou (pp. 369-380)
Yueyang University
Yueyang, Hunan 414000
P. R. China

FOREWORD

This book contains 33 papers from among the 41 papers presented at the Eighth International Conference on Fibonacci Numbers and Their Applications which was held at the Rochester Institute of Technology, Rochester, New York, from June 22 to June 26, 1998. These papers have been selected after a careful review by well known referees in the field, and they range from elementary number theory to probability and statistics. The Fibonacci numbers and recurrence relations are their unifying bond.

It is anticipated that this book, like its seven predecessors, will be useful to research workers and graduate students interested in the Fibonacci numbers and their applications.

June 1, 1999   The Editor

F.T. Howard
Mathematics and Computer Science
Wake Forest University
Box 7388 Reynolda Station
Winston-Salem, NC USA

THE ORGANIZING COMMITTEES

LOCAL COMMITTEE

Anderson, Peter G., Chairman

Arpaya, Pasqual

Biles, John

Orr, Richard

Radziszowski, Stanislaw

Rich, Nelson

INTERNATIONAL COMMITTEE

Horadam, A.F. (Australia), Co-Chair

Philippou, A.N. (Cyprus), Co-Chair

Bergum, G.E. (U.S.A.)

Filipponi, P. (Italy)

Harborth, H. (Germany)

Horibe, Y. (Japan)

Howard, F. (U.S.A.)

Johnson, M. (U.S.A.)

Kiss, P. (Hungary)

Phillips, G.M. (Scotland)

Turner, J. (New Zealand)

Waddill, M.E. (U.S.A.)

Conference Participants

LIST OF CONTRIBUTORS TO THE CONFERENCE

AGRATINI, OCTAVIAN, "Unusual Equations in Study."

*ANDO, SHIRO, (coauthor Daihachiro Sato), "On the Generalized Binomial Coefficients Defined by Strong Divisibility Sequences."

*ANATASSOVA, VASSIA K., (coauthor J.C. Turner), "On Triangles and Squares Marked With Goldpoints — Studies of Golden Tiles."

*ANTZOULAKOS, DEMETRIOS L., (coauthor Andreas N. Philippou), "Multivariate Pascal Polynomials of Order K with Probability Applications."

*ATANASSOV, K.T., (coauthor A.G. Shannon), "Fibonacci Planes and Spaces."

*BICKNELL-JOHNSON, MARJORIE, "The Smallest Positive Integer Having F_k Representations as Sums of Distinct Fibonacci Numbers."

*BICKNELL-JOHNSON, MARJORIE, "The Zeckendorf-Wythoff Array Applied to Counting the Number of Representations of N as Sums of Distinct Fibonacci Numbers."

*BILES, JOHN A., "Composing with Sequences: ...but is it Art?"

*CARAGIU, MIHAI, (coauthor William Webb), "Invariants for Linear Recurrences."

*COOPER, CURTIS, (coauthor Robert E. Kennedy), "Base 10 RATS Cycles and Arbitrarily Long Base 10 RATS Cycles."

DILCHER, KARL, "Nested Squares and Evaluations of Integer Products."

*ELIA, MICHELE, (coauthor Piero Filipponi), "Quintics $x^5 - 5x - k$, The Golden Section, and Square Lucas Numbers."

*ERICKSEN, LARRY, "The Pascal-DeMoivre Moments and Their Generating Functions."

*FIELDER, DANIEL C., (coauthor Cecil O. Alford), "Investigating Special Binary Sequences With Some Computer Help."

FIELDER, DANIEL C., (coauthor Marjorie Bicknell-Johnson), "The 1st 330 terms of Sequence A013583."

*FILIPPONI, PIERO, (coauthor Alwyn F. Horadam), "Integration Sequences of Jacobsthal and Jacobsthal-Lucas Polynomials."

*FREITAG, HERTA T., (coauthors Marjorie Bicknell-Johnson and George M. Phillips), "A Property of the Unit Digits of Recursive Sequences."

*FREITAG, HERTA T., (coauthor Daniel C. Fielder), "On General Divisibility of Sums of Integral Powers of the Golden Ratio."

*FREITAG, HERTA T., (coauthor George M. Phillips), "Sylvester's Algorithm and Fibonacci Numbers."

*GROSSMAN, GEORGE W., (coauthor Sivaram K. Narayan), "On the Characteristic Polynomial of the j-th Order Fibonacci Sequence."

*HORADAM, A.F., "Quasi Morgan-Voyce Polynomials and Pell Convolutions."

*HORIBE, YASUICHI, "On an Asymptotic Maximality of the Fibonacci Tree."

*HOWARD, F.T., "Generalizations of a Fibonacci Identity."

*KIMBALL, WILLIAM A., (coauthor William A. Webb), "Some Generalizations of Wolstenholme's Theorem."

---

*The asterisk indicates that the paper is included in this book.

*KIMBERLING, CLARK, "Card Sorting Related to Fibonacci Numbers."

*LEE, JACK, "On the Inhomogeneous Geometric Line-Sequence."

*LUCA, FLORIAN, "Fibonacci Numbers of the Form $k^2 + k + 2$."

*MELHAM, R.S., "On Certain Polynomials of Even Subscripted Lucas Numbers."

MORGAN, MARK D., "The Distribution of Second Order Linear Recurrence Sequences Mod 2^m."

*MÜLLER, SIGUNA, "On the Rank of Appearance of Lucas Sequences."

*RABINOWITZ, STANLEY, "Algorithmic Simplification of Reciprocal Sums."

RABINOWITZ, STANLEY, "Proving identities by Computer."

*ROTKIEWICZ, ANDRZEJ, "Solved and Unsolved Problems on Pseudoprime Numbers and Their Generalizations."

SANCHIS, LAURA, "On the Occurrence of Certain Fibonacci Numbers in the Zeckendorf Decomposition of nF_n."

*SHANNON, A.G., (coauthor A.F. Horadam), "Some Relationships Among Vieta, Morgan-Voyce and Jacobsthal Polynomials."

*SOMER, LAWRENCE, "Special Multipliers of Lucas Sequences Modulo p^r."

*SZYBIST, JOHN, (coauthor Peter G. Anderson), "Digital Halftoning Using Error Diffusion and Linear Pixel Shuffling."

TAKAHASHI, TOMAKAZU, "Forcasting Method for Stock Prices by Using Elliott Wave Principle and Neural Network."

*TURNER, JOHN C., "On Vector Sequence Recurrence Equations in Fibonacci Vector Geometry."

*ZHOU, CHIZHONG, "Constructing Identities Involving K^{th}-order F-L Numbers by Using the Characteristic Polynomial."

INTRODUCTION

The Fibonacci numbers

$$1, 1, 2, 3, 5, 8, 13, 21, 34, 55, 89, \ldots,$$

were first mentioned in 1202 in the *Liber Abaci*, a book written by Leonardo of Pisa to introduce the Hindu-Arabic numeral system to western Europe. Leonardo, perhaps the greatest mathematician of the Middle Ages, wrote under the name of Fibonacci — a contraction of "filius Bonacci" (son of Bonacci). In *Liber Abaci* the numbers appeared in the famous rabbit problem, but they were not called "Fibonacci numbers" until the nineteenth century, when the French mathematician Edouard Lucas used that term. Lucas studied the Fibonacci numbers extensively, and the simple generalization

$$2, 1, 3, 4, 7, 11, 18, 29, 47, 76, 123, \ldots,$$

bears his name.

The rich and interesting history of the Fibonacci numbers in the eighteenth and nineteenth centuries can be found in L. E. Dickson's *History of the Theory of Numbers, volume 1*. During the twentieth century, interest in Fibonacci numbers and their applications rose rapidly. In 1961 the Soviet mathematician N. Vorobyov published *Fibonacci Numbers*, and Verner E. Hoggatt, Jr., followed in 1969 with his *Fibonacci and Lucas Numbers*. Meanwhile, in 1963, Hoggatt and his associates founded The Fibonacci Association and began publishing *The Fibonacci Quarterly*. They also organized a Fibonacci Conference in California, U.S.A., each year for almost sixteen years until 1979.

In 1984, the First International Conference on Fibonacci Numbers and Their Applications was held in Patras, Greece, and the proceedings from that conference were published. It was anticipated at that time that this conference would be the first of a series of international conferences on the subject to be held every two or three years in different countries. With this intention as a motivating force, The Second, Third, Fourth, Fifth, Sixth and Seventh International Conference on Fibonacci Numbers and Their Applications were respectively held in alternate years at San Jose, California; Pisa, Italy; Winston-Salem, North Carolina; St. Andrews, Scotland; Pullman, Washington and Graz, Austria. The proceedings from these seven conferences have also been published. Because of the continuous success of the preceding seven conferences, The Eighth International Conference on Fibonacci Numbers and Their Applications was held at Rochester, New York, June 22-26, 1998, and a Ninth Conference is scheduled in July 2000 in Luxembourg.

It is impossible to overemphasize the importance and relevance of the Fibonacci numbers to the mathematical and physical sciences as well as other areas of study. The Fibonacci numbers appear in almost every branch of mathematics, including number theory, combinatorics, differential equations, probability, statistics, numerical analysis, and linear algebra. They also occur in physics, biology, chemistry, and electrical engineering.

It is believed that the contents of this book, like its predecessors, will prove useful to everyone interested in this important branch of mathematics and that this material may lead to additional results on Fibonacci numbers both in mathematics and in their applications to science and engineering.

The editor would like to acknowledge The Fibonacci Association and Rochester Institute of Technology for their financial and other assistance in making the conference a success. He would also like to thank Gerald Bergum, Marjorie Bicknell-Johnson and Calvin Long for their advice, assistance and encouragement. Finally, the editor thanks the technical typist, Patricia Solsaa, for her excellent work.

The Editor

ON THE GENERALIZED BINOMIAL COEFFICIENTS DEFINED BY STRONG DIVISIBILITY SEQUENCES

Shiro Ando and Daihachiro Sato

1. STRONG DIVISIBILITY SEQUENCES AND GENERALIZED
BINOMIAL COEFFICIENTS

Given a positive integer sequence $\{a_n\}$, where $n = 1, 2, 3, \cdots$, we replace $n!$ in the defining form

$$\binom{n}{r} = \frac{n!}{r!s!}, \quad r + s = n$$

of the binomial coefficients with $\prod a_n = a_n a_{n-1} \cdots a_2 a_1$, where we define $\prod a_0 = 1$. When the resulting numbers

$$\left[\begin{matrix}n\\r\end{matrix}\right] = \frac{\prod a_n}{\prod a_r \prod a_s}, \quad r + s = n$$

are all integers, we call them the generalized binomial coefficients defined by $\{a_n\}$, and the sequence $\{a_n\}$ a Raney sequence. In particular, the Fibonacci sequence $\{F_n\}$ is a Raney sequence, and the generalized binomial coefficients defined by it are called Fibonomial coefficients.

H.W. Gould stated in [8] that his conjecture on the GCD hexagonal equality would be valid even for Fibonomial coefficients and for Gauss'-binomial coefficients, the generalized binomial coefficients defined by the sequence $\{1 + q + q^2 + + q^{n-1}\}$. A.P. Hillman and V.E. Hoggatt Jr. proved the GCD hexagonal equality in [10], for generalized binomial coefficients defined by the sequences, satisfying the conditions

---

This paper is in final form and no version of it will be submitted for publication elsewhere.

$$\gcd(a_m, a_n) \mid a_{m+n} \text{ and } \gcd(a_m, a_n) \mid a_{n-m} \ (m < n). \tag{1}$$

We noticed in [3] that these conditions are equivalent to

$$a_{(m,n)} = (a_m, a_n), \tag{2}$$

which had been studied in M. Ward [15], as the special case of the divisibility sequences which satisfy the condition

$$m \mid n \Rightarrow a_m \mid a_n. \tag{3}$$

Later in the investigation of divisibility sequences satisfying a linear recurrence, M. Hall [9] and M. Ward himself, adding the initial term a_0, adopted the sequence $\{a_n\}$, where $n = 0, 1, 2, \cdots$. C. Kimberling [13] referred to this sequence satisfying (2) as a strong divisibility sequence. However, we use the initial term a_1 as in [15].

Thus, we define a divisibility sequence to be a positive integer sequence $\{a_n\}$, where $n = 1, 2, 3, \cdots$, satisfying (3) and a strong divisibility sequence (SDS) to be one satisfying (2). Since we are concerned with the strong divisibility sequences, the generalized binomial coefficients always mean the ones defined by a strong divisibility sequence in the following. We call the infinite triangular number array that has these generalized binomial coefficients as its entries, a generalized Pascal triangle.

Since a strong divisibility sequence S is a divisibility sequence, every term is divisible by a_1, and the generalized binomial coefficients defined by S are unchanged if we divide every term of S by a_1. Thus, we may assume $a_1 = 1$ in the following.

Example 1: The Fibonacci sequence $\{F_n\}$ is a strong divisibility sequence. The generalized binomial coefficients defined by $\{F_n\}$ are called Fibonomial coefficients.

Example 2: Let $g > 1$ be an integer. The sequence $\{G_n\}$, where $G_n = 1 + g + g^2 + \cdots + g^{n-1}$ is a strong divisibility sequence. The generalized binomial coefficients defined by $\{G_n\}$ are obtained by substituting g for q in Gauss' q-binomial coefficients.

Example 3: Let $\{a_n\}$ be a strong divisibility sequence and let $P = \{p_i \mid i = 1, 2, \cdots, t\}$ be a finite set of primes. The sequence $\{b_n\}$ defined by

$$b_n = p_1^{v_{p_1}(a_n)} p_2^{v_{p_2}(a_n)} \cdots p_t^{v_{p_t}(a_n)},$$

where $v_p(a_n)$ denotes the p-adic exponential valuation of a_n, is also a strong divisibility sequence.

2. DEFINITIONS AND NOTATIONS

Let $S = \{a_n\}$, where $n = 1$, 2, 3, ···, be a positive integer sequence and p be a prime. Then we call the sequence $S_p = \left\{ p^{v_p(a_n)} \right\}$ the p-factor of S.

If we define the product of two sequences $S = \{a_n\}$ and $T = \{b_n\}$ by $ST = \{a_n b_n\}$, then the product of finite number of sequences will be determined inductively.

Let M be an infinite set of positive integer sequences such that for each n, only finitely many sequences in M have an n-th term not equal to 1. The product of the sequences in M can be defined as the sequence whose n-th term is the product of the n-th terms of the sequences in M. Then the sequence S is represented as

$$S = \prod_p S_p, \tag{4}$$

where p runs all the primes. This formula is called the **prime factoring** of S.

The strong divisibility sequence whose terms are all 1 is denoted by

$$E = \{1, 1, 1, \cdots\},$$

and is called the unit sequence.

The sequence $\{b_n\}$ in Example 3 of the first section is the product of the p-factors of $\{a_n\}$ for p in $P = \{p_1, p_2, \cdots, p_t\}$. It is obtained by replacing all of the p_i-factors in the prime factoring of $\{a_n\}$, except for the p_i in P, with the unit sequence E.

A strong divisibility sequence where every term is a nonnegative power of a given prime p is called a **p-local strong divisibility sequence.**

Let K be a given positive integer. The smallest number n such that $K \mid a_n$ is called the **rank of apparition** of K in the sequence $S = \{a_n\}$ and is denoted by $\rho(K)$. If there is no such number, the rank of apparition of K is infinity. The sequence $T = \{b_n\}$, where $b_n = \min\{a_n, K\}$ is called the **K restriction of** $S = \{a_n\}$.

3. A CHARACTERIZATION OF THE STRONG DIVISIBILITY SEQUENCE

M. Ward [15] showed that following two conditions are equivalent.

(i) $\{a_n\}$ is a strong divisibility sequence.

(ii) For every prime power p^k, $p^k \mid a_n$ if and only if $\rho(p^k) \mid n$.

Theorem 1: A positive integer sequence $S = \{a_n\}$ is a strong divisibility sequence if and only if the p-factor S_p of S is a strong divisibility sequence for every prime p.

Proof: The condition that $(a_m, a_n) = a_d$ for all positive integers m and n, where $d = (m,n)$, is equivalent to $v_p(a_d) = \min\{v_p(a_m), v_p(a_n)\}$ for every prime p. Equivalently, S_p is a strong divisibility sequence for every p.

Theorem 2: A sequence $L_p = \{a_n\}$ where every a_n is a nonnegative power of a given prime p, is a p-local SDS if and only if the p^k restriction of L_p is a periodic with period $\rho(p^k)$ for every $p^k \leq \sup\{a_n\}$.

Proof: If the p^k restriction of L_p has period $\rho(p^k)$ then $p^k \mid a_n$ if and only if $\rho(p^k) \mid n$. Since this holds for all $p^k \leq \sup\{a_n\}$, L_p is an SDS by Ward's results.

Using induction on k it is easy to show that if L_p is an SDS and $\{b_n\}$ is the p^k restriction of L_p then $\{b_n\}$ is periodic with period $\rho(p^k)$ for every $p^k \leq \sup\{a_n\}$.

(i) If $k = 0$, then $\{b_n\} = E$, which is periodic with period 1.

(ii) Now we assume that $1 \leq k \leq \sup\{v_p(a_n)\}$ and the statement is true for the p^{k-1} restriction of L_p. The p^k restriction $\{b_n\}$ of $\{a_n\}$ is periodic with period $\rho(p^k)$ for n such that $\rho(p^k) \nmid n$, since $\{b_n\}$ coincides with the p^{k-1} restriction of $\{a_n\}$ for such n and $\rho(p^{k-1}) \mid \rho(p^k)$. On the other hand, for n such that $\rho(p^k) \mid n$, $b_n = p^k$. Thus, $\{b_n\}$ is periodic with period $\rho(p^k)$.

From (i) and (ii) we complete the proof.

Remark: It is clear by the proof of Theorem 2 that a p-local SDS $\{a_n\}$ is symmetric in the range $1 \leq n \leq \rho(p^k) - 1$ for any positive integer $k \leq \sup\{v_p(a_n)\}$.

Now let $L_p = \{a_n\}$ be a p-local SDS which is not E. We put $c_k = \rho(p^k)/\rho(p^{k-1})$ for k with $1 \leq k \leq \sup\{v_p(a_n)\}$.

If $\{a_n\}$ is unbounded, then there corresponds a positive integer sequence $\{c_n\}$ for which $c_1 > 1$ and there are infinitely many n such that $c_n > 1$.

Conversely, given a positive integer sequence $\{c_n\}$ which satisfies these conditions, we can define a p-local SDS $\{a_n\}$ to which the given sequence $\{c_n\}$ corresponds by putting $a_n = 1$ if $c_1 \nmid n$ and $a_n = p^k$ if $c_1 c_2 \cdots c_k \mid n$ and $c_1 c_2 \cdots c_{k+1} \nmid n$.

If $\{a_n\}$ is bounded and $\max\{a_n\} = p^s$, then the corresponding $\{c_n\}$ is a finite sequence of length s and satisfies $c_1 > 1$. Conversely, if $\{c_n\}$ is such a finite positive integer sequence, then we can find a bounded p-local SDS $\{a_n\}$ to which the given sequence $\{c_n\}$ corresponds in a similar manner stated above. Thus, we have:

Corollary: Let $L_p = \{a_n\}$ be a p-local SDS which is not E. If we put $c_k = \rho(p^k)/\rho(p^{k-1})$ for k with $1 \leq k \leq \sup\{v_p(a_n)\}$, there corresponds a finite or infinite positive integer sequence of period ratios $\{c_n\}$. Conversely, given a finite or infinite positive integer sequence $\{c_n\}$ where $c_1 > 1$ and there are infinitely many n such that $c_n > 1$ if it is infinite, we can find a p-local SDS to which $\{c_n\}$ corresponds and every p-local SDS can be given in this manner.

4. GCD CENTER COVERING STAR ON THE GENERALIZED PASCAL TRIANGLE

In this section, we will apply the results obtained in the previous section to the GCD covering problem on the generalized Pascal triangle. The terminology and the symbols are similar to those used in our previous paper [4], where we discussed the same problem in Pascal's triangle.

$$A_2 \quad A_1$$
$$A_3 \quad X \quad A_6$$
$$A_4 \quad A_5$$

Figure 1

Let X be any element in a generalized Pascal triangle and let $A_1, A_2, \cdots, A_6$ be six points in order surrounding X. We call the set R_i, where $i \in \{1, 2, \cdots, 6\}$ of the successive t_i points on the half line XA_i (but excluding X itself), a ray of length t_i and their union $S = R_1 \cup R_2 \cup \cdots \cup R_6$ (empty rays with length 0 are admitted), a star of type $T = [t_1, t_2, t_3, t_4, t_5, t_6]$ with center X. In the following, we sometimes call a star of type T shortly "a star T" since we are interested in its properties which are independent of its location. We discussed in [4] a necessary and sufficient condition that rays of a star configuration in Pascal's triangle cover its center with respect to GCD and LCM independent of its location in Pascal's triangle. In the case of GCD, which means $\gcd(S \cup \{X\}) = \gcd S$, we had following results.

Proposition 1: Each of five star configurations

$$\{A_1, A_2\}, \ \{A_3, A_4\}, \ \{A_5, A_6\}, \ \{A_1, A_3, A_5\}, \ \{A_2, A_4, A_6\},$$

shown in Figure 2 covers its center with respect to GCD.

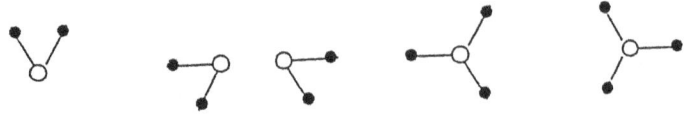

Figure 2

Proposition 2: Let

$$T_1 = [101001], T_2 = [011001], T_3 = [100101], T_4 = [011010], T_5 = [100110], T_6 = [010110]$$

be six types of Star configurations in Pascal's triangle. For $T = [t_1 t_2 t_3 t_4 t_5 t_6]$ and a positive integer t, we define tT as $tT = [tt_1 tt_2 tt_3 tt_4 tt_5 tt_6]$. Then for any positive integer t, each of $tT_i (i = 1, 2, \cdots, 6)$ does not cover its center with respect to GCD.

From these two propositions we concluded that a necessary and sufficient condition that a star configuration S in Pascal's triangle covers its center with respect to GCD independent of its location in Pascal's triangle is that S contains one of five star configurations shown in Figure 2.

It is easy to show that Proposition 1 can be generalized to the case of the generalized Pascal triangle, using the fundamental property of the strong divisibility sequences shown in [3]. For instance, we have

$$a_n \gcd\{A_1, A_2\} = \gcd(a_n A_1, a_n A_2) = \gcd\{a_s X, a_r X\} = X \gcd\{a_s, a_r\}.$$

Since $n = r + s$, $\gcd\{a_s, a_r\} \mid a_n$ so that $\gcd\{A_1, A_2\} \mid X$.

Thus $\{A_1, A_2\}$ covers X, and similarly, $\{A_3, A_4\}$ and $\{A_5, A_6\}$ cover X. Also, from $\gcd\{A_1, A_3 A_5\} \mid \gcd\{a_n A_1, a_{s+1} A_3, a_{r+1} A_5\} = \gcd\{a_s X, a_r X, a_{n+1} X\} = X \gcd\{a_s, a_r, a_{n+1}\}$, and $\gcd\{a_s, a_r, a_{n+1}\} \mid a_{n+1-r-s} = a_1 = 1$, we have $\gcd\{A_1, A_3, A_5\} \mid X$. Thus $\{A_1, A_3, A_5\}$ covers X, and similarly, $\{A_2, A_4, A_6\}$ covers X.

As a trivial counter example, when $\{a_n\}$ is the unit sequence E, Proposition 2 is not valid in the generalized Pascal triangle without an additional condition. The purpose of this section is to investigate such conditions.

Theorem 3: In the generalized Pascal triangle defined by a strong divisibility sequence $\{a_n\}$, we have the following property concerning the six types of star configurations $T_i (i = 1, 2, \cdots, 6)$ given in Proposition 2.

A star configuration of type tT_i does not cover its center for any positive integer t, if and only if:

(i) $\{a_n\}$ has a p-factor whose range is unbounded.

or

(ii) $\{a_n\}$ has infinitely many p-factors whose ranges contain more than two values.

Proof: We prove the theorem for $i = 1$. We can prove the other cases in the same way.

First, we assume that $\{a_n\}$ does not satisfy either of the two conditions (i) and (ii). Let P_1 be the set of all primes such that the p-factors of $\{a_n\}$ have ranges of size larger than two. Then, P_1 must be finite. The range of each p-factor has the largest number K_p since $\{a_n\}$ does not satisfy the condition (i). We denote the rank of apparition of K_p in $\{a_n\}$ by $\rho(K_p)$, and put $\max \{\rho(K_p) \mid p \in P_1\} = t$.

Now, we will show that the star tT_1 covers its center X with respect to GCD. If $X = \prod a_n / \prod a_r \prod a_s$, where $\prod a_n = a_n a_{n-1} \cdots a_2 a_1$, , then tT_1 consists of $3t$ points:

$$A_{1,h} = (a_s a_{s-1} \cdots a_{s-h+1} / a_n a_{n-1} \cdots a_{n-h+1}) X, \quad (h = 1, 2, \cdots, t),$$
$$A_{3,h} = (a_r a_{r-1} \cdots a_{r-h+1} / a_{s+1} a_{s+2} \cdots a_{s+h}) X, \quad (h = 1, 2, \cdots, t), \tag{5}$$
$$A_{6,h} = (a_s a_{s-1} \cdots a_{s-h+1} / a_{r+1} a_{r+2} \cdots a_{r+h}) X, \quad (h = 1, 2, \cdots, t).$$

Fixing a $p \in P_1$, we put $\rho(K_p) = \rho$. Then $\rho \leq t$. Since ρ is a period fo the p-factor of $\{a_n\}$, we have $v_p(a_s a_{s-1} \cdots a_{s-\rho+1}) = v_p(a_n a_{n-1} \cdots a_{n-\rho+1})$ so that $v_p(a_{1,\rho}) = v_p(X)$.

For $p \notin P_1$, we can assume that $p \mid a_r$ and $p \mid a_s$, since otherwise p does not divide a_r or a_s, which give $v_p(A_{3,1}) \leq v_p(X)$ or $v_p(A_{1,1}) \leq v_p(X)$, respectively. Using the property of the strong divisibility sequence we get $p \mid a_n$ from $n = r + s$. By the assumption $p \notin P_1$, the range of the p-factor of $\{a_n\}$ consists of two values: 1 and a positive power of p since we assumed $a_1 = 1$ (see Page 3). Since $v_p(a_s) > 0$ and $v_p(a_n) > 0$, we must have $v_p(a_s) = v_p(a_n)$ an then $v_p(A_{1,1}) = v_p(X)$. This proves the necessity of the conditions.

In order to prove the sufficiency of the conditions, assume that $\{a_n\}$ satisfies the conditions (i) or (ii). Let t be a positive integer. Since the number of prime powers whose ranks of apparitions do not exceed t is finite, in either case, there exist a prime p and different powers p^j and p^k such that the ranks of apparitions r, s of p^j and p^k satisfy $t < r < s$ and $0 < v_p(a_r) = j < v_p(a_s) = k$, nd there are not terms between a_r and a_s whose p-adic valuations are larger than j. Fixing such p, j and k, we put

$$X = \prod a_n / \prod a_r \prod a_s, \text{ where } n = r + s.$$

Then we have $r \mid s$, $r \mid n$ and $v_p(a_n) = j$. Since the p^j restriction of the p-factor of $\{a_m\}$ has period r and is symmetric in $1 \leq m \leq r - 1$, we have

$$v_p(a_{r-h}) = v_p(a_{s-h}) = v_p(a_{n-h}) = v_p(a_{r+h}) = v_p(a_{s+h}) = v_p(a_{n+h}) \text{ for } h = 1, 2, \cdots, r - 1.$$

Concerning $3t$ points (6) belonging to tT_1, the term numbers m of a_m which appear in the coefficients of X satisfy $1 < m < 2s$. In this range, the p-factor of $\{a_m\}$ is periodic with period s, and $v_p(a_m) < k$ except for $m = s$, when $v_p(a_s) = k$. Then, we have $v_p(A_{1,h}) > v_p(x)$ since $v_p(a_s) > v_p(a_n)$, $v_p(a_{s-1}\cdots a_{s-h+1}) = v_p(a_{n-1}\cdots a_{n-h+1})$. We also have $v_p(A_{3,h}) > v_p(x)$ from $v_p(a_r) > v_p(a_{s+h})$, $v_p(a_{r-1}\cdots a_{r-h+1}) = v_p(a_{s+1}\cdots a_{s+h-1})$, and $v_p(A_{6,h}) > v_p(X)$ from $v_p(a_s) > v_p(a_{r+h})$, $v_p(a_{s-1}\cdots a_{s-h+1}) = v_p(a_{r+1}\cdots a_{r+h-1})$, so that tT_1 does not cover its center.

Corollary: On the generalized Pascal triangle defined by strong divisibility sequence satisfying the conditions (i) or (ii) in Theorem 3, a necessary and sufficient condition that a star configuration S covers its center with respect to GCD is that S contains one of the five stars given in figure 2.

It is well known (See V.E. Hoggatt, Jr. [12]) that the rank of apparition of any positive integer m in the Fibonacci sequence is less than m^2. Therefore, the Fibonacci sequence satisfies both of the conditions (i) and (ii) of Theorem 3. Thus, the Corollary above is applicable to the Fibonomial triangle.

REFERENCES

[1]   Ando, S. and Sato, D. "A GCD Property on Pascal's Pyramid and the Corresponding LCM Property of the Modified Pascal Pyramid." <u>Applications</u> <u>of</u> <u>Fibonacci</u> <u>Numbers</u>, Volume 3. Edited by G.E. Bergum, A.F. Horadam and A.N. Philippou. Kluwer Academic Publishers, Dordrecht, The Netherlands, 1990: pp. 7-14.

[2]   Ando, S. and Sato, D. "Translatable and Rotatable Configurations Which Give Equal Product, Equal GCD and Equal LCM Properties Simultaneously." <u>Applications</u> <u>of</u> <u>Fibonacci</u> <u>Numbers</u>, Volume 3. Edited by G.E. Bergum, A.F. Horadam and A.N. Philippou. Kluwer Academic Publishers, Dordrecht, The Netherlands, 1990: pp. 15-26.

[3]   Ando, S. and Sato, D. "On the Proof of GCD and LCM Equalities Concerning the Generalized Binomial and Multinomial Coefficients." <u>Applications</u> <u>of</u> <u>Fibonacci</u> <u>Numbers</u>, Volume 4. Edited by G.E. Bergum, A.F. Horadam and A.N. Philippou. Kluwer Academic Publishers, Dordrecht, The Netherlands, 1991: pp. 9-16.

[4]   Ando, S. and Sato, S. "A Necessary and Sufficient Condition that Rays of a Star
        Configuration on Pascal's Triangle Cover Its Center with Respect to GCD and LCM."
        Applications of Fibonacci Numbers, Volume 5. Edited by G.E. Bergum, A.F. Horadam
        and A.N. Philippou. Kluwer Academic Publishers, Dordrecht, The Netherlands, 1993:
        pp. 11-36.

[5]   Ando, S. and Sato, D. "On the Minimal Center Covering Stars with Respect to GCD in
        Pascal's Pyramid and Its Generalizations." Applications of Fibonacci Numbers, Volume
        5. Edited by G.E. Bergum, A.F. Horadam and A.N. Philippou. Kluwer Academic
        Publishers, Dordrecht, The Netherlands, 1993: pp. 37-43.

[6]   Ando, S. and Sato, D. "Minimal Center Covering Stars with Respect to LCM in Pascal's
        Pyramid and its Generalizations." Applications of Fibonacci Numbers, Volume 6.
        Edited by G.E. Bergum, A.F. Horadam and A.N. Philippou. Kluwer Academic
        Publishers, Dordrecht, The Netherlands, 1996: pp. 23-30.

[7]   Ando, S. and Sato, D. "Multiple Color Version of the Star of David Theorems on
        Pascal's Triangle and Related Arrays of Numbers." Applications of Fibonacci Numbers,
        Volume 6. Edited by G.E. Bergum, A.F. Horadam, A.N. Philippou. Kluwer Academic
        Publishers, Dordrecht, The Netherlands, 1996: pp. 31-45.

[8]   Gould, H.W. "A new Greatest Common Divisor Property of the Binomial Coefficients."
        The Fibonacci Quarterly, Vol. 10.6 (1972): pp. 579-584, 628.

[9]   Hall, M. "Divisibility Sequences of Third Order." American Journal of Mathematics,
        Vol. 58 (1936): pp. 577-584.

[10]   Hillman, A.P. and Hoggatt, V.E. Jr. "A Proof of Gould's Pascal Hexagon Conjecture."
        The Fibonacci Quarterly, Vol. 10.6 (1972): pp. 565-568, 598.

[11]   Hitotumatu, S. and Sato, D. "Star of David Theorem (I)." The Fibonacci Quarterly,
        Vol. 13.1 (1975): p. 70.

[12]   Hoggatt, V.E. Jr. Fibonacci and Lucas Numbers. Houghton Mifflin Company, Boston,
        1969.

[13]   Kimberling, C. "Strong Divisibility Sequences with Nonzero Initial Term." The
        Fibonacci Quarterly, Vol. 16.6 (1978): pp. 541-544.

[14]   Sato, D. and Hitotumatu, S. "Simple Proof that a p-adic Pascal's Triangle is 120 Degree
        Rotatable." Proceedings of the American Mathematical Society, Vol. 59 (1976): pp. 406-
        407.

S. ANDO AND D. SATO

[15]   Ward, M. "Note on Divisibility Sequences." *Bulletin of the American Mathematical Society*, Vol. *42* (1936): pp. 843-845.

AMS Classification Numbers: 11A05, 11B65, 11B83

ON TRIANGLES AND SQUARES MARKED WITH GOLDPOINTS—STUDIES OF GOLDEN TILES

Vassia K. Anatassova and J.C. Turner

INTRODUCTION

In late 1996, Turner sent the Atanassov family a Christmas card, on which he had inscribed the star diagram shown below and had set a Christmas puzzle for them to attempt. He included also a set of equilateral triangles, made from card and marked with golden cut points; various jig-saw problems with these cards were suggested. The Atanassovs became interested in these problems, and subsequent developments with related geometric and combinatoric themes led to the results described in this paper.

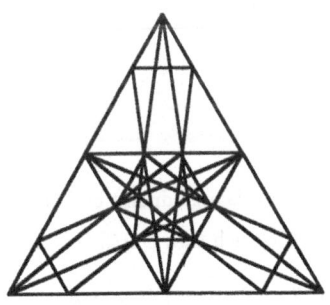

The Fibonacci Star

Christmas puzzle: Find how many points in this Star are interior golden cut points of line segments, given that the figure is based on four equilateral triangles, and that the two boundary points immediately below the apex are golden cut points of the top triangle's sides.

This paper is in final form and no version of it will be submitted for publication elsewhere.

1. EQUILATERAL TRIANGLES AND GOLDPOINTS

In [1] Turner described the Fibonacci vector polygon, which is a plot of the Fibonacci position vectors $F_n \equiv (F_{n-1},\ F_n,\ F_{n+1})$, the points being 'joined-up' by straight lines. The polygon lies in the plane $x + y - z = 0$, all of whose integer-points (i.e. points with integer coordinates) have six regularly spaced nearest integer-point neighbours; for that reason, he called the plane the *honeycomb plane*: it can be tessellated by regular hexagons.

He showed that the vectors F_n tend, as $n \to \infty$, to lie on the line $x/1 = y/\alpha = z/\alpha^2$ where α denotes the golden mean. He was intrigued to observe how this line, as it leaves the origin and moved upwards, cuts sides of the hexagons (and hence equilateral triangles) in ratios which involve α.

The simplest case is the triangle closest to the origin: its upper side is cut in the ratio $\alpha:1$. This happens in the next higher triangle too, with the order of the ratio reversed. Thereafter, moving upwards through adjacent triangles the ratios change, and they all involve powers of the golden mean α. The following two theorems reflect the observations made upon the first two triangles; these results constitute the *raison d'être* of the triangular tiles to be described below.

<u>Definition:</u> A *goldpoint* is an internal point of a line-segment, which divides the segment length in the ratio $\alpha:1$ or $1:\alpha$, where α is the golden mean $(1 + \sqrt{5})/2$.

<u>Theorem 1:</u> Let $ABDC$ be a rhombus, with angle θ at A and D. Let AR be a ray through A which cuts CB in P and CD in Q.

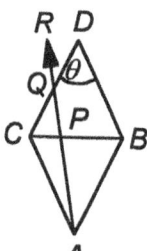

Then P is a goldpoint of CB if and only if Q is the goldpoint of CD such that $CQ/QD = \alpha$.

If $\theta = 60°$, then $PC = QD$.

<u>Proof:</u>   Since $AB \parallel CD$, $\angle CQP = \angle PAB$ and $\angle QCP = \angle PBA$; and $\angle QPC = \angle APB$ (vertically opposite angles).   Hence $\triangle PQC$ is similar to $\triangle PAB$.   Therefore $PB/PC = AB/QC = CD/CQ$ (since $AB = CD$).   If P is a goldpoint of CB (and by construction $CP < PB$), then $PB/PC = \alpha$. Hence $CD/CQ = \alpha$, so Q is a goldpoint of CD: note that $CD/CQ = (CQ + QD)/CQ = 1 + QD/CQ$. Therefore $QD/CQ = \alpha - 1 = 1/\alpha$, hence

$CQ/QD = \alpha$.   For   the   converse,   suppose   that $CQ/QD = \alpha$.   Then $PB/BC = CD/CQ = 1+1/\alpha$, hence P is a goldpoint of CB. Finally, if $\theta = 60°$ then $\triangle BCD$ is equilateral, so $CB = CD$. Then since $CB/PB = CD/CQ = \alpha$, we have $PB = CQ$ and hence $PC = QD$.   □

The following theorem shows how the 60°-rhombus can be set in the honeycomb plane, and $3D$ coordinate geometry used to determine the line through AR.

Theorem 2: Let the rhombus of Theorem 1, with $\theta = 60°$, be set in the plane $x + y - z = 0$ with points A, B, C, D being respectively $(0,0,0)$, $(1,0,1)$, $(0,1,1)$, $(1,1,2)$. Then if P is a goldpoint of CB, the ray AR lies in the line $x/1 = y/\alpha = z/\alpha^2$, which has d.c.s $(1, \alpha, \alpha^2)/2\alpha$.

Proof: In the accompanying diagram, $u \in \left(0, \frac{1}{2}\right)$ is the fractional distance of P from C to B.

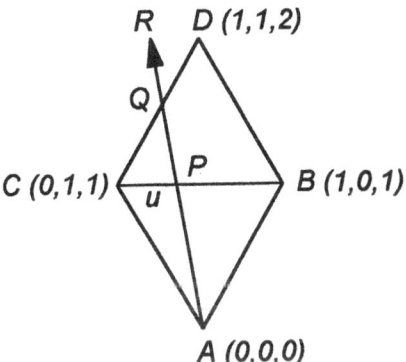

Figure 1. A 60°-rhombus in the plane $x + y - z = 0$

Since P is a goldpoint, and AR is to the left of AD, $PB/PC = \alpha$.

Therefore:

$$PB^2 = \alpha^2 PC^2$$

$$\text{so } (1-u)^2 + (u-1)^2 = \alpha^2(u^2 + u^2 + 0)$$

$$\Rightarrow \quad u - 1 = \pm u\alpha \quad \text{(reject +, since } u > 0)$$

$$\Rightarrow \quad u = 1/\alpha^2.$$

Hence point P has coordinates $(1/\alpha^2, 1 - 1/\alpha^2, 1) = 1/\alpha^2(1, \alpha, \alpha^2)$.

Therefore the line through AR has direction ratios $1 : \alpha : \alpha^2$; and the equation of the line follows immediately. The direction cosines are $(1, \alpha, \alpha^2)/\sqrt{1 + \alpha^2 + \alpha^4}$; and $1 + \alpha^2 + \alpha^4 = 4\alpha^2$, hence the required formula.   □

2. GOLDEN TILES AND COMBINATION RULES

The above theorems show that the adjacent equilateral triangles, with goldpoints marked on their sides, can give rise to interesting geometry involving the golden mean α. Having discovered this, Turner made up (in cardboard) a set of equilateral triangles, and he marked upon each side of every triangle one of its goldpoints. He proposed the following rules of equivalence and combination for the triangles (and, by extension, for polygons):

Rule 1. Two marked triangles in a plane are *equivalent* if one triangle can be rotated, and translated, until it and its three side-marks* coincide with the other triangle and its side-marks.

Rule 2. Two marked triangles can be *combined* if they can be placed in a plane with two sides coinciding and with the goldpoints in those two sides coinciding too. (Note: This is the *jigsaw combination move*; we shall speak of *jigging* two triangles together.) If we wish to indicate that two triangles, say T_1 and T_2, can be jigged, we shall write T_1*T_2. These same rules are to apply generally, to polygons.

Definition of names:

Golden tiles: Polygons with sides each marked with a goldpoint will be called *golden tiles*. The abbreviations TGP and SGP will be used for golden tiles made respectively from equilateral triangles and squares. These are abbreviations for 'triangles with gold-points' and 'squares with goldpoints' respectively.

Tile figure: The *tile figure* of a golden tile with one mark per side is the polygon which is obtained by drawing on the tile straight lines with join up all pairs of marks occurring on adjacent sides of the tile.

Jigsaw: When two or more golden tiles are combined by jigsaw combination moves, the result will be called a *jigsaw* and the pattern displayed upon it, resulting from all the tile figures, will be called the *jigsaw pattern*.

Problems:

Many types of problem can be posed about golden tiles and jigsaws. Combinatorial and geometric ones include: How many different (i.e. inequivalent) tiles can be derived from a given polygonal shape? How many different ways can 2 (or 3, or 4 etc.) be combined to form a jigsaw in a plane (or on the surface of a given polyhedral solid)? What kinds of tile figures are there:

---

*[In this paper, only one goldpoint per side will be marked. In general, however, both goldpoints may be marked on any side of a triangle, or polygon. Rules 1 and 2 cover the general case.]

and what are their geometric properties (for example, what are their symmetries)? What kinds of jigsaw patterns can be made with a given set of golden tiles? The list could go on and on.

In this paper we have space only to show a few of the results we have obtained, and those only for TGPs and SGPs. We begin by showing the tile figures for TGPs, and discussing a few of their properties.

3. THE TGPs, AND SOME BASIC PROPERTIES

If one thinks of marking one goldpoint on each side of an equilateral triangle, since there are two different goldpoints on each side, there are 2^3 ways of placing the marks. However, the eight resulting TGPs are not all inequivalent. In fact, there are four equivalence classes, each with two TGPs in them. The set of four inequivalent TGPs is shown in Figure 2.

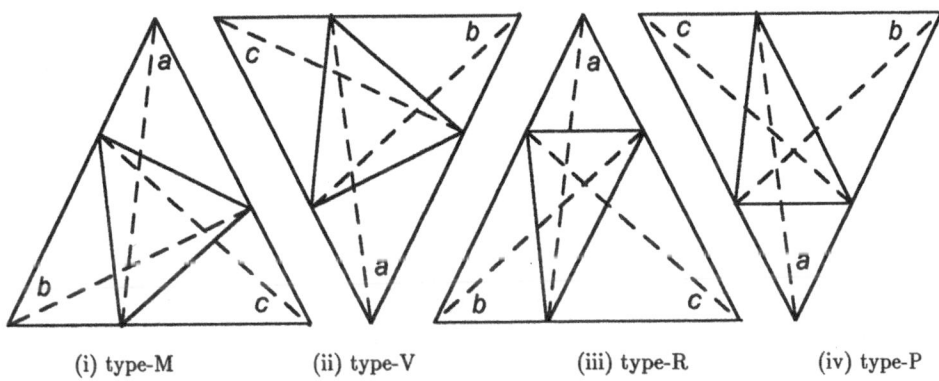

(i) type-M   (ii) type-V   (iii) type-R   (iv) type-P

Figure 2. The four types of TGP

It will be noted that the four tile figures in the TGPs are of two types, namely: an equilateral triangle, and a scalene triangle. The two tiles with equilateral triangles are inequivalent, and likewise are the two with scalene triangles.

Rotations: When studying tile figures and jigsaw patterns, it will sometimes be necessary to specify rotations of TGPs in the plane. We define a positive rotation to be an anti-clockwise motion, and a negative rotation to be a clockwise one. Sometimes the vertices of a golden tile will be labeled, so that rotations and combinations can be explicitly defined.

Turner coloured his original golden tiles Mauve (M), Violet (V), Red (R) and Pink (P); hence the letters that we use to name these TGP types. He also drew dotted lines on them, joining all vertices to the goldpoints on sides opposite them. These added to the geometric complexity of the tile figure, of course: the next diagram shows one of these figures, and a theorem about its properties follows.

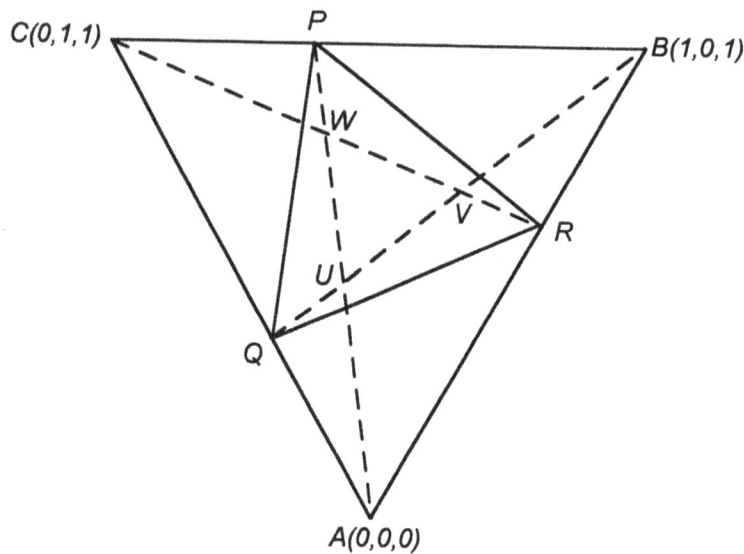

Figure 3. A TGP with full and dotted lines, labeled for Theorem 3

Theorem 3: Let ABC be an equilateral triangle, with P, Q, R goldpoints of the sides, marked such that $\triangle PQR$ is equilateral. Let AP meet BQ in U, and CR in W; and let CR meet BQ in V. Then:

    (i) U, V, W are the mid-points of AP, BQ, CR respectively.

    (ii) U, V, W are goldpoints of AW, BU, CV respectively.

    (iii)   Triangle areas: $\triangle UVW = \frac{1}{4}\triangle PQR = \frac{1}{2\alpha^4}\triangle ABC.$

Proof: The $3D$ coordinate system and lower triangle, explained for Theorem 2, is used again here. We study only ratios, or relative figures, so that the results are general (i.e. independent of the choice of side length ($\sqrt{2}$) for the triangle). We could use well-known vector geometry techniques to find the coordinates of P, Q, R, U, and V, and by these means effect proofs directly. However, we shall use a mixture of Euclidean and vector geometry, to demonstrate interesting ways for studying tile figures.

    (i)   Using $3D$ vector geometry, we find the points P, U and W, thus: $P = (\alpha C + B)/(\alpha + 1) = (1/\alpha^2, 1/\alpha, 1)$; similarly $Q = 1/\alpha^2(0, 1, 1)$. Then $U = \frac{1}{2}(1/\alpha^2, 1/\alpha, 1)$ is found from intersection of lines AP, QB, by standard procedures with coordinates. Hence U is the mid-point of AP, since A is $(0,0,0)$. Similarly, V and W are mid-points of BQ and CR.

(ii) From their coordinates, we see immediately that $W = \alpha U$. Thus $AW : AU = \alpha : 1$, hence U is a goldpoint of AW. Similarly V, W are goldpoints of BU, CV respectively.

(iii) The following argument leads to the ratio of areas of triangles PQR and ABC.

$$\Delta PQR = \Delta ABC - 3\Delta AQR, \quad (\text{since } QPC \equiv PBR \equiv AQR).$$

And $\Delta AQR = (1/\alpha)AQB = (1/\alpha)((1/\alpha^2)\Delta ABC)$.

Therefore $\Delta PQR = (1 - 3/\alpha^3)\Delta ABC = (2/\alpha^4)\Delta ABC$.

Next we find triangle UVW as a ratio of triangle ABC, thus:

$$\Delta UVW = \Delta PQR - 3\Delta QVR.$$

Now $\Delta QVR = (1/2)\Delta QBR$ (V is the mid-point of QB)

$$= (1/2\alpha^2)\Delta AQB = (1/2\alpha^2)(1/\alpha^2)\Delta ABC.$$

Therefore $\Delta UVW = (1/2\alpha^4)\Delta ABC$.

Putting the results for ΔPQR and ΔUVW together, the proof of (iii) is completed. □

It is of interest to note that we can cast part (i) of Theorem 3 in a way similar to that of Theorem 2, where a ray AR was involved. Thus:

Theorem 3'(i): Let ΔABC be equilateral; and let AR_1, BR_2 be rays rotating about A and B respectively, with AD always kept equal to CE. Then when their intersection X is mid-point of AE, or of BD, E is a goldpoint of BC and D is a goldpoint of AC.

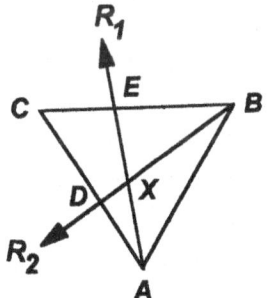

4. FORMING JIGSAWS, AND TILING WITH TGPs

We shall now discuss various results about how the TGPs can be combined using the jigsaw rule. We shall first treat combinations of two tiles: and then give examples of how TGPs can be used to tile the plane. Later several examples of jigsaw tiling of polyhedra surfaces by TGPs will be given. Because of lack of space, we can only give diagrams of these tilings, and not discuss the many aspects of their jigsaw patterns which beg to be studied and described.

We shall, however, begin with a few definitions of concepts which are involved in jigsaw formations, so that the reader will appreciate how rapidly the complexity of these combined objects develops. At every step in the development, new combinatorial problems of increasing difficulty suggest themselves.

Orientation and labeling of TGPs

Any jigsaw move in a plane can be carried out by a rotation, followed by a translation without rotation. In order to specify rotations of a TGP, it is necessary to have a fixed line in the plane (say H, which we shall say is 'horizontal') and refer the TGP to that line in a well-defined manner.

Before saying how the fixed line becomes useful, we shall define an 'n-jigsaw' and a 'connected n-jigsaw', thus:

Definitions: An *n-jigsaw* is a jigsaw composed of n tiles, combined by jigsaw moves. It is a *connected n-jigsaw* if every tile in it has at least one side (or point) coinciding with a side (or point) of another of the n tiles.

Labeling of TGPs: In Figure 2 we showed the four possible TGPs, and identified each by a letter (standing for a colour). We also labeled the three vertices of each tile, using the letter a, b and c, with anti-clockwise alphabetic ordering around the triangle in each case.

It may be noted that we always drew a triangle with one side horizontal (parallel to line H); theorems 1 and 2 below tell us that with any n-jigsaw of TGPs we an always rotate it in the plane until every one of its tiles has a horizontal side. Then, in each tile there will be one vertex either above its horizontal side or below it. Suppose in a particular tile T (where T is one of the four types M, V, R or P) this vertex has label l (where l is a, b, or c). Then we propose the following notation: T_l means that the l-vertex points downwards in T (it is below the horizontal side); whereas $T_{l'}$ means that the l-vertex points upwards in T (it is above the horizontal side). The two diagrams in Figure 4 illustrate this convention. They should convince the reader that the notation used, together with the labelings of TGPs given in Figure 2 above, specifies the orientation of tile T relative to line H precisely.

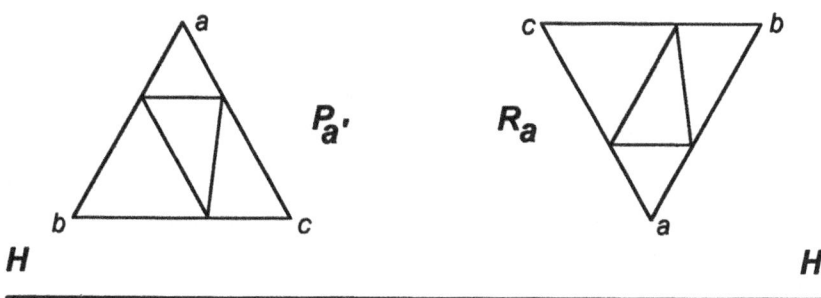

Figure 4. Two TGPs, labeled and oriented relative to line H

Theorem 1: If a tile T_1 has a horizontal side, and T_1*T_2 is a 2-jigsaw, then tile T_2 has a horizontal side.

Proof: If T_2 is jigged to T_1 on the horizontal side of T_1, then T_2 has a horizontal side which coincides with that. If T_2 is jigged to one of T_1's sloping sides, then the 2-jigsaw is in the form of a rhombus: this is a parallelogram, one of whose sides is horizontal and *not* in T_1. Hence this horizontal side is in T_2. Hence proof of Theorem 1.

Theorem 2: Let a connected jigsaw (say J_n) be composed of n TGP tiles. Then if any one tile (say T_1) has a horizontal side, each of the tiles in the n-jigsaw has a horizontal side.

Proof: Proof is easily established by induction, using Theorem 1.

Linear TGP jigsaws

A jigsaw composed of TGPs and which takes one of the following four linear forms will be called a linear TGP jigsaw, and designated an $LTGP$. Its *length* will be the number of TGPs used in its formation.

Linear forms: (i) △ ▽ △···▽   (ii) △ ▽ △···△

                       (iii) ▽ △ ▽···▽   (iv) ▽ △ ▽···△.

Example

The following $LTGP$ of length 4 is composed of one TGP of each type:

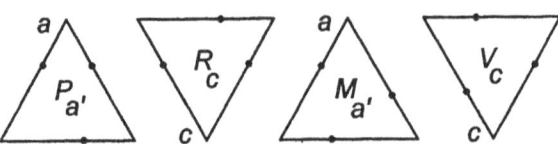

The symbol string $P_{a'} * R_c * M_{a'} * V_c$ gives a proper definition of one of these, and shows that one such actually exists. Moreover, since $V_c * P_{a'}$ is a valid combination, it follows that this $LTGP$ can be extended indefinitely (in both directions). Hence we can say that this 4-jigsaw, of type defined by its symbol string, *tiles the jigsaw line.*

We can now ask the question: Does this $LTGP$ tile the plane? The answer is: Yes. Two of these tiles may be placed end-to-end, or on top of or under one another, in suitably staggered positions. And the process may be continued indefinitely. These remarks could be made precise, using the symbols developed above; but we will leave it to the reader's visual imagination to confirm them.

Jigsaws with two tiles

Having studied single $TGPs$, and linear combinations of them, albeit most briefly, we can look at 2-jigsaws. Two jigged equilateral triangles form a 60° rhombus. We shall say that two of these rhombuses, each formed from two $TGPs$, are equivalent if their overall geometric patterns are identical (can be placed exactly over each other).

We ask how many inequivalent rhombus tiles there are, and the answer is 16. We have not space to give diagrams of all these; we present only two, below, to demonstrate how varieties of geometric patterns are presented on these rhombuses. Under the diagram we give a table which shows how often the various V, M, P and R tiles can be jigged together in pairs, in a total of 16 inequivalent ways.

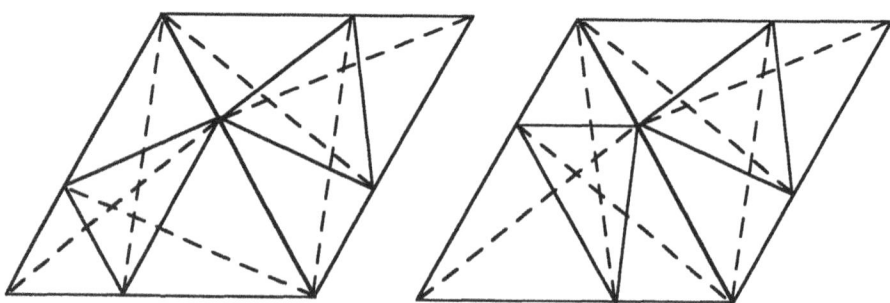

Figure 5. The two (M, P)-rhombuses

No. Pairs	V	M	P	R	Totals
V	0	1	1	2	4
M		0	2	1	3
P			2	5	7
R				2	2
Total					16

Table: Numbers of possible rhombuses, from TGP-pairs

As final comments on tiling the plane, we note that neither the V- nor the M-type tiles combine with themselves in pairs, so they cannot tile the plane. Although there are two types of (R, R) combinations, it can be shown that the R-tiles cannot tile the plane. Similarly, the pink tiles cannot tile the plane. Some of the rhombuses will tile the plane (e.g. $P_b * R_{a'}$ will do so); the (V, P) rhombus will not.

Tiling surfaces of solids

The first-named author of this paper has discovered many examples of tilings of polyhedra. A small selection of her findings are presented below. There is no space to discuss her work on the geometric patterns, and various equivalence classes which she discovered: some of her findings will be given in [1].

Each of the following diagrams is a developed, planar view of a surface tiling, for the solid which is named above it. If these were to be cut out and folded, all the boundary edges could be matched in pairs, with goldpoints coinciding on these edges; and the solid's surface would result.

(A) Two pyramids, each using four TGPs of different types

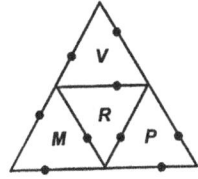 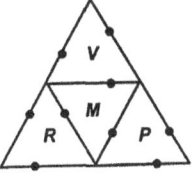

(B) Two pyramids, using two type-P and two type-R TGPs

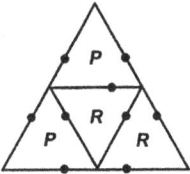 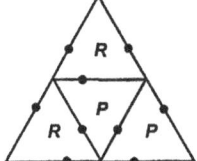

(C) A pyramid using four *TGPs* of each type

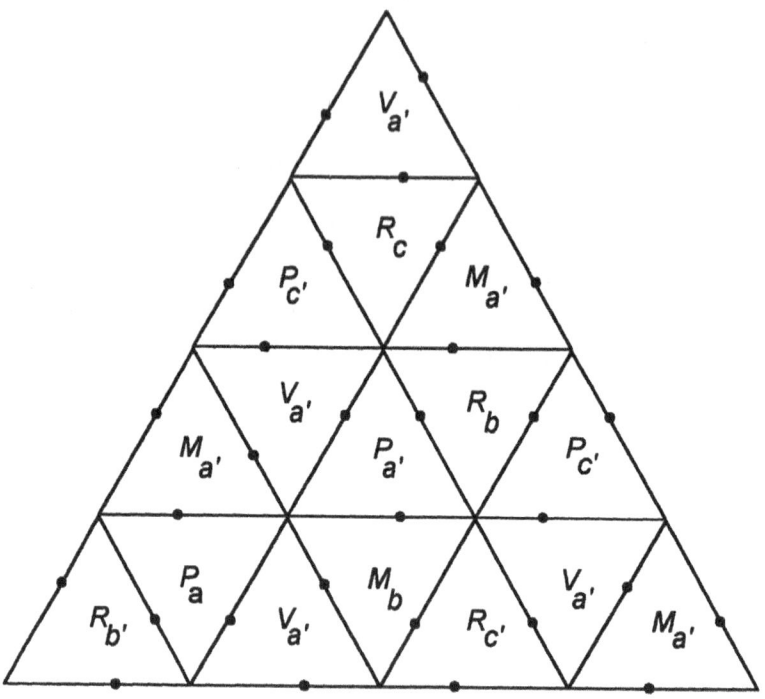

(D) An octahedron using two *TGPs* of each type

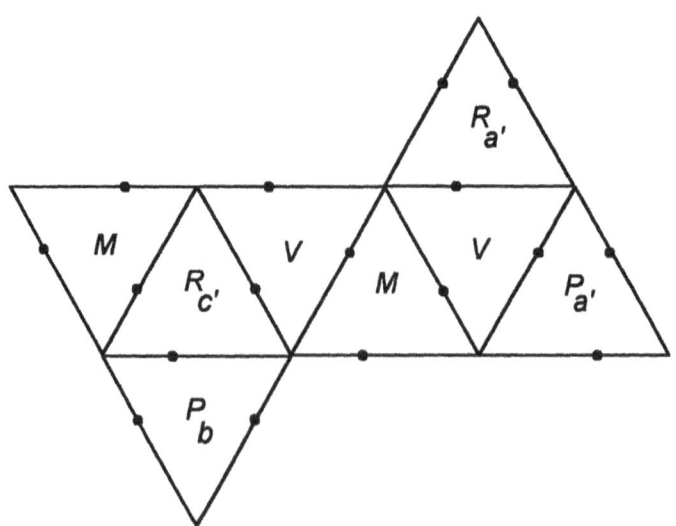

(E) A regular icosahedron using five TGPs of each type.

[N.B. neighbouring faces are everywhere coloured differently.]

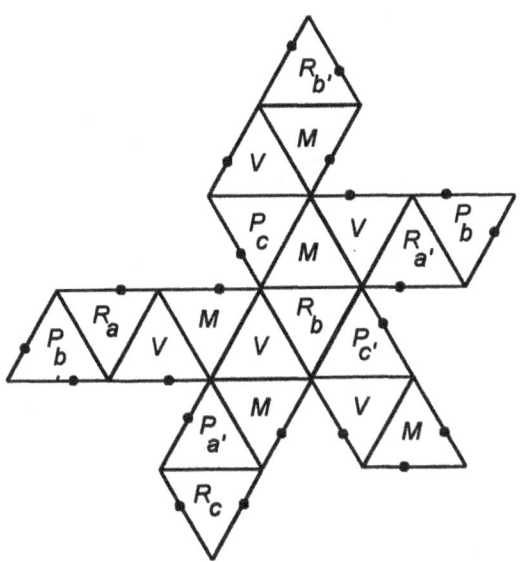

5. ON SQUARES WITH GOLDPOINTS (SGPs)

Vassia Atanassova, the first-named author, extended the idea of golden tiles to squares marked with goldpoints (SGPs), and she has discovered much about them. Not only has she studied some of their combinatorial and geometric problems, but she has also invented intriguing games which make use of them. For example, in [1] is described a kind of chess game, using SGPs.

We have only space here to show the different types of SGP that can be drawn; and to demonstrate a few solutions to the problem of tiling the plane with them.

The six different (inequivalent) SGPs

The four goldpoints of a square may be marked in $2^4 = 16$ ways, since there are two ways to place a goldpoint on each of its sides. Allowing for plane rotations, there are six equivalence classes, which may be described in terms of the tile figures. These figures are: square (2 types), rectangle (1 type), trapezium (1 type) and quadrangle (2 types). A set of distinct representatives of these types is shown in Figure 6. (Note, incidentally, that they are arranged so that they jig-tile a 2×3-rectangle: and observe the centrally placed kite and diamond shapes, which play roles in SGP jigsaw patterns.)

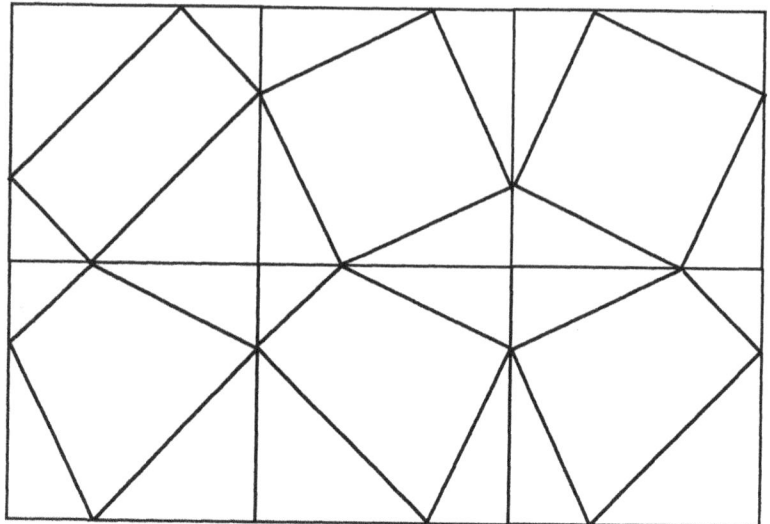

Figure 6. The six possible types of SGP

<u>On tiling the plane</u>

The four diagrams, each of four SGPs arranged in a 2×2 square, which are presented in Figure 7, demonstrate how the plane may be tiled by SGPs in the following ways:

(i)   Using only the rectangle tile;

(ii)   Using only the trapezium tile;

(iii)   Using both types of quadrangle;

(iv)   Using both types of square and both types of quadrangle.

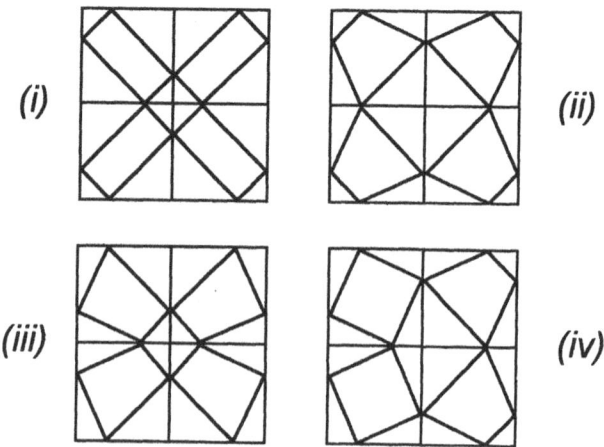

Figure 7. Four ways to tile the plane with SGPs

In each case, it is obvious how each 2×2 jig-tile can be jigged with a copy of itself, on either side, or on top or bottom; and thence how the plane can be tiled by further copies.

There are other solutions for jigging SGPs into 2×2 squares, some of which tile the plane, and some which do not. (N. B. Another plane 2×2 tiling solution may be found on the right side of Figure 6.)

Further work on plane tiling, and tiling of polyhedra surfaces, is described in [1].

6. SUMMARY

The first part of the paper was concerned with two theorems about rhombuses with a line being drawn from a vertex to cut two sides in goldpoints. Then definitions of golden tiles, tile figures and jigsaws were given, followed by a theorem on properties of a type-V tile figure.

The rest of the paper dealt with golden tiles which were either equilateral triangles or squares marked with one goldpoint on each of their sides. A variety of jigsaw problems were described, and solutions given.

The construction of golden tiles, and their applications to jigsaw tiling of plane and solid figures, provides a potentially endless source of fascinating problems in combinatorics and geometry. All results from these are of necessity, in view of the manner of tile markings, directly related to the golden mean.

In his book *The Divine Proportion*, Dover (1970), H.E. Huntley has given many geometric instances of what he calls the 'ubiquity of the golden mean'. Several of them relate to the angle 36° or multiples of that, for example when α arises in the geometry of a pentagram star (p. 28). Our relations of α to the angle 60° can hardly be new, as this kind of trigonometry has been studied at least since the Golden Age of classical Greece. However, we believe that our treatment of them, followed by their use in the study of golden tiles, is new.

REFERENCES

[1]   Atanassov, K. and V., Shannon, A.G. and Turner, J. C. Visual Perspectives on Number Sequences. In preparation, 1998.

[2]   Turner, J. C. and Shannon, A.G. "Introduction to a Fibonacci Geometry". Applications of Fibonacci Numbers, Volume 7. Edited by G. E. Bergum, A. F. Horadam and A. N. Philippou. Kluwer Academic Publishers, Dordrecht, The Netherlands, 1998: pp. 435-448.

AMS Classification Numbers: 11B37, 11B39, 10A35

MULTIVARIATE PASCAL POLYNOMIALS OF ORDER K WITH PROBABILITY APPLICATIONS

Demetrios L. Antzoulakos and Andreas N. Philippou

1. INTRODUCTION

For any fixed positive integer k, let $C_k(n,m)$ be the entry at the intersection of row n ($n \geq 0$) and column m ($m \geq 0$) in the Pascal triangle of order k, viz., T_k. Then

$$C_1(n,0) = 1 \text{ for } n \geq 0 \text{ and } C_1(n,m) = 0 \text{ for } m \geq 1,$$

and for $k \geq 2$, $C_k(0,0) = 1$, $C_k(0,m) = 0$ for $m \geq 1$, and

$$C_k(n,m) = \begin{cases} \sum\limits_{i=0}^{m} C_k(n-1,m-i), & 0 \leq m \leq k-1 \text{ and } n \geq 1 \\ \sum\limits_{i=0}^{k-1} C_k(n-1,m-i), & m \geq k \text{ and } n \geq 1. \end{cases} \tag{1}$$

This particular definition is due to Philippou and Georghiou [22] (see also Turner [26] and Bollinger [6]). Many theoretical properties of the numbers $C_k(n,m)$ have been established by Bollinger [6, 7, 8, 9], and Bollinger and Burchard [10]. For easy reference, some of them are listed below:

$$C_k(n,m) = C_k(n, n(k-1)-m), \; 0 \leq m \leq n(k-1); \tag{2}$$

$$\sum_{m=0}^{n(k-1)} C_k(n,m)t^m = (1+t+t^2+\cdots+t^{k-1})^n; \tag{3}$$

$$C_k(n,m) = \sum_{j=0}^{n} \binom{n}{j} C_{k-1}(j,m-j); \tag{4}$$

This paper is in final form and no version of it will be submitted for publication elsewhere.

27

$$C_k(n_1 + n_2, m) = \sum_{m_1 + m_2 = m} C_k(n_1, m_1) C_k(n_2, m_2); \tag{5}$$

$$C_k(n, m) = \sum_{\substack{n_0 + n_1 + \cdots + n_{k-1} = n \\ n_1 + 2n_2 + \cdots + (k-1)n_{k-1} = m}} \binom{n}{n_0, n_1, \cdots, n_{k-1}}; \tag{6}$$

$$C_k(n, m) = \sum_j (-1)^j \binom{n}{j} \binom{n-1+m-kj}{n-1}. \tag{7}$$

Freund [12] showed that $C_k(n, m)$ is the number of ways of putting m identical objects into n urns so that no urn contains more than $k - 1$ objects. Bollinger [6] discussed the appearance of these numbers in k-in-a-row problems, and Balakrishnan, Balasurbramanian and Viveros [4] showed their role in describing the characteristics of a sampling plan based on the theory of runs. Recently, Balasubramanian, Viveros and Balakrishnan [5], studied special cases of the multiparameter binomial and multiparameter negative binomial distributions of order k, even though they call them extended, and expressed their probability mass functions in terms of these numbers.

In the present paper we introduce and study a generalized Pascal triangle of order k $(k \geq 1)$ to be denoted by $T_k(\alpha_0, \alpha_1, \cdots, \alpha_{k=1}) \equiv T_k(\alpha)$. This is a (left-justified) rectangular array whose rows are indexed by $n = 0, 1, 2, \cdots$ and columns by $m = 0, 1, 2, \cdots$, and its entry at the intersection of row n and column m is a polynomial with real variables $\alpha_0, \alpha_1, \cdots, \alpha_{k-1}$. We denote these entries by $A_k(n, m; \alpha_0, \alpha_1, \cdots, \alpha_{k-1}) \equiv A_k(n, m; \alpha)$ and we call them multivariate Pascal polynomials of order k. In Section 2 the definition of $A_k(n, m; \alpha)$ is given and we derive appropriate analogues of (2)-(6) (see Lemma 2.1 and Theorems 2.1 and 2.2). Also, their relation with the multivariate Fibonacci polynomials of the same order of Philippou and Antzoulakos [19] is established (see Theorem 2.3). In Section 3 we show that the probability mass function of the multiparameter distributions of order k of Philippou [18] can be expressed in terms of $A_k(n, m; \alpha)$ (see Theorem 3.1). As by-products we obtain analogous expressions in terms of $A_k(n, m; \alpha)$ and $C_k(n, m)$ for a number of distributions of order k (see Proposition 3.2 and relations (36)-(45)). Furthermore, a probability result regarding the length of the longest success run in n Bernoulli trials is given in terms of $C_k(n, m)$ (see Proposition 3.1).

2. MULTIVARIATE PASCAL POLYNOMIALS OF ORDER K

In this section we introduce and study a generalized Pascal triangle of order k $(k \geq 1)$ to be denoted by $T_k(\alpha_0, \alpha_1, \cdots, \alpha_{k-1}) \equiv T_k(\alpha)$. The entries of $T_k(\alpha)$ are polynomials with (real) variables $\alpha_0, \alpha_1, \cdots, \alpha_{k-1}$. We denote these entries by $A_k(n, m; \alpha_0, \alpha_1, \cdots, \alpha_{k-1}) \equiv A_k(n, m; \alpha)$

and call them multivariate Pascal polynomials of order k with variables $\alpha_0, \alpha_1, \cdots, \alpha_{k-1}$. These polynomials are defined as follows:

Definition 2.1: For any fixed positive integer k the polynomials $A_k(n, m; \alpha), n, m \geq 0$, are said to be the multivariate Pascal polynomials of order k or the entries in $T_k(\alpha)$, if

$$A_1(n, 0; \alpha_0) = \alpha_0^n, \quad A_1(n, m; \alpha_0) = 0 \text{ for } m \geq 1,$$

and for $k \geq 2$, $A_k(0, 0; \alpha) = 1$, $A_k(0, m; \alpha) = 0$ for $m \geq 1$, and

$$A_k(n, m; \alpha) = \begin{cases} \displaystyle\sum_{i=0}^{m} \alpha_i A_k(n-1, m-i; \alpha), & 0 \leq m \leq k-1 \text{ and } n \geq 1 \\[2mm] \displaystyle\sum_{i=0}^{k-1} \alpha_i A_k(n-1, m-i; \alpha), & m \geq k \text{ and } n \geq 1. \end{cases}$$

As an example, here are the first few rows of $T_3(\alpha_0, \alpha_1, \alpha_2)$:

Rows n	\multicolumn{7}{c}{Columns m}
	0	1	2	3	4	5	6
0	1	0	0	0	0	0	0
1	α_0	α_1	α_2	0	0	0	0
2	α_0^2	$2\alpha_0\alpha_1$	$\alpha_1^2 + 2\alpha_0\alpha_2$	$2\alpha_1\alpha_2$	α_2^2	0	0
3	α_0^3	$3\alpha_0^2\alpha_1$	$3\alpha_0\alpha_1^2 + 3\alpha_0^2\alpha_2$	$6\alpha_0\alpha_1\alpha_2 + \alpha_1^3$	$3\alpha_0\alpha_2^2 + 3\alpha_1^2\alpha_2$	$3\alpha_1\alpha_2^2$	α_2^3

It follows from Definition 2.1 that there are $n(k-1)+1$ nonzero entries in row n $(n \geq 0)$ and

$$A_k(n, \ m; \ 1, \ 1, \cdots, \ 1) = C_k(n, m), \tag{8}$$

by means of (1). Also, it can be easily seen by induction on n that

$$A_k(n, \ m; \ \alpha_0, \ \alpha_1, \cdots, \ \alpha_{k-1}) = A_k(n, \ n \ (k-1) - m; \ \alpha_{k-1}, \ \alpha_{k-2}, \cdots, \ \alpha_0), \tag{9}$$

which reduces to (2) for $a_i = 1$ $(0 \leq i \leq k-1)$ by means of (8). We proceed now to derive the analogue of (3).

Lemma 2.1: Let $A_k(n, m; \alpha)$ be the entries of the n-th row in $T_k(\alpha)$. Then

$$\sum_{m=0}^{n(k-1)} A_k(n, m; \alpha) t^m = (\alpha_0 + \alpha_1 t + \alpha_2 t^2 + \cdots + \alpha_{k-1} t^{k-1})^n, \quad n \geq 0. \tag{10}$$

Proof: Note that (10) is true for $n = 0, 1$. Next, assume that (10) is true for $n = r$. Then

D.L. ANTZOULAKOS AND A.N. PHILIPPOU

$$(\alpha_0 + \alpha_1 t + \alpha_2 t^2 + \cdots + \alpha_{k-1} t^{k-1})^{r+1}$$

$$= (\alpha_0 + \alpha_1 t + \alpha_2 t^2 + \cdots + \alpha_{k-1} t^{k-1}) \sum_{m=0}^{r(k-1)} A_k(r, m; \alpha) t^m.$$

Definition 2.1 implies that the coefficient of t^i, $0 \le i \le (r+1)(k-1)$, in the right-hand side of the equality is $A_k(r+1, i; \alpha)$. Therefore

$$(\alpha_0 + \alpha_1 t + \alpha_2 t^2 + \cdots + \alpha_{k-1} t^{k-1})^{r+1} = \sum_{m=0}^{(r+1)(k-1)} A_k(r+1, m; \alpha) t^m,$$

and this completes the proof of the lemma by induction.

In the following theorem we derive the analogues of (4) and (5).

Theorem 2.1: Let $A_k(n, m; \alpha)$ be the (n, m)-entry in $T_k(\alpha)$. Then

(a) $\quad A_k(n, m; \alpha_0 \alpha_1, \cdots, \alpha_{k-1}) = \sum_{j=0}^{\min\{n, m\}} \binom{n}{j} \alpha_0^{n-j} A_{k-1}(j, m-j; \alpha_1, \alpha_2, \cdots, \alpha_{k-1});$

(b) $\quad A_k(n_1 + n_2, m; \alpha) = \sum_{m_1 + m_2 = m} A_k(n_1, m_1; \alpha) A_k(n_2, m_2; \alpha).$

Proof: Using Lemma 2.1 and the binomial theorem we get

$$\sum_{m=0}^{n(k-1)} A_k(n, m; \alpha) t^m = \left[\alpha_0 + (\alpha_1 t + \alpha_2 t^2 + \cdots + \alpha_{k-1} t^{k-1}) \right]^n$$

$$= \sum_{j=0}^{n} \binom{n}{j} \alpha_0^{n-j} t^j (\alpha_1 + \alpha_2 t + \cdots + \alpha_{k-1} t^{k-2})^j$$

$$= \sum_{j=0}^{n} \binom{n}{j} \alpha_0^{n-j} t^j \sum_{i=0}^{j(k-2)} A_{k-1}(j, i; \alpha_1, \alpha_2, \cdots, \alpha_{k-1}) t^i.$$

The proof of part (a) follows by equating the coefficients of identical powers of t on both sides of the equality. The proof of part (b) follows by the same reasoning since

$$\sum_{m=0}^{(n_1+n_2)(k-1)} A_k(n_1 + n_2, m; \alpha) t^m = \sum_{m_1=0}^{n_1(k-1)} A_k(n_1, m_1; \alpha) t^{m_1} \sum_{m_2=0}^{n_2(k-1)} A_k(n_2, m_2; \alpha) t^{m_2},$$

as Lemma 2.1 implies.

The procedure of bracketing off one term of a multinomial used in the proof of Theorem 2.1(a) can be repeated with the remaining multinomial part. This offers the opportunity to express $A_k(n, m; \alpha)$ as a sum of products of binomial coefficients. For instance, in the case where $k = 3$, we get

$$A_3(n, m; \alpha_0 \alpha_1, \alpha_2) = \sum_{j=SIGE(\frac{m}{2})}^{\min\{n, m\}} \binom{n}{j} \binom{j}{m-j} \alpha_0^{n-j} \alpha_1^{2j-m} \alpha_2^{m-j},$$

where $SIGE(x)$ denotes the smallest integer that is greater than or equal to x, also called the "ceiling" of x.

Theorem 2.2: Let $A_k(n, m; \alpha)$ be the (n, m)-entry in $T_k(\alpha)$. Then

(a) $A_k(n, m; \alpha) = \displaystyle\sum_{\substack{n_0 + n_1 + \cdots + n_{k-1} = n \\ n_1 + 2n_2 + \cdots + (k-1)n_{k-1} = m}} \binom{n}{n_0, n_1, \cdots, n_{k-1}} \prod_{i=0}^{k-1} \alpha_i^{n_i};$

(b) $A_k(n, m; \alpha) = \displaystyle\sum_{n_1 + 2n_2 + \cdots + (k-1)n_{k-1} = m} \binom{n}{n - \sum_{i=1}^{k-1} n_i, n_1, \cdots, n_{k-1}} \alpha_0^{n - \sum_{i=1}^{k-1} n_i} \prod_{i=1}^{k-1} \alpha_i^{n_i}.$

Proof: By the multinomial theorem we get

$$(\alpha_0 + \alpha_1 t + \alpha_2 t^2 + \cdots + \alpha_{k-1} t^{k-1})^n$$

$$= \sum_{n_0 + n_1 + \cdots + n_{k-1} = n} \binom{n}{n_0, n_1, \cdots, n_{k-1}} t^{n_1 + 2n_2 + \cdots + (k-1)n_{k-1}} \prod_{i=0}^{k-1} \alpha_i^{n_i}. \tag{11}$$

Equating the coefficients of identical powers of t in relations (10) and (11), we get

$$A_k(n, m; \alpha) = \sum_{\substack{n_0 + n_1 + \cdots + n_{k-1} = n \\ n_1 + 2n_2 + \cdots + (k-1)n_{k-1} = m}} \binom{n}{n_0, n_1, \cdots, n_{k-1}} \prod_{i=0}^{k-1} \alpha_i^{n_i}. \tag{12}$$

Relation (12) shows part (a) of the theorem. The proof of part (b) then follows by noting that

$$\binom{n}{n - \sum_{i=1}^{k-1} n_i, n_1, \cdots, n_{k-1}} = 0,$$

if $\sum_{i=1}^{k-1} n_i > n$.

We mention that part (a) of Theorem 2.2 is the analogue of (6), while part (b) of Theorem 2.2 for $\alpha_i = 1$ $(0 \le i \le k-1)$ reduces to a new expression of $C_k(n, m)$.

Upon using relations (6) and (12) we get the following identities:

$$A_k(n, m; \alpha, \alpha\beta, \cdots, \alpha\beta^{k-1}) = \alpha^n \beta^m C_k(n, m); \tag{13}$$

$$A_k(n, m; \alpha, \alpha, \cdots, \alpha) = \alpha^n C_k(n, m); \tag{14}$$

$$A_k(n, m; \alpha\beta^{k-1}, \alpha^2\beta^{k-2}, \cdots, \alpha^k) = \alpha^{n+m}\beta^{n(k-1)-m} C_k(n, m); \tag{15}$$

$$A_k(n, m; \alpha^{k-1}, \alpha^{k-2}, \cdots, 1) = \alpha^{n(k-1)-m} C_k(n, m); \tag{16}$$

$$A_{k+1}(n,m;\beta^k,\alpha\beta^{k-1},\alpha^2\beta^{k-2},\cdots,\alpha^k) = \alpha^m\beta^{nk-m}C_{k+1}(n,m). \qquad (17)$$

In the following theorem we relate the multivariate Pascal polynomials of order k ($k \geq 2$) with the multivariate Fibonacci polynomials of the same order of Philippou and Antzoulakos [19], to be denoted here by $H_n^{(k)}(\alpha_0,\alpha_1,\cdots,\alpha_{k-1}) \equiv H_n^{(k)}(\alpha)$. We recall from [19] that

$$\sum_{n=0}^{\infty} H_{n+1}^{(k)}(\alpha)t^n = \left(1 - \sum_{i=0}^{k-1} a_i t^{i+1}\right)^{-1}. \qquad (18)$$

Theorem 2.3: Let $\left\{H_n^{(k)}(\alpha)\right\}_{n=0}^{\infty}$ be the sequence of multivariate Fibonacci polynomials of order k. Then

$$H_{n+1}^{(k)}(\alpha) = \sum_{i=SIGE(\frac{n}{k})}^{n} A_k(i,n-i;\alpha), \quad n \geq 0.$$

Proof: Using (10) and (18) we get

$$\sum_{n=0}^{\infty} H_{n+1}^{(k)}(\alpha)t^n = \sum_{r=0}^{\infty} (a_0 t + a_1 t^2 + \cdots + a_{k-1}t^k)^r$$

$$= \sum_{r=0}^{\infty} t^r (a_0 + a_1 t + \cdots + a_{k-1}t^{k-1})^r$$

$$= \sum_{r=0}^{\infty} \sum_{m=0}^{r(k-1)} A_k(r,m;\alpha)t^{r+m}.$$

Setting $r+m=l$ we have that $0 \leq l \leq \infty$, $SIGE\left(\frac{l}{k}\right) \leq r \leq l$ and $m = l-r$. Therefore

$$\sum_{n=0}^{\infty} H_{n+1}^{(k)}(\alpha)t^n = \sum_{l=0}^{\infty} \sum_{r=SIGE(\frac{l}{k})}^{l} A_k(r,l-r;\alpha)t^l,$$

from which the proof of the theorem follows.

Philippou and Antzoulakos [19] noted that

$$H_n^{(k)}(x^{k-1},x^{k-2},\cdots,1) = f_n^{(k)}(x) \text{ and } H_n^{(k)}(x,x,\cdots,x) = F_n^{(k)}(x), \quad n \geq 0, \qquad (19)$$

where $\{f_n^{(k)}(x)\}_{n=0}^{\infty}$ and $\{F_n^{(k)}(x)\}_{n=0}^{\infty}$ is the sequence of Fibonacci and Fibonacci-type polynomials of order k of Philippou, Georghiou and Philippou [23] and [25], respectively. Theorem 2.3 and relations (14), (16) and (19) imply the following corollary.

Corollary 2.1: Let $\left\{f_n^{(k)}(x)\right\}_{n=0}^{\infty}$ and $\left\{F_n^{(k)}(x)\right\}_{n=0}^{\infty}$ be the sequence of Fibonacci and Fibonacci-type polynomials of order k, respectively. Then for $n \geq 0$

(a) $\quad f_{n+1}^{(k)}(x) = \sum_{i=SIGE(\frac{n}{k})}^{n} x^{ki-n} C_k(i, n-i);$ $\qquad\qquad$ (20)

(b) $\quad F_{n+1}^{(k)}(x) = \sum_{i=SIGE(\frac{n}{k})}^{n} x^{i} C_k(i, n-i).$ $\qquad\qquad$ (21)

Since $SIGE(\frac{n}{k}) \leq i \leq n$ is equivalent with $0 \leq n-i \leq \left[n-\frac{n}{k}\right]$, where $[x]$ denotes the greatest integer in x, and by setting $n-i=j$, we have that (21) reduces to

$$F_{n+1}^{(k)}(x) = \sum_{j=0}^{\left[n-\frac{n}{k}\right]} x^{n-j} C_k(n-j, j), \quad n \geq 0,$$

which is the main result of Philippou and Georghiou [22].

Remark 2.1: Let $P_n^{(r)}(x_1, x_2, \cdots, x_n)$ and $B_{n,r}(x_1, x_2, \cdots)$, respectively, denote the potential polynomials and the partial Bell polynomials (cf. Chapter III in Comtet [11]), and set

$$P_{n;k-1}^{(r)}(x_1, x_2, \cdots, x_{k-1}) \equiv P_n^{(r)}(x_1, x_2, \cdots, x_n), \quad \text{for } x_j = 0, \ j = k, k+1, \cdots, n,$$

and

$$B_{n,r;k}(x_1, x_2, \cdots, x_k) \equiv B_{n,r}(x_1, x_2, \cdots), \quad \text{for } x_j = 0, \ j = k+1, k+2, \cdots.$$

It can be easily deduced then, that

$$A_k(n, m; \alpha_0, \alpha_1, \cdots, \alpha_{k-1}) = \frac{a_0^n}{m!} P_{m;k-1}^{(n)}(x_1, x_2, \cdots, x_{k-1}), \quad x_j = j! a_j / a_0,$$

and

$$A_k(n, m; \alpha_0, \alpha_1, \cdots, \alpha_{k-1}) = \frac{n!}{(m+n)!} B_{m+n,n;k}(x_1, x_2, \cdots, x_k), \quad x_j = j! a_{j-1}.$$

Utilizing the above remark, several properties of the potential and partial Bell polynomials may be expressed in terms of the multivariate Pascal polynomials of order k, which would be useful in studying the distributions of order k. The authors intend to investgate this track in a forthcoming paper.

3. PROBABILITY APPLICATIONS

Consider a die with $k+1$ faces marked $\{0, 1, 2, \cdots, k\}$ and let p_i $(0 \leq p_i < 1, \ 0 \leq i \leq k,$ and $p_0 + p_1 + \cdots + p_k = 1)$ be the turn-up side probabilities. Let X be a random variable denoting the total score in the necessary number of rolls of the die until face marked 0 appears r times. Philippou [18] derived the distribution of X, called it negative binomial distribution of order k with parameters $p_1, p_2 \cdots, p_k$, and denoted it by $NB_k(r; p_1, p_2, \cdots, p_k)$. He also derived this distribution by compounding the extended (or multiparameter) Poisson distribution of order k of Aki [1], which he denoted by $P_k(\lambda_1, \lambda_2, \cdots, \lambda_k)$. He also obtained a logarithmic series distribution of the same order and denoted it by $LS_k(p_1, p_2, \cdots, p_k)$, as the zero truncated limit of the distribution of X. Philippou, Antzoulakos and Tripsiannis [21] called the distribution of the total score in n rolls of the die binomial distribution of order k with parameters $n, p_1, p_2, \cdots, p_k$, and denoted it by $B_k(n; p_1, p_2, \cdots, p_k)$ (see also Panaretos and Xekalaki [15]). Using the symbol " $\sim$ " as shorthand for the phrase "is distributed as", we recall from the above authors that:

$$\text{If } X \sim NB_k(r; p_1, p_2, \cdots, p_k) \text{ then } \sum_{m=0}^{\infty} P(X = m)t^m = p_0^r\left(1 - \sum_{i=1}^{k} p_i t^i\right)^{-r}; \qquad (22)$$

$$\text{If } X \sim LS_k(p_1, p_2, \cdots, p_k) \text{ then } \sum_{m=1}^{\infty} P(X = m)t^m = \frac{1}{\log p_0} \log\left(1 - \sum_{i=1}^{k} p_i t^i\right); \qquad (23)$$

$$\text{If } X \sim P_k(\lambda_1, \lambda_2, \cdots, \lambda_k) \text{ then } \sum_{m=0}^{\infty} P(X = m)t^m = \exp\left[\sum_{i=1}^{k} \lambda_i(t^i - 1)\right]; \qquad (24)$$

$$\text{If } X \sim B_k(n; p_1, p_2, \cdots, p_k) \text{ then } \sum_{m=0}^{nk} P(X = m)t^m = \left(p_0 + \sum_{i=1}^{k} p_i t^i\right)^n. \qquad (25)$$

Now, consider the following transformations of the parameters $p_1, p_2, \cdots, p_k$:

(A) $p_i = P^{i-1}Q \ (1 \le i \le k)$ so that $p_0 = P^k$;

(B) $p_i = Q/k \ (1 \le i \le k)$ so that $p_0 = P$;

(C) $p_i = Q_1$ and $p_i = P_1 P_2 \cdots P_{i-1} Q_i \ (2 \le i \le k)$ so that $p_0 = P_1 P_2 \cdots P_k$;

(D) $p_i = P^i Q^{k-i} \ (1 \le i \le k, \ 0 \le P, \ Q \le 1 \text{ and } \dfrac{Q^{k+1} - P^{k+1}}{Q - P} = 1)$ so that $p_0 = Q^k$.

It was noted by Philippou [18] that

$$NB_k(r; p_1, p_2, \cdots, p_k) \text{ reduces to } NB_{k,I}(r, P) \text{ under } (A); \qquad (26)$$

$$NB_k(r; p_1, p_2, \cdots, p_k) \text{ reduces to } NB_{k,II}(r, P) \text{ under } (B); \qquad (27)$$

$$NB_k(r; p_1, p_2, \cdots, p_k) \text{ reduces to } ENB_k(r; P_1, P_2, \cdots, P_k) \text{ under } (C); \qquad (28)$$

$$LS_k(p_1, p_2, \cdots, p_k) \text{ reduces to } LS_{k,I}(P) \text{ under } (A); \qquad (29)$$

$$LS_k(p_1, p_2, \cdots, p_k) \text{ reduces to } LS_{k,II}(P) \text{ under } (B); \qquad (30)$$

$$LS_k(p_1, p_2, \cdots, p_k) \text{ reduces to } ELS_k(P_1, P_2, \cdots, P_k) \text{ under } (C); \qquad (31)$$

$$P_k(\lambda_1, \lambda_2, \cdots, \lambda_k) \text{ reduces to } P_k(\lambda) \text{ if } \lambda_1 = \lambda_2 = \cdots = \lambda_k = \lambda. \qquad (32)$$

Here and in the sequel, (a) $NB_{k,I}(r, P)$ and $P_k(\lambda)$ denote, respectively, the negative binomial distribution of order k with support $\{0, 1, \cdots\}$ and the Poisson distribution of the same order of Philippou, Georghiou and Philippou [24]; (b) $NB_{k,II}(r, P)$ denotes the compound

Poisson (or negative binomial) distribution of order k of Philippou [16], with $P = \alpha/(\alpha + k)$; (c) $LS_{k,I}(P)$ denotes the logarithmic series distribution of order k of Aki, Kuboki and Hirono [2]; (d) $LS_{k,II}(P)$ denotes the logarithmic series distribution of order k, type II of Philippou [18]; (e) $ENB_k(r; P_1, P_2, \cdots, P_k)$ and $ELS_k(p_1, p_2, \cdots, P_k)$ denote, respectively, the extended negative binomial distribution of order k with support $\{0, 1, \cdots\}$ and the extended logarithmic series distribution of the same order of Aki [1].

Furthermore, we note presently that

$$NB_k(r; p_1, p_2, \cdots, p_k) \text{ reduces to } ENB(k + 1, r, P) \text{ under } (D); \tag{33}$$

$$B_k(n; p_1, p_2, \cdots, p_k) \text{ reduces to } EB(k + 1, n, P) \text{ under } (D); \tag{34}$$

$$B_k(n; p_1, p_2, \cdots, p_k) \text{ reduces to } \tilde{B}_k(n, Q) \text{ under } (B). \tag{35}$$

Here and in the sequel, $ENB(k + 1, r, P)$ and $EB(k + 1, n, P)$ denote, respectively, the distributions of the random variables $Z_r^{(k+1)}$ and $X_n^{(k+1)}$ of Balasubramanian, Viveros and Balakrishnan [5], and $\tilde{B}_k(n, Q)$ denotes a distribution studied by Philippou [17].

Theorem 3.1: Let $A_k(n, m; \alpha)$ be the (n, m)-entry in $T_k(\alpha)$. Then the following hold true.

(a) If $X \sim NB_k(r; p_1, p_2, \cdots, p_k)$, then, for $m \geq 0$

$$P(X = m) = p_0^r \sum_{j=0}^{m} \binom{r + j - 1}{j} A_k(j, m - j; p_1, p_2, \cdots, p_k);$$

(b) If $X \sim LS_k(p_1, p_2, \cdots, p_k)$, then, for $m \geq 1$

$$P(X = m) = (-\log p_0)^{-1} \sum_{j=1}^{m} \frac{1}{j} A_k(j, m - j; p_1, p_2, \cdots, p_k);$$

(c) If $X \sim P_k(\lambda_1, \lambda_2, \cdots, \lambda_k)$, then, for $m \geq 0$

$$P(X = m) = \exp\left(-\sum_{i=1}^{k} \lambda_i\right) \sum_{j=0}^{m} \frac{1}{j!} A_k(j, m - j; \lambda_1, \lambda_2, \cdots, \lambda_k);$$

(d) If $X \sim B_k(n; p_1, p_2, \cdots, p_k)$, then, for $0 \leq m \leq nk$

$$P(X = m) = A_{k+1}(n, m; p_0, p_1, \cdots, p_k).$$

Proof: Let $X \sim NB_k(r; p_1, p_2, \cdots, p_k)$. Using relations (10) and (22) we get

$$\sum_{m=0}^{\infty} P(X = m) t^m = p_0^r \left(1 - \sum_{i=1}^{k} p_i t^i\right)^{-r}$$

$$= p_0^r \sum_{j=0}^{\infty} \binom{r + j - 1}{j} t^j \left(p_1 + p_2 t + \cdots + p_k t^{k-1}\right)^j$$

$$= p_0^r \sum_{j=0}^{\infty} \binom{r + j - 1}{j} t^j \sum_{i=0}^{j(k-1)} A_k(j, i; p_1, p_2, \cdots, p_k) t^i.$$

The proof of part (a) follows by matching the coefficients of identical powers of t on both sides of the equality. The proof of parts (b) and (c) follow by the same reasoning, on the basis of the following: (i) If $X \sim LS_k(p_1, p_2, \cdots, p_k)$, then relations (10) and (23) imply that

$$\sum_{m=1}^{\infty} P(X=m)t^m = (\log p_0)^{-1} \log \left(1 - \sum_{i=1}^{k} p_i t^i \right)$$

$$= (-\log p_0)^{-1} \sum_{j=1}^{\infty} \frac{1}{j} t^j \left(p_1 + p_2 t + \cdots + p_k t^{k-1} \right)^j$$

$$= (-\log p_0)^{-1} \sum_{j=1}^{\infty} \frac{1}{j} t^j \sum_{i=0}^{j(k-1)} A_k(j, i; p_1, p_2, \cdots, p_k) t^i.$$

(ii) If $X \sim P_k(\lambda_1, \lambda_2, \cdots, \lambda_k)$, then relations (10) and (24) imply that

$$\sum_{m=0}^{\infty} P(X=m)t^m = \exp \left(-\sum_{i=1}^{k} \lambda_i \right) \exp \left(\sum_{i=1}^{k} \lambda_i t^i \right)$$

$$= \exp \left(-\sum_{i=1}^{k} \lambda_i \right) \sum_{j=0}^{\infty} \frac{1}{j!} t^j (\lambda_1 + \lambda_2 t + \cdots + \lambda_k t^{k-1})^j$$

$$= \exp \left(-\sum_{i=1}^{k} \lambda_i \right) \sum_{j=0}^{\infty} \frac{1}{j!} t^j \sum_{i=0}^{j(k-1)} A_k(j, i; \lambda_1, \lambda_2, \cdots, \lambda_k) t^i.$$

The proof of part (d) follows directly by relations (10) and (25).

Theorem 3.1 and relations (26)-(35), in conjunction with relations (13)-(17) imply the following expressions:

Let $X \sim NB_{k,I}(r, P)$. Then for $m \geq 0$

$$P(X=m) = P^{kr+m} \sum_{j=0}^{m} \binom{r+j-1}{j} \left(\frac{Q}{P} \right)^j C_k(j, m-j). \tag{36}$$

Let $X \sim NB_{k,II}(r, P)$. Then for $m \geq 0$

$$P(X=m) = P^r \sum_{j=0}^{m} \binom{r+j-1}{j} \left(\frac{Q}{P} \right)^j C_k(j, m-j). \tag{37}$$

Let $X \sim ENB_k(r; P_1, P_2, \cdots, P_k)$. Then for $m \geq 0$

$$P(X=m) = (P_1 P_2 \cdots P_k)^r \sum_{j=0}^{m} \binom{r+j-1}{j} A_k(j, m-j; Q_1, P_1 Q_2, \cdots, P_1 P_2 \cdots P_{k-1} Q_k). \tag{38}$$

Let $X \sim LS_{k,I}(P)$. Then for $m \geq 1$

$$P(X = m) = P^m(-k\log P)^{-1} \sum_{j=1}^{m} \frac{1}{j} \left(\frac{Q}{P}\right)^j C_k(j, m-j). \tag{39}$$

Let $X \sim LS_{k,II}(P)$. Then for $m \geq 1$

$$P(X = m) = (-\log P)^{-1} \sum_{j=1}^{m} \frac{1}{j} \left(\frac{Q}{P}\right)^j C_k(j, m-j). \tag{40}$$

Let $X \sim ELS_k(P_1, P_2, \cdots, P_k)$. Then for $m \geq 1$

$$P(X = m) = [-\log(P_1 P_2 \cdots P_k)]^{-1} \sum_{j=1}^{m} \frac{1}{j} A_k(j, m-j; Q_1, P_1 Q_2, \cdots, P_1 P_2 \cdots P_{k-1} Q_k). \tag{41}$$

Let $X \sim P_k(\lambda)$. Then for $m \geq 0$

$$P(X = m) = \exp(-k\lambda) \sum_{j=0}^{m} \frac{\lambda^j}{j!} C_k(j, m-j). \tag{42}$$

Let $X \sim ENB(k+1, r, P)$. Then for $m \geq 0$

$$P(X = m) = \left(\frac{P}{Q}\right)^m \sum_{j=0}^{m} \binom{r+j-1}{j} Q^{k(j+r)} C_k(j, m-j). \tag{43}$$

Let $X \sim EB(k+1, n, P)$. Then for $0 \leq m \leq nk$

$$P(X = m) = \left(\frac{P}{Q}\right)^m Q^{kn} C_{k+1}(n, m). \tag{44}$$

Let $X \sim \tilde{B}_k(n, Q)$. Then for $0 \leq m \leq nk$

$$P(X = m) = A_{k+1}\left(n, m; P, \frac{Q}{k}, \cdots, \frac{Q}{k}\right). \tag{45}$$

It is noted that relations (43) and (44) are two of the main results of Balasubramanian, Viveros and Balakrishnan [5].

Proposition 3.1: Let S_n, L_n and X_n be random variables denoting, respectively, the number of successes, the length of the longest success run and the outcome of the nth trial in n (≥ 1) independent Bernoulli trials with success probability p ($0 < p < 1$). Then for $0 \leq k \leq r \leq n$

(a) $P(L_n \leq k, S_n = r) = p^r q^{n-r} C_{k+1}(n-r+1, r);$

(b) $P(L_n \le k, S_n = r, X_n = F) = p^r q^{n-r} C_{k+1}(n-r,r).$

Proof: A careful look at the proof of Lemma 3.1 of Philippou and Makri [20] reveals that

$$P(L_n \le k, S_n = r) = p^r q^{n-r} \sum_{i=0}^{k} \sum_{\substack{n_0+n_1+\cdots+n_k = n-r \\ n_1+2n_2+\cdots+kn_k = r-i}} \binom{n}{n_0, n_1, \cdots, n_k}$$

$$= p^r q^{n-r} \sum_{i=0}^{k} C_{k+1}(n-r, r-i), \tag{46}$$

which along with relation (1) shows part (a) of the proposition. Part (b) follows from (46) for $i = 0$.

Part (a) of Proposition 3.1 reveals that the number of binary numbers of length n that have a total of r ones, no k consecutive is given by $C_k(n-r+1, r)$, and it is Theorem 3.3 of Bollinger [6].

In the following proposition we relate the probability mass function of the binomial distribution of order k, $B_k(n, p)$, of Philippou and Makri [20] (see also Hirano [14]), with the entries of the Pascal triangle of order k.

Proposition 3.2: Let X be a random variable distributed as $B_k(n, p)$. Then for $\alpha = n - km$ and $0 \le m \le [n/k]$ we have that

$$P(X = m) = p^n \sum_{y=[\alpha/k]}^{\alpha} \binom{y+m}{m} \left(\frac{q}{p}\right)^y C_k(y+1, \alpha-y).$$

Proof: Godbole [13] showed that

$$P(X = m) = p^{km} \sum_{y=[\alpha/k]}^{\alpha} \binom{y+m}{m} \sum_{i=0}^{k-1} p^i P(L_{r_i} \le k-1, S_{r_i} = r_i - y, X_{r_i} = F), \tag{47}$$

where $r_i = n - km - i$ ($0 \le i \le k-1$). By using part (b) of Proposition 3.1 we have that (47) reduces to

$$P(X = m) = \sum_{y=[\alpha/k]}^{\alpha} \binom{y+m}{m} q^y p^{n-y} \sum_{i=0}^{k-1} C_k(y, \alpha-y-i),$$

which along with (1) establishes the proposition.

Remark 3.1: Using relation (7) to replace $C_k(n, m)$ in Proposition 3.2 and in the expressions (36), (37), (39), (40) and (42), we readily rederive several results of Godbole [13] and Antzoulakos and Philippou [3].

ACKNOWLEDGEMENT

The authors should like to thank an anonymous referee for his specific comments which led to Remark 2.1.

REFERENCES

[1]   Aki, S. "Discrete Distributions of Order k on a Binary Sequence." *Ann. Inst. Statist. Math*, Vol. *37* (1985): pp. 205-224.

[2]   Aki, S., Kuboki, H. and Hirano, K. "On Discrete Distributions of Order k." *Ann. Inst. Statist. Math.*, Vol. *36* (1984): pp. 431-440.

[3]   Antzoulakos, D.L. and Philippou, A.N. "Expressions in Terms of Binomial Coefficients for Some Multivariate Distributions of Order k." Runs and Patterns in Probability. Edited by A.P. Godbole and S.G. Papastavridis. Dordrecht: Kluwer, 1994: pp. 1-14.

[4]   Balakrishnan, N., Balasubramanian, K. and Viveros, R. "On sampling Inspection Plans Based on the Theory of Runs." *The Math. Scientist*, Vol. *18* (1993): pp. 113-126.

[5]   Balasubramanian, K., Viveros, R. and Balakrishnan, N. "Some Discrete Distributions Related to Extended Pascal Triangles." *The Fibonacci Quarterly*, Vol. *35.5* (1995): pp. 415-425.

[6]   Bollinger, R.C. "Fibonacci k-Sequences, Pascal-T Triangles, and k-In-A-Row Problems." *The Fibonacci Quarterly*, Vol. *22.2* (1984): pp. 146-151.

[7]   Bollinger, R.C. "A Note on Pascal-T Triangles, Multinomial Coefficients, and Pascal Pyramids." *The Fibonacci Quarterly*, Vol. *24.2* (1986): pp. 140-144.

[8]   Bollinger, R.C. "The Mann-Shanks Primality Criterion in the Pascal-T Triangle T_3." *The Fibonacci Quarterly*, Vol. *27.3* (1989): pp. 272-275.

[9]   Bollinger, R.C. "Extended Pascal Triangles." *Math. Magazine*, Vol. *66* (1993): pp. 87-94.

[10]   Bollinger, R.C. and Burchard, Ch. L. "Lucas's Theorem and Some Related Results for Extended Pascal Triangles." *Amer. Math. Monthly*, Vol. *97* (1990): pp. 198-204.

[11]   Comtet, L. Advanced Combinatorics. Reidel, Dordrecht: Holland, 1974.

[12]   Freund, J.E. "Restricted Occupancy Theory-A Generalization of Pascal's Triangle." *Amer. Math. Monthly*, Vol. *63* (1956): pp. 20-27.

[13]   Godbole, A.P. "Specific Formulae for Some Success Run Distributions." *Statist. Probab. Lett.*, Vol. *10* (1990): pp. 199-124.

[14]   Hirano, K. "Some Properties of the Distributions of Order k." Fibonacci Numbers and
       Their Applications, Volume 1. Edited by G.E. Bergum, A.F. Horadam and A.N.
       Philippou. Kluwer Academic Publishers, Dordrecht, The Netherlands, 1986: pp. 43-53.

[15]   Panaretos, J. and Xekalaki, E. "On Generalized Binomial and Multinomial Distribution
       and Their Relation to Generalized Poisson Distributions." Ann. Inst. Statist. Math, Vol.
       38 (1986): pp. 223-231.

[16]   Philippou, A.N. "Poisson and Compound Poisson Distributions of Order k and Some of
       their Properties." Zapiski Nauchnykh Seminarov Leningradskogo Otdeleniya
       Matematicheskogo Instituta im V.A. Steklova An SSSR, Vol. 130 (1983): pp. 175-180 (In
       Russian, English summary).

[17]   Philippou, A.N. "Distributions and Fibonacci Polynomials of Order k, Longest Runs,
       and Reliability of Consecutive-k-Out-of-n: F Systems." Fibonacci Numbers and Their
       Applications, Volume 1. Edited by G.E. Bergum, A.F. Horadam, and A.N. Philippou.
       Kluwer Academic Publishers, Dordrecht, The Netherlands, 1986: pp. 203-227.

[18]   Philippou, A.N. "On Multiparameter Distributions of Order k." Ann. Inst. Statist.
       Math., Vol. 40 (1988): pp. 467-475.

[19]   Philippou, A.N. and Antzoulakos, D.L. "Multivariate Fibonacci Polynomials of Order k
       and the Multiparameter Negative Binomial Distribution of the Same Order." Fibonacci
       Numbers and Their Applications, Volume 3. Edited by G.E. Bergum, A.F. Horadam and
       A.N. Philippou. Kluwer Academic Publishers, Dordrecht, The Netherlands, 1990:
       pp. 273-279.

[20]   Philippou, A.N. and Makri, F.S. "Successes, Runs and Longest Runs." Statist. Probab.
       Lett., Vol. 4 (1986): pp. 211-215.

[21]   Philippou, A.N., Antzoulakos, D.L. and Tripsiannis, G.A. "Multivariate Distributions of
       order k, Part II." Statist. Probab. Lett, Vol. 10 (1990): pp. 29-35.

[22]   Philippou, G.N. and Georghiou, C. "Fibonacci-Type Polynomials and Pascal Triangles
       of Order k." Fibonacci Numbers and Their Applications, Volume 1. Edited by G.E.
       Bergum, A.F. Horadam and A.N. Philippou, Dordrecht. Kluwer Academic Publishers,
       Dordrecht, The Netherlands, 1990: pp. 229-233.

[23]   Philippou, A.N., Georghiou, C. and Philippou, G.N. "Fibonacci Polynomials of Order k,
       Multinomial Expansions and Probability." Internat. J. Math. & Math. Sci., Vol. 6.3
       (1983): pp. 545-550.

[24]   Philippou, A.N., Georghiou, C. and Philippou, G.N. "A Generalized Geometric Distribution and Some of Its Properties." *Statist. Probab. Lett.*, Vol. *1* (1983): pp. 171-175.

[25]   Philippou, A.N., Georghiou, C. and Philippou, G.N. "Fibonacci-Type Polynomials of Order k with Probability Applications." *The Fibonacci Quarterly*, Vol. *23.2* (1985): pp. 100-105.

[26]   Turner, S.J. "Probability via the Nth Order Fibonacci-T-Sequence." *The Fibonacci Quarterly*, Vol. *17.1* (1979): pp. 23-28.

AMS Classification Numbers: 11B65, 62E15, 11B39

FIBONACCI PLANES AND SPACES

K.T. Atanassov and A.G. Shannon

1. INTRODUCTION

Two types of Fibonacci plane were introduced in [1]. They were based on the idea of Tirman and Jablinski [4], where the infinite Fibonacci square generalizes a fourth Cartesian quadrant. The purpose of this paper is to describe a third Fibonacci plane and to outline the associated algebra.

Fibonacci planes can be classified as positive, negative and standard. A positive Fibonacci plane has the form in Table 1 [1], whereas the negative and standard Fibonacci planes are shown in Tables 2 and 3.

...	...	...	...	...	...	...	...	...
...	8	5	3	2	1	1	0	...
...	5	3	2	1	1	0	1	...
...	3	2	1	1	0	1	1	...
...	2	1	1	0	1	1	2	...
...	1	1	0	1	1	2	3	...
...	1	0	1	1	2	3	5	...
...	0	1	1	2	3	5	8	...
...	...	...	...	...	...	...	...	...

Table 1: Positive Fibonacci plane

...	...	...	...	...	...	...	...	...
...	−8	−5	−3	−2	−1	−1	0	...
...	−5	−3	−2	−1	−1	0	1	...
...	−3	−2	−1	−1	0	1	1	...
...	−2	−2	−1	0	1	1	2	...
...	−1	−1	0	1	1	2	3	...
...	−1	0	1	1	2	3	5	...
...	0	1	1	2	3	5	8	...
...	...	...	...	...	...	...	...	...

Table 2: Negative Fibonacci plane

This paper is in final form and no version of it will be submitted for publication elsewhere.

43

...	...	...	...	...	...	...	...	...
...	-8	5	-3	2	-1	1	0	...
...	5	-3	2	-1	1	0	1	...
...	-3	2	-1	1	0	1	1	...
...	2	-1	1	0	1	1	2	...
...	-1	1	0	1	1	2	3	...
...	1	0	1	1	2	3	5	...
...	0	1	1	2	3	5	8	...
...	...	...	...	...	...	...	...	...

Table 3: Standard Fibonacci plane

We can see that the diagonal with "0" is the general line in the plane, and the square from [3] is, for example, the marked part of the plane.

2. GENERAL FIBONACCI PLANES

This construction can be extended directly for the plane which corresponds to the generalized Fibonacci sequence with initial values a and b [2]. The construction can also be generalized to a Fibonacci 3-dimensional or n-dimensional space $(n \geq 4)$, so that by analogy with the concept of the Fibonacci plane we can also construct the Fibonacci cubes and Fibonacci spaces. Let a, b, c, d be fixed integers. Following the idea of Lee and Lee [3], we can construct the square in Table 4 which we call a Fibonacci square.

a	b	$a+b$	$a+2b$	...
c	d	$c+d$	$c+2d$	...
$a+c$	$b+d$	$a+b+c+d$	$a+2b+c+2d$	...
$a+2c$	$b+2d$	$a+b+2c+2d$	$a+2b+2c+4d$	...
$2a+3c$	$2b+3d$	$2a+2b+3c+3d$	$2a+4b+3c+6d$	...
...	...	...	...	...

Table 4: Fibonacci square

Clearly every row and every column of this square is a generalized Fibonacci sequence. By analogy, a Tribonacci sequence can be constructed where every row and every column is a Tribonacci sequence as in Table 5.

a	b	c	$a+b+c$
d	e	f	$d+e+f$
g	h	i	$g+h+i$
$a+d+g$	$b+e+h$	$c+f+i$	...

Table 5: Tribonacci square

The Fibonacci cube will have contiguous planes of the form in Figure 1, and so the element of the Fibonacci cube in cell (i,j,k) $(0 \leq i,j,k \leq \infty)$ will have the value $F_{i+j+k-3}$.

$k = 0$

3	2	1
2	1	1
1	1	0

$j = 0$

1	2	3
1	1	2
0	1	1

$i = 0$

3	5	8
2	3	5
1	2	3

Figure 1: Contiguous faces of Fibonacci cube

Obviously every sub-plane in the Fibonacci cube can be interpreted as a Fibonacci square in the sense of [3]. We can now construct a Fibonacci space. In practice, every one of the above forms of the Fibonacci plane can generate a corresponding form for a Fibonacci space. Hence, we can have positive, negative and standard Fibonacci spaces. Furthermore, the above constructions can be generalized in a geometric sense to an n-dimensional Fibonacci space.

3. 3-DIMENSIONAL SPACE

Algebraically, the initial four cells of a generalized 3-dimensional Fibonacci space with elements $H_{i,j,k}$ can be taken as

$$H_{i,j,0} = \begin{array}{|c|c|c|} \hline a & b & \cdot \\ \hline c & d & \cdot \\ \hline \cdot & \cdot & \cdot \\ \hline \end{array} \quad H_{i,0,k} = \begin{array}{|c|c|c|} \hline a & e & \cdot \\ \hline c & g & \cdot \\ \hline \cdot & \cdot & \cdot \\ \hline \end{array} \quad H_{0,j,k} = \begin{array}{|c|c|c|} \hline a & b & \cdot \\ \hline e & f & \cdot \\ \hline \cdot & \cdot & \cdot \\ \hline \end{array}$$

so that in general

$$H_{i,j,k} = F_{i-1}F_{j-1}F_{k-1}a + F_{i-1}F_jF_{k-1}b + F_iF_{j-1}F_{k-1}c +$$

$$F_iF_jF_{k-1}d + F_{i-1}F_{j-1}F_ke + F_{i-1}F_jF_kf + F_iF_{j-1}F_kg$$

since, in each cross-sectional plane, the elements (or faces of each 'atomic' cube) must satisfy the Fibonacci recurrence relation. For instance,

$$H_{i,j,0} = F_{i-1}F_{j-1}a + F_{i-1}F_jb + F_iF_{j-1}c + F_iF_jd,$$

or, for example,

$$H_{3,2,0} = F_2F_1a + F_2F_2b + F_3F_1c + F_3F_2b,$$

$$= a + b + 2c + 2d,$$

as can be seen in Table 4. Furthermore,

$$H_{i,0,0} = F_{i-1}a + F_ic,$$

$$H_{0,j,0} = F_{j-1}a + F_jb,$$

$$H_{0,0,k} = F_{k-1}a + F_ke,$$

which yield the particular cases

$$H_{0,0,0} = a,$$

$$H_{1,0,0} = c, \quad H_{1,1,0} = d,$$

$$H_{0,1,0} = b, \quad H_{0,1,1} = f,$$

$$H_{0,0,1} = e, \quad H_{1,0,1} = g.$$

By way of conclusion we note that ways of carrying out geometric operations in a Fibonacci space are outlined in [5]. Acknowledgement is made of corrections and suggestions of an anonymous reference towards the improvement of this paper.

REFERENCES

[1]   Atanassov, K.T. "A remark on a Fibonacci plane." *Bulletin of Number Theory &* *Related Topics*, Vol. *13*, (1989): pp. 69-71.

[2]   Horadam, A.F. "A generalized Fibonacci sequence." *American Mathematical Monthly*, Vol. *68.5* (1961): pp. 455-459.

[3]   Lee, J.Z. and Lee, J.S. "Some properties of the generalization of the Fibonacci Sequence." *The Fibonacci Quarterly*, Vol. *25.2* (1987): pp. 111-117.

[4]   Tirman, A. and Jablinski, T. Jr. "Identities derived on a Fibonacci multiplication table." *The Fibonacci Quarterly*, Vol. *26.4* (1988): pp. 328-331.

[5]   Turner, J.C. and Shannon, A.G. "Introduction to a Fibonacci geometry." Applications of Fibonacci Numbers, Volume 7. Edited by G.E. Bergum, A.F. Horadam and A.N. Philippou, Kluwer Academic Publishing, Dordrecht, The Netherlands, 1998: pp. 435-448.

AMS Classification Numbers: 11B39, 51M20

THE SMALLEST POSITIVE INTEGER HAVING F_k REPRESENTATIONS AS SUMS OF DISTINCT FIBONACCI NUMBERS

Marjorie Bicknell-Johnson

Let $R(N)$ be the number of representations of the nonnegative integer N as a sum of distinct Fibonacci numbers. Consider the specialized and related sequence 1, 3, 8, 16, 24, $\cdots$, A_n, whose nth term is the least positive integer N such that $n = R(N)$. Fielder [3] has calculated $\{A_n\}$ for $1 \leq n \leq 330$ and for $1 \leq N \leq F_{28} - 1$. This paper exhibits formulas for $\{A_n\}$ such that $n = F_k$, L_k, or $2F_k$.

1. BASIC RECURSION PROPERTIES OF $R(N)$

The most obvious property in a table of $R(N)$ is the palindromic subsequences it contains, beginning and ending with 1, for N in the interval F_n $1 \leq N \leq F_{n+1} - 1$, as stated in Theorem 1. By symmetry, all possible values for $R(N)$ occur for N in the first half of the interval, or for $F_n \leq N \leq F_n + F_{n-1}/2 < F_n + F_{n-2}$. Theorems 1   through 4 restate Theorems 1 and 3, and Lemmas 5 and 8, from [1].

Theorem 1: When $0 \leq M \leq F_{n-1}$,

$$R(F_{n+1} - 1 - M) = R(F_n - 1 + M), \quad n \geq 3. \tag{1}$$

Theorem 2: when $0 \leq M \leq F_{n-1}$,

$$R(F_n + M) = R(F_{n+1} - 2 - M), \quad n \geq 3. \tag{2}$$

Theorem 3: $R(N)$ for the interval $F_n \leq N \leq F_{n+1} - 1$ is given by

$$R(F_n + K) \quad = R(F_{n-2} + K) + R(K), \quad 0 \leq K \leq F_{n-3} - 1; \tag{3}$$

This paper is in final form and no version of it will be submitted for publication elsewhere.

$$R(F_n + K) \quad = 2R(K), \qquad\qquad F_{n-3} \le K \le F_{n-2} - 1; \qquad (4)$$

$$R(F_n + K) \quad = R(F_{n+1} - 2 - K), \qquad F_{n-2} \le K \le F_{n-1} - 1. \qquad (5)$$

Theorem 4: Let $F_n \le N \le F_{n+1} - 2$. Then

$$R(F_{n+1} + N) \qquad = R(N) + R(N - F_n); \qquad\qquad\qquad (6)$$

$$R(F_{n+2k+1} + N) \quad = (k+1)R(N), \quad k \ge 1;$$

$$R(F_{n+2k} + N) \qquad = kR(N) + R(F_{n+1} - 2 - N), \, k \ge 1. \qquad (7)$$

Theorem 5: Given $F_n \le N \le F_{n+1} - 2$,

$$R(F_{n+4} + N) = R(F_{n+2} + N) + R(N), \, n \ge 3. \qquad (8)$$

Proof: Taking $k = 2$ in (7), $R(F_{n+4} + N) = 2R(N) + R(F_{n+1} - 2 - N)$, while $R(F_{n+2} + N) = R(N) + R(F_{n+1} - 2 - N)$ by taking $k = 1$ in (7).   •

2. MAXIMUM VALUES FOR $R(N)$ BY INTERVAL

Theorem 6:

$$\text{If } F_{2k} \le N < F_{2k} + F_{2k-2}, \text{ then } R(N) \le F_{k+1}, \quad k \ge 1. \qquad (9)$$

Proof: Theorem 6 is true for $1 \le k \le 13$ by calculation [3].

Assume that Theorem 6 holds for $k = n - 1$ and $k = n$; that is,

$$\text{when} \quad F_{2n-2} \le N < F_{2n-2} + F_{2n-4}, \qquad R(N) \le F_n; \qquad (10)$$

$$\text{when} \quad F_{2n} \le M < F_{2n} + F_{2n-2}, \qquad R(M) \le F_{n+1}. \qquad (11)$$

We will show that

$$\text{when} \quad F_{2n+2} \le P < F_{2n+2} + F_{2n}, \qquad R(P) \le F_{n+2}, \qquad (12)$$

which is (9) with $k = n + 1$. We can use (10) to derive (13):

$$F_{2n} \le F_{2n+1} - 2 - N < F_{2n} + F_{2n-2}. \qquad (13)$$

Replacing M by $F_{2n+1} - 2 - N$ in (11), $R(F_{2n+1} - 2 - N) \le F_{n+1}$ when $F_{2n-2} \le N < F_{2n-2} + F_{2n-4}$. Next, take $m = 2n - 2$ in (8):

$$R(F_{2n+2} + N) = R(F_{2n} + N) + R(N). \qquad (14)$$

From (10), $R(N) \le F_n$.   Replacing n by $2n$ in (2), when $0 \le N \le F_{2n-1}$, $R(F_{2n+1} - 2 - N) = R(F_{2n} + N)$, so $R(F_{2n} + N) \le F_{n+1}$. Putting these values in (14),

$$R(F_{2n+2} + N) \le F_{n+1} + F_n = F_{n+2}. \qquad (15)$$

Adding F_{2n+2} to each member of inequality (10),

$$F_{2n-2} + F_{2n+2} \leq N + F_{2n+2} \qquad < F_{2n-2} + F_{2n-4} + F_{2n+2};$$

$$F_{2n+2} \leq (N + F_{2n+2}) \qquad < F_{2n+2} + F_{2n}. \qquad (16)$$

Thus, (15) and (16) give (12) with $P = N + F_{2n+2}$, finishing a proof by induction.   •

Theorem 7: If $F_{2k+1} \leq N < F_{2k+1} + F_{2k-1}$, then $R(N) \leq 2F_k$, $k \geq 1$.

Theorem 8: If $F_{2k} \leq N < F_{2k+1}$, then $R(N) \leq F_{k+1}$, $k \geq 1$; if $F_{2k+1} \leq N < F_{2k+2}$, then $R(N) \leq 2F_k$, $k \geq 1$.

Theorem 7 holds for $1 \leq k \leq 13$ by calculation using data from [3] and the proof is similar to that of Theorem 6. Theorem 8 follows from Theorems 6 and 7 by applying (1) and examining endpoints.

Theorem 9: If $R(M) = F_k$, then $M > F_{2k-2}$; if $M < F_{2k-2}$, then $R(M) < F_k$, $k \geq 5$.

Proof: Let $M < F_{2k-2}$. By Theorem 8, if $F_{2k-3} \leq M < F_{2k-2}$, $R(M) \leq 2F_{k-2}$; if $F_{2k-4} \leq M < F_{2k-3}$, $R(M) \leq F_{k-1}$. In either case, $R(M) < F_k$. If $M < F_{2k-4}$, then $R(M) \leq 2F_{k-3} < F_k$ or $R(M) \leq F_{k-2} < F_k$. If $R(M) = F_k$, then we must have $M \geq F_{2k-2}$, but $R(F_{2k-2}) = k-1 < F_k$, making $M > F_{2k-2}$.   •

Theorem 10:

$$F_n^2 \qquad\qquad = F_{2n-1} - F_{n-1}^2; \qquad (17)$$

$$F_{n+1}^2 - 1 \qquad = F_{2n} + F_{n-1}^2 - 1; \qquad (18)$$

$$R(F_{n+1}^2 - 1) \qquad \leq F_{n+1}. \qquad (19)$$

Proof: (17) and (18) are minor variations of (I_{11}) and (I_{10}) given by Hoggatt in [4]. Since $F_{2n} < F_{2n} + F_{n-1}^2 - 1 < F_{2n+1}$, we may apply Theorem 8 to (18) to obtain (19).   •

3. RESULTS REGARDING A_n FOR $n = F_k$, L_k, OR $2F_k$

Theorem 11: $R(F_k^2 - 1) = F_k$, $k \geq 1$. $\qquad\qquad\qquad\qquad\qquad\qquad (20)$

Proof: Combine Equations (6.20) and (6.25) from Carlitz [2].

Theorem 12: $N = F_k^2 - 1$ is the smallest positive integer N such that $R(N) = F_k$, $k \geq 3$.

Proof: Theorem 12 is true for $3 \leq k \leq 14$ by calculation using data from [3]. Let M be an integer such that $M < F_n^2 - 1$ and $R(M) = F_n$ for some $n > 14$. Since $R(M) = F_n$, $M > F_{2n-2}$ by Theorem 9. By applying (18) with $n+1$ replaced by n,

$M < F_n^2 - 1 = F_{2n-2} + F_{n-2}^2 - 1,$ or $M = F_{2n-2} + K$ where

$0 \le K < F_{n-2}^2 - 1 = F_{2n-6} + F_{n-4}^2 - 1,$ so that $R(K) < F_{n-2}$ by Theorem 9. Note also that $R(F_{2n-4} + K) < F_{n-1}$ by Theorem 9. Replacing n by $(2n-2)$ in (3),

$$R(M) = R(F_{2n-2} + K) = R(F_{2n-4} + K) + R(K) < F_{n-1} + F_{n-2} = F_n,$$

a contradiction, since we took $R(M) = F_n$. Thus, Theorem 12. ●

Theorem 13: If $N < F_k^2 - 1,$ the $R(N) < F_k,$ $k \ge 5.$

Theorem 13 is a consequence of the proof of Theorem 12, where we found that $R(M) < F_n$ when $M < F_n^2 - 1.$ We can restate Theorem 11 using (18) as $R(F_{2k} + F_{k-1}^2 - 1) = F_{k+1},$ $k \ge 1;$ compare with $R(N) = L_k$ in Theorem 14.

Theorem 14: $R(F_{2k+1} + F_{k+1}^2 - 1) = L_k,$ $k \ge 2.$ (21)

Proof: Using (18),

$$N = F_{2k+1} + F_{k+1}^2 - 1 = F_{2k+1} + (F_{2k} + F_{k-1}^2 - 1) = F_{2k+1} + K.$$

Apply (6) with $K = N$ and $2k = n$ to write

$$R(F_{2k+1} + K) = R(K) + R(K - F_{2k}) = R(F_{k+1}^2 - 1) + R(F_{k-1}^2 - 1)$$

$$= F_{k+1} + F_{k-1} = L_k,$$

making use of Theorem 11. ●

Theorem 15: $N = F_{2k+1} + F_{k+1}^2 - 1$ is the smallest positive integer such that $R(N) = L_k,$ $k \ge 4.$

Proof: Theorem 15 is true for $4 \le k \le 12$ by calculation using data from [3]. Suppose $R(M) = L_n$ for some $n > 12$ such that $M < F_{2n+1} + F_{n+1}^2 - 1.$ Then $M = F_{2n+1} + (F_{2n} + K)$ where $K < F_{n-1}^2 - 1$ and $R(K) < F_{n-1},$ and $(F_{2n} + K) < F_{n+1}^2 - 1$ so that $R(F_{2n} + K) < F_{n+1}$ by Theorem 13. Using (6), with $N = F_{2n} + K,$ and n replaced by $2n,$

$$R(M) = R(F_{2n} + K) + R(F_{2n} + K - F_{2n}) < F_{n+1} + F_{n-1} < L_n,$$

a contradiction to $R(M) = L_n.$ Thus, N is the smallest positive integer such that $R(N) = L_k,$ finishing Theorem 15. ●

Theorem 16: $R(F_{2n+3} F_{2n} - 2)$ $= 2F_{2n},$ $n \ge 1$ (22)

$R(F_{2n+1} F_{2n+4})$ $= 2F_{2n+1},$ $n \ge 1.$ (23)

Proof: To prove (23), take $n = (m+2)$ in Eq. (6.37) given by Carlitz in [2]. By calculation, (22) is true for $1 \leq n \leq 5$. Assume that (22) holds for all $n \leq k-1$. Derive (24) by replacing n by $2k$ in (I_{11}) as given by Hoggatt [4]:

$$F_{4k+1} = F_{2k}^2 + F_{2k+1}^2 \qquad = F_{2k}(F_{2k+3} - 2F_{2k+1}) + F_{2k+1}(2F_{2k} - F_{2k-2})$$

$$= F_{2k}F_{2k+3} - F_{2k+1}F_{2k-2}$$

which becomes (24) upon reorganization:

$$F_{2k+3}F_{2k} - 2 = F_{4k+1} + F_{2k+1}F_{2k-2} - 2 = F_{4k+1} + K. \tag{24}$$

Notice that, replacing k by $(k-1)$ in (24),

$$K = F_{2k+1}F_{2k-2} - 2 = F_{4k-3} + F_{2k-1}F_{2k-4} - 2. \tag{25}$$

Taking $n = (4k-3)$ and $K = N$ in (8),

$$R(F_{4k+1} + K) = R(F_{4k-1} + K) + R(K). \tag{26}$$

Next, by letting $n = (4k-1)$ in (2),

$$R(F_{4k-1} + K) = R(F_{4k} - 2 - K)$$

$$= R(F_{4k} - F_{2k+1}F_{2k-2})$$

$$= R(F_{2k}(F_{2k+1} + F_{2k-1}) - F_{2k+1}F_{2k-2})$$

$$= R(F_{2k-1}F_{2k+2})$$

$$= 2F_{2k-1} \tag{27}$$

by taking $n = (k-1)$ in (23). $R(K)$ as needed for (26) is

$$R(K) = R(F_{2k+1}F_{2k-2} - 2) = 2F_{2k-2} \tag{28}$$

which is (22) with $n = (k-1)$. Going back to (24) and (26),

$$R(F_{2k+3}F_{2k} - 2) = R(F_{4k-1} + K) + R(K) = 2F_{k-1} + 2F_{2k-2} = 2F_{2k}$$

which is (22) when $n = k$, proving (22) by mathematical induction. $\bullet$

Theorem 17: $R(F_{k+3}F_k - 1 + (-1)^{k+3}) = 2F_k$, $k \geq 1$.   (29)

Proof: Combine (22) and (23) from Theorem 16. $\bullet$

Theorem 18: The smallest positive integer N such that $R(N) = 2F_k$ is given by $N = F_{k+3}F_k - 1 + (-1)^{k+3}$, $k \geq 1$.

Proof: Theorem 18 is true by calculation for $1 \leq k \leq 13$, using data from [3]. Replace $2k$ by k in (24) and (25) to write (30):

$$F_{k+3}F_k = F_{2k+1} + F_{k+1}F_{k-2} \quad = F_{2k+1} + F_{2k-3} + F_{k-1}F_{k-4}$$

$$= F_{2k+1} + F_{2k-3} + F_{2k-7} + F_{k-3}F_{k-6}. \tag{30}$$

For convenience, let $w = -1 + (-1)^{k+3}$ in the rest of this discussion. Suppose that $M < N = F_{k+3}F_k + w$ and $R(M) = 2F_k$ for some $k > 13$. Then $F_{2k+1} < M = F_{2k+1} + K < F_{2k+1} + F_{k+1}F_{k-2} + w$ where $K < F_{k+1}F_{k-2} + w = F_{2k-3} + F_{k-1}F_{k-4} + w$ by (30) so that $R(K) \leq F_{k-1}$ if $F_{2k-4} < K < F_{2k-3}$ by Theorem 8 or $R(K) < F_{k-1}$ if $K < F_{2k-4}$ by Theorem 9. From (30),

$$F_{2k-2} - 2 - K \quad = F_{2k-2} - 2 - (F_{2k-3} + F_{2k-7} + F_{k-3}F_{k-6} + w)$$

$$= F_{2k-5} + F_{2k-8} - F_{k-3}F_{k-6} - 2 - w < F_{2k-4}$$

and $R(F_{2k-2} - 2 - K) < F_{k-1}$ by Theorem 9. Applying (7), with $N = K$ and $n = 2k - 3$,

$$R(M) = R(F_{2k+1} + K) \quad = R(F_{(2k-3)+2(2)} + K)$$

$$= 2R(K) + R(F_{2k-2} - 2 - K) \leq 2F_{k-1} + F_{k-1} < 2F_k,$$

a contradiction to $R(M) = 2F_k$. Thus, $N = F_{k+3}F_k + w$ is the smallest positive integer such that $R(N) = 2F_k$. ●

REFERENCES

[1]   Bicknell-Johnson, M. and Fielder, D.C. "The Number of Representations of N Using distinct Fibonacci Numbers, Counted by Recursive Formulas." Accepted for publication by *The Fibonacci Quarterly*.

[2]   Carlitz, L. "Fibonacci Representations." *The Fibonacci Quarterly*, Vol. *6.4* (1968): pp. 193-220.

[3]   Fielder, D.C. Private communications and tables calculated using "Mathematica."

[4]   Hoggatt, V.E., Jr. Fibonacci and Lucas Numbers. Boston: Houghton Mifflin, 1969.

AMS Classification Numbers: 11B39, 11B37, 11Y55

THE ZECKENDORF-WYTHOFF ARRAY APPLIED TO COUNTING THE NUMBER OF REPRESENTATIONS OF N AS SUMS OF DISTINCT FIBONACCI NUMBERS

Marjorie Bicknell-Johnson

Let $R(N)$ be the number of representations of the non-negative integer N as a sum of distinct Fibonacci numbers. The Zeckendorf representation of N is the unique representation of N as the sum of distinct Fibonacci numbers, using no two consecutive Fibonacci numbers. Recursive relationships for computing $R(N)$ from the Zeckendorf representation of N appear in [2]. C. Kimberling [8] showed that an array of integers sorted by Zeckendorf representations is the same as an array of Wythoff pairs introduced by D. Morrison [10]. W. Lang [9] derives the Zeckendorf representation of an integer from its Wythoff representation and shows the equivalence of the two representations.

Each row in the Zeckendorf-Wythoff array is a Fibonacci sequence. We shall prove that successive values of $R(N)$ for integers N in the odd columns of a row form an arithmetic progression, and $R(N-1)$ is a constant. Then we shall apply properties of Zeckendorf representations and of Wythoff pairs to count $R(N)$.

1. THE ZECKENDORF-WYTHOFF ARRAY WITH ITS ZECKENDORF PROPERTIES

The Zeckendorf-Wythoff array partitions the positive integers into columns, sorted by their Zeckendorf representations. We will say that N begins (ends) with F_k if the largest (smallest) Fibonacci number used in the Zeckendorf representation of N is F_k. Column k in the Zeckendorf-Wythoff array consists of integers N, taken in increasing order, whose Zeckendorf representations end with F_{k+1}. Each row of the array consists of integers N having Zeckendorf representations of the same form; i.e., any two successive entries in the row have the

---

This paper is in final form and no version of it will be submitted for publication elsewhere.

corresponding subscripts used in their Zeckendorf representations differing by one. The 10×10 northwest corner of the Zeckendorf-Wythoff array appears in Table 1.

1	2	3	5	8	13	21	34	55	89
4	7	11	18	29	47	76	123	199	322
6	10	16	26	42	68	110	178	288	466
9	15	24	39	63	102	165	267	432	699
12	20	32	52	84	136	220	356	576	932
14	23	37	60	97	157	254	411	665	1076
17	28	45	73	118	191	309	500	809	1309
19	31	50	81	131	212	343	555	898	1453
22	36	58	94	152	246	398	644	1042	1686
25	41	66	107	173	280	453	733	1186	1919

Table 1: The Zeckendorf-Wythoff Array

Some general properties of the array follow from the Zeckendorf representations themselves. Let $N = w_{ij}$ be the element in the ith row and jth column of the Zeckendorf-Wythoff array. The jth column has F_{j+1} as its first entry, and the Zeckendorf representation of every integer listed in the jth column ends in F_{j+1}; further, the indices of the Fibonacci numbers used in the Zeckendorf representation of $w_{i,j+1}$ are each one more than the corresponding indices in the Zeckendorf representation of w_{ij}. Each row is a generalized Fibonacci sequence, and every positive Fibonacci sequence of integers (with a possible shift of indices) appears as a row in the array [8, 10]. Every positive integer occurs exactly once, and each successive element in the first column is the smallest integer not yet used.

If $F_n \leq N \leq F_{n+1} - 1$ is an element in a given column in the Zeckendorf-Wythoff array, then $N + F_{n+k}$, $k \geq 2$, is an element in the same column, and $R(N)$ satisfies recursions, as proved in Theorem 3 of [2] and restated here as Theorems 1.1 and 1.2, within that column. Eq. (1.1) arises by taking $k = 1$ and $k = 2$ in Eq. (10) of [2].

Theorem 1.1. If the Zeckendorf representation of N begins with F_n, then elements of the column containing N in the Zeckendorf-Wythoff array satisfy

$$R(F_{n+4} + N) = R(F_{n+2} + N) + R(N). \tag{1.1}$$

Theorem 1.2. If N begins with F_n, then

$$R(F_{n+2k-1} + N) = kR(N), \quad k \geq 2. \tag{1.2}$$

Since each row contains a generalized Fibonacci sequence of positive integers, $R(N-1) = R(w_{ij}-1)$ will be a constant for integers N within any given row for j sufficiently large, as proved in [1]. Further, the number of representations of N as sums of distinct Fibonacci numbers can be written from the Zeckendorf representation of N, as proved as Theorem 4 in [2] and restated as Theorem 1.3 below, where $[x]$ is the greatest integer in x.

Theorem 1.3. If the Zeckendorf representation of N ends with F_k, $k \geq 2$, then there exists a constant q,

$$R(N) = R(N-1)R(F_k) - q, \quad 0 \leq q \leq R(N-1), \tag{1.3}$$

where $R(F_k) = [k/2]$ and $R(F_k - 1) = 1$, $k \geq 1$.

From Theorem 1.3, we expect successive values of $R(N)$ for integers N in the odd columns in a given row to form an arithmetic progression if the column number j is sufficiently large. Tables 2 and 3 list $R(N)$ and $R(N-1)$ for N in Table 1, showing $R(N-1)$ a constant for the first ten N in each of the first ten rows of the Zeckendorf-Wythoff array.

1	1	2	2	3	3	4	4	5	5
1	1	3	3	5	5	7	7	9	9
2	2	4	4	6	6	8	8	10	10
2	2	5	5	8	8	11	11	14	14
1	1	4	4	7	7	10	10	13	13
3	3	6	6	9	9	12	12	15	15
2	2	6	6	10	10	14	14	18	18
3	3	6	6	9	9	12	12	15	15
3	3	7	7	11	11	15	15	19	19
2	2	7	7	12	12	17	17	22	22

Table 2: $R(N)$ for N in the Zeckendorf-Wythoff Array

1	1	1	1	1	1	1	1	1	1
2	2	2	2	2	2	2	2	2	2
2	2	2	2	2	2	2	2	2	2
3	3	3	3	3	3	3	3	3	3
3	3	3	3	3	3	3	3	3	3
3	3	3	3	3	3	3	3	3	3
4	4	4	4	4	4	4	4	4	4
3	3	3	3	3	3	3	3	3	3
4	4	4	4	4	4	4	4	4	4
5	5	5	5	5	5	5	5	5	5

Table 3: $R(N-1)$ for N in the Zeckendorf-Wythoff Array

2. THE ZECKENDORF-WYTHOFF ARRAY WITH ITS WYTHOFF PROPERTIES

Next we explore Wythoff properties of the Zeckendorf-Wythoff array, in which every positive integer appears exactly once as does every Wythoff pair. We let $\alpha = (1 + \sqrt{5})/2$ be the golden ratio, and let $[x]$ be the greatest integer function. The Wythoff pairs are pairs of numbers $([n\alpha], [n\alpha^2])$ closely related to the Fibonacci numbers and which give the winning positions in Wythoff's game (see [4] or [7], for example). It is known [10] that every natural number n is of the form $[n\alpha]$ or $[n\alpha^2]$ but not both. A common notation for a Wythoff pair is (a_n, b_n), and the first forty Wythoff pairs and some of their properties are given in [7]. The subscript notation leads to subscripted subscripts which are difficult to read. Therefore, let us define

$$a(n) = a_n = [n\alpha], \quad b(n) = b_n = [n\alpha^2] \tag{2.1}$$

and we take $aa(n) = a(a(n))$ to mean "a with the subscript a_n." Table 4 gives a very short table of Wythoff pairs.

n	1	2	3	4	5	6	7	8	9	10
a_n	1	3	4	6	8	9	11	12	14	16
b_n	2	5	7	10	13	15	18	20	23	26

Table 4: The first ten Wythoff Pairs

Note that $a_7 = a(7) = 11$, which also can be written $ab(3)$ since $b_3 = 7$. We can write any element of the Zeckendorf-Wythoff array in terms of Wythoff pairs. For example, the Zeckendorf representation of the numbers in the 5th column of the Zeckendorf-Wythoff array all end with F_6 and are of the form $abba(m)$, where m is the row number. Some useful properties from [7] include

$$n + a(n) = b(n), \quad a(n) + b(n) = ab(n), \quad aa(n) + 1 = b(n). \qquad (2.2)$$

Also, $a(n)$ is the least integer not yet used in building Table 4.

The first column in the Zeckendorf-Wythoff array is given by

$$w_{m1} = aa(m) = a(m) + m - 1 = [[m\alpha]\alpha]. \qquad (2.3)$$

Notice that, since each row is a generalized Fibonacci sequence, one can obtain the row number by successively subtracting terms until reaching $\cdots$, $aa(m)$, $a(m)$, $(m-1)$. This can also be done when the Zeckendorf representation of an element N in the mth row is known, since $aa(m)$ ends in F_2 and N ends in F_{m+1}, and the terms of the Zeckendorf representation of N will each have a subscript which is $(m-1)$ more than the corresponding subscript in the Zeckendorf representation of $aa(m)$. Further, since $aa(m) = [\alpha a(m)]$, the mth row contains a generalized Fibonacci sequence $\{H_n\}$, $H_{n+2} = H_{n+1} + H_n$ for $H_2 = aa(m)$, $H_1 = a(m)$ such that $H_2 = [\alpha H_1]$ so that $R(H_n - 1)$ is a constant for $n \geq 2$, as proved in [1]. Thus, $R(N-1)$ is a constant for all N in any given row in the Zeckendorf-Wythoff array. That means that successive values of $R(N)$ for all N in that row in odd (or in even) columns form an arithmetic progression, and we can write Theorems 2.1, 2.2, and 2.3. We let $N = w_{mk}$, the element in the mth row, kth column in the Zeckendorf-Wythoff array, and apply Theorem 1.3, noting that N ends with F_{k+1} and $q = R(w_{m1} - 1) - R(w_{m1})$.

Theorem 2.1. $R(w_{mk} - 1) = R(w_{m1} - 1)$.

Theorem 2.2. $R(w_{mk}) = R(w_{m1} - 1)[(k-1)/2] + R(w_{m1})$.

Theorem 2.3. $R(w_{m, 2n-1}) = R(w_{m, 2n})$, $\quad n = 1, 2, \cdots$.

Corollary 2.3.1. Let A_n be the least positive integer N such that $R(N) = n$, so that $\{A_1, A_2, A_3, \cdots, \} = \{1, 3, 8, 16, 24, \cdots\}$. Then A_n ends with F_{2j}, $j \geq 2$, $n \geq 2$.

We note that A_n cannot end with an odd-subscripted Fibonacci number, since $R(N)$ for N in an even column of the array equals $R(M)$ for a smaller integer M in the preceding column. The first appearance of a particular value for $R(N)$ occurs as $R(aa(m) - 1)$ for the integer $aa(m)$ in the first column. It is interesting that the only $A_n < 12000$ whose Zeckendorf representation ends with F_8 is $A_{61} = 7218$, and all others in that range [5] end with F_4 or F_6.

We return to Wythoff properties of the Zeckendorf-Wythoff array. As proved in [6], Wythoff pairs are related to Fibonacci numbers by

$$a(F_{2k}) = F_{2k+1} - 1, \quad \text{and} \quad a(F_{2k+1}) = F_{2k+2};$$
$$b(F_{2k}) = F_{2k+2} - 1, \quad \text{and} \quad b(F_{2k+1}) = F_{2k+3}. \tag{2.4}$$

Successive entries in the mth row of the Zeckendorf-Wythoff array are given by $aa(m)$, $ba(m)$, $aba(m)$, $bba(m)$, $abba(m)$, $bbba(m)$, $\cdots$,

$$w_{m,2k-1} = ab^{k-1}a(m), \qquad w_{m,2k} = bb^{k-1}a(m), \quad k \geq 1. \tag{2.5}$$

Returning to Theorem 2.2 and using the Wythoff pairs notation,

$$R(ab^{k-1}a(m)) = (k-1)R(aa(m) - 1) + R(aa(m)), \quad k \geq 1. \tag{2.6}$$

3. IDENTITIES FOR $R(N)$ WITHIN THE ZECKENDORF-WYTHOFF ARRAY

With sequences which are Fibonacci sequences, it is very easy to use a spreadsheet program such as *Excel* to make large tables of the Zeckendorf-Wythoff array and to extract interesting rows.

$aa(m)$	$ba(m)$	$aba(m)$	$bba(m)$	$abba(m)$	$bbba(m)$
1	2	3	5	8	13
9	15	24	39	63	102
64	104	168	272	440	712
441	714	1155	1869	3024	4893
3025	4895	7920	12815	20735	33550
20736	33552	54288	87840	142128	229968
142129	229970	372099	602069	974168	1576237
974169	15766239	2550408	4126647	6677055	10803702
6677056	10803704	17480760	28284464	45765224	74049688
N: F_{2k}^2	$F_{2k}F_{2k+1}$	$F_{2k+1}^2 - 1$	$F_{2k}F_{2k+3}$	$F_{2k+2}^2 - 1$	$F_{2k}F_{2k+5}$

Table 5

Table 5 has rows extracted from the Zeckendorf-Wythoff array, where each row number is the square of an odd-subscripted Fibonacci number. The form of each column is listed below the array, which gives rows for $m = F_{2k-1}^2$, $k = 1, 2, \cdots, 9$. Apparently,

$$aa(F_{2k-1}^2) = F_{2k}^2;$$
$$ba(F_{2k-1}^2) = F_{2k}F_{2k+1};$$
$$aba(F_{2k-1}^2) = F_{2k+1}^2 - 1. \tag{2.7}$$

Of particular interest are the columns headed $aba(m)$ and $abba(m)$, since every entry in those columns is a member of the sequence $\{A_n\}$, where A_n is the smallest positive integer such that $R(N) = n$, for the particularly pleasing special case $n = F_r$.

$R(aa(m))$	$R(ba(m))$	$R(aba(m))$	$R(bba(m))$	$R(abba(m))$	$R(bbba(m))$
1	1	2	2	3	3
2	2	5	5	8	8
5	5	13	13	21	21
13	13	34	34	55	55
34	34	89	89	144	144
89	89	233	233	377	377
233	233	610	610	987	987
610	610	1597	1597	2584	2584
1597	1597	4181	4181	6765	6765
F_{2k-1}	F_{2k-1}	F_{2k+1}	F_{2k+1}	F_{2k+2}	F_{2k+2}

Table 6: $R(N)$ for Table 5

Table 6 list $R(N)$ for the entries of Table 5, with the form of each column given beneath it. Taking N in successive rows of Table 5, $R(N-1) = 1, 3, 8, 21, \cdots, F_{2k}$. Comparing tables, it appears that

$$R(aa(F_{2k-1}^2)) = R(F_{2k}^2) = F_{2k-1} = R(F_{2k}F_{2k+1}); \qquad (2.8)$$

$$R(aba(F_{2k-1}^2)) = R(F_{2k+1}^2 - 1) = F_{2k+1} = R(F_{2k}F_{2k+3}); \qquad (2.9)$$

$$R(abba(F_{2k-1}^2)) = R(F_{2k+2}^2 - 1) = F_{2k+2} = R(F_{2k}F_{2k+5}). \qquad (2.10)$$

Lastly, computing $R(N-1)$ for entries in table 5,

$$R(aa(F_{2k-1}^2)-1) = R(aba(F_{2k-1}^2)-1) = R(abba(F_{2k-1}^2)-1) = F_{2k},$$

which leads to

$$R(F_{2k}^2 - 1) = R(F_{2k}F_{2k+1} - 1) = R(F_{2k+1}^2 - 2) = R(F_{2k+2}^2 - 2) = F_{2k}. \qquad (2.11)$$

Identities (2.8) through (2.11) can be proved by mathematical induction. Carlitz [3] gives $R(F_k^2 - 1) = F_k$, as well as the identities listed in (2.11). Except for the relationships to Tables 5 and 6, all of the identities above can be derived from identities stated in [3].

REFERENCES

[1]   Bicknell-Johnson, M. "A Note on a Representation Conjecture by Hoggatt."
       Applications of Fibonacci Numbers. Volume 7. Edited by G.E. Bergum, A.F. Horadam
       and A.N. Philippou. Kluwer Academic Publishers, Dordrecht, The Netherlands, 1997:
       pp. 39-42.

[2]   Bicknell-Johnson, M. and Fielder, D.C. "The Number of Representations of N Using
       Distinct Fibonacci Numbers, Counted by Recursive Formulas." *The Fibonacci Quarterly.*
       To appear.

[3]   Carlitz, L. "Fibonacci Representations." *The Fibonacci Quarterly*, Vol. *6.4* (1968): pp.
       193-220.

[4]   Coxeter, H.S.M. "The Golden Section, Phyllotaxis, and Wythoff's Game." *Scripta
       Mathematica*, Vol. *19* (1953): pp. 135-143.

[5]   Fielder, D.C. Private communication and tables calculated using "Mathematica".

[6]   Hoggatt, V.E. Jr., and Bicknell-Johnson, M. "Representations of Integers in Terms of
       Greatest Integer Functions and the Golden Section Ratio." *The Fibonacci Quarterly*,
       Vol. *17.4* (1979): pp. 306-317.

[7]   Hoggatt, V.E., Jr., Bicknell-Johnson, M. and Sarsfield, R. "A Generalization of
       Wythoff's Game." *The Fibonacci Quarterly*, Vol. *17.3* (1979): pp. 198-211.

[8]   Kimberling, C. "The Zeckendorf Array Equals the Wythoff Array." *The Fibonacci
       Quarterly*, Vol. *33.1* (1995): pp. 3-8.

[9]   Lang, W. "The Wythoff and the Zeckendorf Representations of Numbers are
       Equivalent." Applications of Fibonacci Numbers. Volume 6. Edited by G.E. Bergum,
       A.F. Horadam and A.N. Philippou. Kluwer Academic Publishers, Dordrecht, The
       Netherlands, 1996: pp. 321-337.

[10]   Morrison, D.R. "A Stolarsky Array of Wythoff Pairs." A Collection of Manuscripts
       Related to the Fibonacci Sequence. Santa Clara, California: The Fibonacci Association,
       1980: pp. 134-136.

AMS Classification Numbers: 11B39, 11B37, 11Y55

COMPOSING WITH SEQUENCES: ...BUT IS IT ART?

John A. Biles

INTRODUCTION

Algorithmic composition dates to around 1026, when Guido d'Arezzo used vowels in the text of a choral piece to determine the pitches in the melody. In the 1400's Guillaume Dufay experimented with "formal processes" and even composed a piece using the golden mean. Mozart, whose parlor game *Musikalisches Würfelspiele* allows players to compose minuets with the aid of a pair of dice, is a slightly more recent example of a composer who was a least intrigued by the idea of algorithmic composition. In the twentieth century, serial composers have openly employed algorithmic techniques, and many composers, for example Cage and Xenakis, have composed aleatoric or "chance" music using random number generators [1]. Part of the appeal of algorithmic techniques is the possibility that the composer can focus on the essence of a piece, its form and deep structure, while the algorithm takes care of surface details like the actual notes. Clearly not all composers have such a carefree attitude about choosing "the actural notes," but the use of algorithms as a composer's assistant is gaining acceptance in many musical circles.

Algorithmic techniques can be categorized in several dimensions. One important dimension is stochastic vs. deterministic. Mozart's dice music is an example of a stochastic technique; a random process controls some aspect of the selection of melodic material, and the repeated invocation of the algorithm will yield different results. Deterministic techniques, on the other hand, yield the same sequence of events on repeated invocations, assuming the parameters controlling the process are the same. The use of Fibonacci sequences falls into the deterministic category [3].

This paper is in final form and no version of it will be submitted for publication elsewhere.

Another dimension of algorithmic techniques is how comprehensively the algorithm is used. Some composers use algorithmic techniques only for generating low-level details, like pitches and/or lengths of specific notes, within tight constraints that they set. These composers determine the deep structure of the piece, make most of the larger-scale decisions, and use the algorithm only to generate the surface structure. Other composers may use algorithms that generate sequences that exhibit their own deep structure, which can allow the algorithm a more comprehensive role in creating the composition. The composer still makes plenty of compositional decisions, but these decisions are more collaborative in nature and often serve to emphasize the deep structure inherent in the sequence produced by the algorithm.

Fibonacci sequences can be used comprehensively due to the important property of self-similarity. Many composers in the last 20 years have pointed to the self-similarity of fractals as a justification for using fractal generators to compose music [4]. The contention is that the scale invariance of embedded structures commonly found in fractal sequences provides an "automatic" deep structure, which can serve to provide the underlying form for a composition based on such a sequence. This certainly is an appealing notion because music is widely recognized to be hierarchical in nature, and the recursive nature of scale-invariant structures certainly provides a hierarchy, at least in theory.

Unfortunately, theory doesn't always make it into practice. The problem with much fractal music is that the self-similar deep structure of the sequence often does not translate readily to a musically interesting deep structure. The structures that are so obvious when looking at a plot of a fractal sequence often go unnoticed when the sequence is heard. This is likely due to the temporal nature of music. An image can be viewed all at once, in its entirety. Viewers of an image can shift their focus from "the big picture" to the tiniest of details whenever and as often as they choose. A piece of music, on the other hand, must be listened to from beginning to end, at the appropriate tempo, without lingering over one portion or ignoring another. Certainly the human perceptual system allows the listener to focus attention on different levels from specific sounds to the overall "texture" of the piece; however, we simply cannot hear a piece of music "all at once" the way we can view an image, and we cannot skip around to arbitrary sections of a piece during its performance.

This means that the deep structure of music must unfold linearly over time, which puts pressure on the composer to make the form of a piece apparent. Furthermore, the composer must present the form of the piece while walking the fine line between expectation and surprise. Listeners tend to form mental models of a piece as they hear it. This model sets up harmonic, rhythmic and melodic expectations of what will happen in the piece. Tapping one's foot to the beat of a tune manifests a common rhythmic expectation. Hearing a V chord resolve to a I chord meets a harmonic expectation. Hearing the melody end on the tonic note meets a melodic expectation.

Surprise happens when expectations are not met. Shifting rhythms, dissonant chords, and angular melodies are examples of surprise. The composer must meet enough expectations to make the piece accessible to the audience, while providing enough surprises to make it interesting. A piece with too few surprises is usually termed "boring," while one with too many surprises is usually greeted by comments like, "That's not music--it's noise!". Clearly different listeners bring different musical experiences, attitudes, and knowledge to a given listening experience, and they can form very different mental models while listening to a given piece. In an effort to help listeners form expectations that can be met, the author sometimes provides the audience with a graphical representation of the underlying sequence for a piece. This image serves a function similar to that of the score that a veteran symphony patron often brings to a concert.

The role of the composer, then, is multifaceted, even when composing with Fibonacci sequences. The composer must convey enough of the deep structure of the sequence, meet enough expectations, spring enough surprises, and in general engage enough listeners to make the piece a success. To explore the algorithmic composer's role, we will describe the development of a modest piece, called PGA-1, which is based on a Fibonacci sequence. The development will proceed from a definition of the sequence, its visualization as a graph, and a description of the properties that make it an appealing choice as a compositional device, to the choices made by the composer in evolving the piece from a simple auralization of the sequence to a (hopeful) successful composition.

The conference presentation for this paper includes sound samples that obviously cannot be included in the medium of a written paper. However, those sound samples will be made available on the Internet from the author's home page, http://www.it.rit.edu/~jab/.

THE SEQUENCE

The primary sequence used for PGA-1 was suggested by Peter G. Anderson, hence the rather unimaginative use of his initials in the piece's title. The sequence has been referred to as the Fibonacci partition function, v_j, which counts the number of ways a non-negative integer j can be represented as a sum of distinct Fibonacci numbers (excluding F_1) [2]. The sequence is defined as the coefficients:

$$v_0, \ v_1, \ \cdots, \ v_{F_{N+2}-2}$$

such that

$$\sum_{j=0}^{F_{N+2}-2} v_j x^j = \prod_{k=2}^{N} \left(1 + x^{F_k}\right)$$

where F_n is a Fibonacci number and

$$F_{N+2} - 2 = F_2 + F_3 + \cdots + F_N$$

is an easily derived identity. A plot of this sequence over its first 1291 members appears in Figure 1.

This sequence has several interesting properties that make it appealing as a compositional device. First, it is self-similar. As can be seen in Figure 1, there is a structure, bounded by 1's, that is repeated and elaborated as the sequence progresses. Incidentally, the portion of the sequence used for PGA-1 ends at the midpoint of the 14th elaboration of this structure.

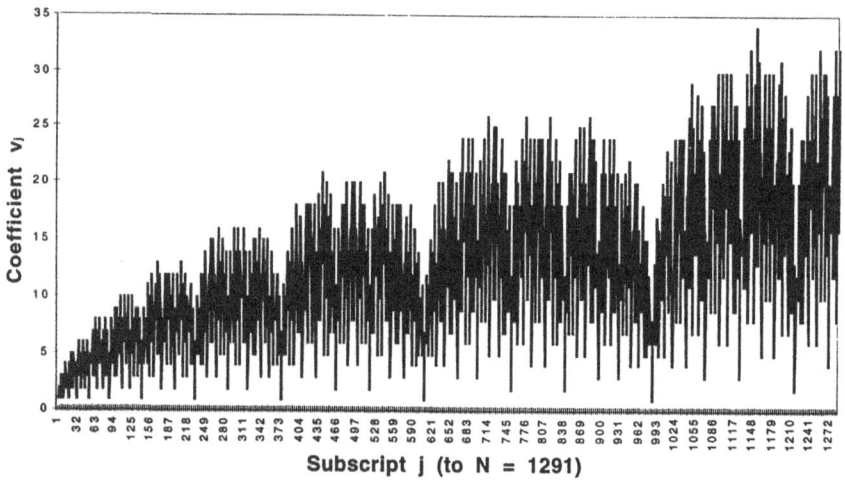

Figure 1. Visualization of Primary Sequence Used for PGA-1

Another interesting property of the sequence is that the repeated structure is symmetric about its midpoint, a fact that is distorted somewhat by the inadequate resolution of Figure 1. Symmetric structures result in a compositional technique called retrograde motion, where a melodic motif is developed by playing it in reverse. The elaborated structure in our sequence divides itself into three sub-structures, the middle one being symmetric (a palindrome) and the right one being the retrograde of the left one. This can be seen clearly in the last complete elaboration, which happens to be the 13th elaboration and runs from element 609 to 986.

The lengths of successive elaborations of the repeated structure are also interesting. The beginning of the sequence, with bars alternating above and below the first six elaborations, which include the first 33 elements, is:

1 1̄ 2 1̄ 2 2 1̄ 3 2 2 3 1̄ 3 3 2 4 2 3 3 1̄ 4 3 3 5 2 4 4 2 5 3 3 4 1̄

The endpoints of the elaborations are shared 1's. If each elaboration counts only one of its two endpoints, the lengths of the successive elaborations form the Fibonacci sequence. Within each elaboration, the lengths of the substructures are similarly related. The three largest substructures in each elaboration, which also share endpoints (1-2, 2-2, 2-1), have Fibonacci lengths if only one endpoint is counted. Furthermore, each substructure can be broken down successively in the same way to the limit of resolution (substructure lengths of 2).

Finally, one can see each elaboration in every successor elaboration, only at a larger scale. For example, the sixth elaboration above is clearly visible as the "low points" of the 13th elaboration (the last complete one) as seen in Figure 1, with the 5's and the outer 4's doubled. Each elaboration, then, retains all the previous elaborations and adds detail with higher values. If the numbers are mapped to pitches, this means that the new details will be heard as higher notes and the established structures as lower notes. Clearly, there is a great deal of structure in this sequence to exploit.

A variant of this sequence was also used in developing PGA-1. This secondary sequence is defined identically to the primary one except that F_1 is also included, which means there are two ones to work with in creating the partitions. In other words, k in the power series starts at 1 instead of 2, and j in the summation runs from 0 to the sum of F_1 to F_N, not F_2 to F_N. A plot of this sequence appears in Figure 2. The secondary sequence shares the major structural properties of the primary sequence, but its elaborations do not descend all the way back to 1 at each endpoint, and its details ascend to higher numbers (note that the vertical scale is 0-60 for Figure 2 versus 0-35 for Figure 1). The secondary sequence was used in PGA-1 to provide counterpoint (a counter-melody) midway through the piece, and to provide velocity information for the primary sequence throughout the piece. Its use will be detailed in the discussion of "The Piece" below.

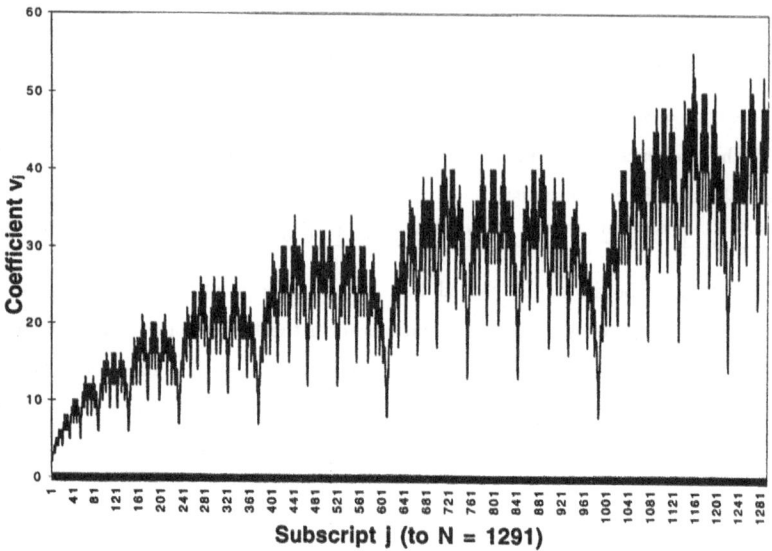

Figure 2. Visualization of Secondary Sequence used for PGA-1

One goal of the algorithmic composer is to make mathematical self-similarity musically meaningful. It is seldom musically satisfying to simply auralize a sequence, but that is a good place to start exploring the compositional potential of a sequence. Such explorations require tools for hearing a sequence. The primary tool used for PGA-1 was a program written by the author, originally for "playing" sequences of stock quotions. Hence, the program is called "Dow" by the author and has been used to generate a series of experiments he shamelessly calls "stock arrangements." Since the Dow program has not been documented elsewhere, the next section describes its capabilities and limitations to set the stage for explaining its use in realizing PGA-1.

THE TOOL

The Dow program reads one or more sequences of numbers from data files and plays each sequence on a separate channel of a MIDI synthesizer. The MIDI (Musical Instrument Digital Interface) standard has emerged as an accessible format for storing and playing back musical information. Note level information is represented by note-on and note-off events. A note-on event contains a pitch (note on a standard keyboard) and a velocity (how hard the key is struck). A note-off event is analogous to releasing a key. The duration of a note is the latency between a note-on event for a given key and the succeeding note-off event. This latency is controlled by the computer that is driving the synthesizer.

The Dow program reads two data elements for each note-level event, a pitch and a velocity. These values are stored in arrays, and the maximum and minimum pitch and velocity are noted. These maxima and minima are used to scale the pitch and velocity data to ranges set by the composer. The scaled pitch data points are mapped to actual notes by using them as indices into a table of 49 pitches comprising a few octaves of a musical scale selected by the composer. For example, the table for a chromatic scale, which uses all 12 tones in each of four octaves, stores a different pitch in each location in the table. Major and minor scales, which use only six or seven notes per octave, store the same pitch in consecutive table locations so that the resulting scale still covers the same four-octave range as the chromatic scale.

If consecutive pitch data points happen to map to the same actual pitch, they will be treated as one note that is held through the duration of those events, which provides some degree of rhythm. The chromatic scale, then, will sound rhythmically "busier" than a major or minor scale because there are more pitches to choose from and there is less chance that consecutive data elements will map to the same pitch.

A piece can consist of up to 16 tracks, each of which is driven by a separate sequence, allowing for counterpoint when multiple sequences are used. Each sequence is played on a separate MIDI channel, and each channel can have its own settings for a host of parameters like tempo, timbre (instrument), loudness, pan (stereo location), key signature, register (octave), and musical scale. The musical scales are particularly important because they control the piece's tonality, or lack thereof. Some of the available scales are traditional ones, like major, minor, whole tone, and chromatic. Other scales are what the author calls progressive, meaning that the key signature for the scale progresses as the scale ascends into higher octaves. We have developed three progressive scales, major, minor and pentatonic, to use with long-term sequences that "start low and end high," and they provide a subtle progression (probably too subtle) among related keys over the course of the piece.

For example, the progressive major scale is based on a hexatonic major scale (six notes per octave, skipping the fourth scale degree). Each half octave of the scale is a sequence of three notes, each a major second apart. The half octaves are separated by a minor third, which occurs where the missing fourth would be. This proceeds for nearly four octaves. The C progressive major scale, then, would be:

C D E   G A B   D E F#   A B C#   E F# G#   B C# D#   F# G# A#

with extra space separating the seven half octaves. Notice that any consecutive pair of half octaves forms a hexatonic major scale, and that 11 of the set of 12 possible pitch classes appear in the entire scale, although in any given octave, only 6 pitches appear. This scale is very consonant, partly because dissonant intervals (like the minor second, tri-tone, augmented fifth, and flatted ninth) do not occur. This means that random notes played simultaneously will sound "nice" to most ears.

An important capability of the Dow program is the ability to (re)set any parameters at arbitrary times in the piece. This allows the composer to synchronize changes in tempo, tonality, instruments, etc., with structural elements of the sequence, and it provides a rich set of tools with which the composer can cross the bridge from auralization to composition. The remainder of the paper will discuss the compositional implications of various parameters and describe the choices made for PGA-1.

THE PIECE

The following is a portion of the description of PGA-1 written for the program of a computer music recital at which the piece was played:

> PGA-1 begins with a very slow, sparse statement of the simplest elaboration of its form in the low register, emanating from a distance at the far left. The form repeats several times, each time somewhat longer, faster and more elaborate, as it approaches the listener. When the sound reaches front and center, the form is stated in a major tonality. It then splits into two counter-melodies, which separate gradually to the left and right as the tempo continues to accelerate and the tonality evolves through minor and pentatonic scales. This continues until the resulting textures become atonal and somewhat tense. Finally, the tension is released by returning to a major tonality for the final elaboration of the form.

This musical overview is intended for an audience about to hear the piece. For a compositional overview of the piece, we refer to Table 1, which summarizes some of the parameters manipulated and serves as a score of sorts. We will discuss the compositional role of each parameter in the table and explain the setting used in PGA-1.

Table 1. Parameter Map for PGA-1

Each row of Table 1 represents a section of PGA-1. The sections were defined to coincide with the major structures of the sequence as discussed above. We have identified the sections by elaboration numbers (the Elab column in Table 1). The Leng column gives the length of each elaboration as the number of elements including only one endpoint. By the 12th elaboration, the sections become lengthy enough to warrant splitting them, giving us sections 12a, 12b, 13a and 13b. The splits were made between the second and third subsection in the 12th elaboration and between the first and second subsections in the 13th elaboration. Section 14 is only the first half of that elaboration and was not split further.

The Tempo column shows the gradual acceleration from a very slow 10 beats per minute (BPM) to a moderately fast 180 BPM, where a beat is defined by the Dow program to be two events (two eighth notes). In general, tempi below about 15 BPM (30 events per minute) are so slow that the listener cannot perceive much in the way of melodic content; each note is usually perceived in isolation rather than as part of a phrase. At tempi from about 30 to maybe 300 BPM, the events tend to be perceived as melodic fragments or phrases made up of distinct notes. This is the range of tempi that we tend to associate with traditional music. Tempi beyond about 300 BPM begin to be perceived as textures, with individual notes no longer distinguishable.

The gradual acceleration in PGA-1 was designed so that the length of time for each elaboration was increased at roughly half the rate at which the number of notes for each elaboration increased. This relationship holds for roughly the first half of the sections, after which the acceleration is more gradual. As the Time column in Table 1 shows, the result is a gradual increase in the amount of time for each section up to the 12th elaboration, where the elaborations split into two sections. After numerous experiments with different tempi, the tempo of 170 BPM was deemed aesthetically pleasing for listening to the 13th elaboration, so the tempi for the intervening sections were scaled to arrive at that point smoothly.

The choice of scale(s) is a critical decision in hearing a sequence. In the initial auralization of a sequence, we usually use the four-octave chromatic scale described above, which maps the elements of the sequence to a range of 49 different notes, the maximum level of detail available in the Dow program. One problem with using the same scale for an entire piece, though, is that there usually will be no changes in tonality. A chromatic scale is, by definition, atonal (no tonal center), and while this is the most "accurate" mapping of a sequence to pitches, many listeners regard the result as less accessible because harmonic and melodic expectations are difficult to form. Listening successfully to atonal music takes practice, and many listeners aren't willing to make that effort.

Nonetheless, we used the four-octave chromatic scale for the first eight elaborations in PGA-1. This insured that the individual elements of the sequence would be heard as distinct notes and that the only held notes would be repeated numbers in the sequence and not different numbers that "rounded" to the same pitch. By the ninth elaboration, the sequence becomes interesting enough to serve as a viable melody, and we make a transition to the previously described progressive major scale. This releases some of the tension built up by the slowly accelerating atonal material of the first eight elaborations. Since the progressive major scale has only 21 different pitches for the 49 table elements, the number of held notes increases as well, making the result less active rhythmically.

In the 10th elaboration, the secondary sequence joins the fray to provide counterpoint (a counter melody). The tonality stays the same in order to make the new sequence's entrance more subtle. In the sections derived from the 11th, 12th and 13th elaborations, the tonality shifts to related keys, using progressive minor and petatonic scales, culminating in a return to the chromatic scale for the last part of the 13th elaboration. This atonal section is fairly active, which induces a degree of tension. That tension is released in the beginning of the 14th elaboration by once again returning to the progressive major scale for the remainder of the piece.

The piece finally ends on a high note, so to speak, in the center of the 14th elaboration.

As a brief illustration of how the elements of the sequence map to pitches, Figure 3 shows four "measures" from elaboration 13b, specifically $j = 897$-928. This section was chosen because the author finds it the most compelling moment in the piece. The elements from the primary and secondary sequences are listed in groups of eight to correspond to the eight eighth notes in each measure. The bold elements form palindromes, with the italic bold elements denoting the center points of the palindromes.

Secondary Sequence (upper stave)							
17	18	27	24	33	30	27	36
27	33	39	30	36	27	24	36
30	*36*	*36*	30	36	24	27	36
30	39	33	27	39	27	30	33

Primary Sequence (lower stave)							
3	15	12	12	21	9	18	18
9	24	15	15	21	6	18	18
12	*24*	12	18	18	6	21	15
15	24	9	18	18	0	21	12

Figure 3. Elements 897-928 with secondary sequence on upper stave, primary on lower

Three other MIDI parameters, velocity (loudness), pan (left-right stereo location), and reverb (echo) are manipulated in the piece, the first algorithmically, and the other two directly by the composer. The velocity for each event was determined by using one sequence as the velocity settings for the other. In other words, the velocities for the primary sequence were derived from the secondary sequence, and the velocities for the secondary sequence were derived from the primary sequence. This served to accentuate the elaboration/section boundaries and provided a gradual increase in loudness over the course of the entire piece. In addition, since the two sequences are not totally aligned at an element-to-element level, this provides subtle changes in loudness that are not totally correlated with changes in pitch. In other words, it sounded nice!

The pan and reverb parameters were manipulated to achieve the movement described in the musical overview above. In Table 1, a pan value of 0 means completely left, 127 means completely right, and 64 is in the center. A reverb value of 0 is no reverb effect (totally dry), 127 is maximum (totally wet), and 40 is a typical default value. By starting the piece far left, very wet, and relatively quiet, the sound appears to be off in the distance. By gradually shifting to the center, reducing the reverb, and getting louder, the sound seems to move closer to the listener. When the secondary sequence makes its entrance, the two parts begin relatively close together and spread farther apart. Multiple sounds that come from the same location tend to be perceived as a single voice, even if different instruments are being played. By separating the two voices somewhat gradually, they emerge as different parts, which adds to the complexity and the tension.

The identity of voices brings us to the last major parameter manipulated in PGA-1, the timbres (instruments) used to realize the sequence. Modern MIDI synthesizers have hundreds of preset instruments, ranging from traditional acoustic instruments to classic electronic sounds, with all manner of combinations in between. For example, the tone generator used in realizing PGA-1 boasts 792 different instruments from which to choose, clearly a large and varied palette of timbres. In dealing with rapid sequences of short notes, which is what the sequences for this piece map to, it is advisable to use timbres with crisp attacks and relatively rapid releases like percussive or plucked instruments. Timbres with gradual attacks and releases lead to indistinct or muddy textures, which may be desirable in some situations but would obscure individual notes. Therefore, we decided to use instruments with crisp attacks, and after experimenting with bell-like sounds, we selected a family of plucked instruments for PGA-1.

As the primary instrument, we chose an instrument called HarpVox, which is a harp sound overlayed with a subtle vocal sustain. The instrument sounds reasonable throughout the three octave range need for the piece, and the vocal sustained layer fills the space between notes and provides a gradual release, which at fast tempi results in a chord-like effect as new notes begin while old notes release. To add variety in the later sections, other plucked instruments were used, including pizzicato strings and kalimba (thumb piano). In the final two sections, the piece returns to the HarpVox. The trick in selecting timbres is to find instruments that complement each other when heard concurrently, while providing transitions that offer an appropriate level of surprise. The somewhat limited variation in timbres chosen for this piece fits our desired aesthetic, which is gradual, subtle changes rather than abrupt shifts.

So there you have it--an alleged piece of music derived from Fibonacci-related sequences! Hopefully, PGA-1 transcends mere auralization and succeeds as a composition for many listeners. However, a paper is not the preferred way to experience a piece of music; it must be heard. The conference presentation of this paper included a "live performance" of PGA-1, but for readers of the proceedings, a sound file can be downloaded from the author's web site at http://www.it.rit.edu/~jab. We invite comments!

REFERENCES

[1]   Roads, Curtis. The Computer Music Tutorial. MIT Press, 1996.

[2]   Robbins, Neville. "Fibonacci Partitions." *The Fibonacci Quarterly*, Vol. *34* (1996): pp. 306-313.

[3]   Schroeder, Manfred. Fractals, Chaos, Power Laws: Minutes from an Infinite Paradise. W.H. Freeman and Company, 1991.

[4]   Schulz, Claus-Dieter. *The Fractal Music Project* web site. Viewed most recently in April, 1998. URL is http://www-ks.rus.uni-stuttgart.de/people/schulz/fmusic/.

AMS Classification Numbers: 11B39

INVARIANTS FOR LINEAR RECURRENCES

Mihai Caragiu and William Webb

1. INTRODUCTION

There is no homogeneous polynomial of degree two, $\phi(x,y) = ax^2 + bxy + cy^2$ for which the classical Fibonacci sequence $\{F_n\}$ satisfies an identity of the form

$$\phi(F_n, F_{n-1}) = constant.$$

However, the related sequence of integers $\{0, 1, 3, 8, 21, 55, \cdots\}$ defined by $u_0 = 0$, $u_1 = 1$ and $u_{n+2} = 3u_{n+1} - u_n$ for $n \geq 0$ satisfies the identity

$$u_n^2 - 3u_n u_{n-1} + u_{n-1}^2 - 1.$$

This leads naturally to the following question: how can we determine which homogeneous polynomial identities, if any, are satisfied by a given n-th order linear recurrence? We will treat this question in the following more general setting.

Let F be a field and $A \in M_{n \times n}(F)$ be a $n \times n$ matrix with coefficients in F. Then A introduces a linear transformation of F^n (which we identify with the set of $n \times 1$ column matrices with entries in F) sending a vector $X \in F^n$ into

$$Y = AX. \tag{1}$$

We may view (1) as a discrete dynamics on F^n. An interesting case is that in which $A \in GL(n, F)$, where A induces a linear automorphism of F^n so that (1) defines actually a $\mathbb{Z}$-action. A natural way to look at this dynamical system is to study the linear recurrences

---

This paper is in final form and no version of it will be submitted for publication elsewhere.

75

induced by (1). Thus, any initial vector $X = X_0 \in F^n$ defines a linear recurring sequence $\{X_k\}_{k \geq 0}$ satisfying

$$X_{k+1} = AX_k. \tag{2}$$

In a sense (2) represents a 'movement' in F^n, in which the time variable k is discrete, while X_k is the position of a moving point at the moment k. A legitimate problem arising in this situation is that of finding invariants (or conservation laws, one may say) for such a movement. In this paper we will be concerned with homogeneous polynomials taking a constant value along any orbit of the discrete dynamical system induced by A. As we will see this involves the notion of symmetric product of matrices.

2. INVARIANTS AND SYMMETRIC POWERS

Definition 1: An invariant k-form form A is a homogeneous polynomial ϕ of degree $k \geq 1$ such that for any vector $X = [x_1, \cdots, x_n]^t \in F^n$ we have

$$\phi(y_1, \cdots, y_n) = \phi(x_1, \cdots, x_n) \tag{3}$$

where $Y = [y_1, \cdots, y_n]^t = AX$.

The following theorem characterizes the number of linearly independent invariant k-forms for A (in the case in which F is an infinite field) and the proof shows how these k-forms can be explicitly calculated.

Theorem 1: If F is infinite, the number of linearly independent invariant k-forms for the linear recurring sequences defined by A equals the dimension of the kernel of $[S^k(A)]^t - I_N$ where $S^k(A)$ is the k-th symmetric power of A and $N = \binom{n+k-1}{k}$. Equivalently, the number of linearly independent k-forms for A equals

$$\binom{n+k-1}{k} - R$$

where R is the rank of $S^k(A) - I_N$.

Proof: First, suppose ϕ is linear:

$$\phi(x_1, \cdots, x_n) = h_1 x_1 + \cdots + h_n x_n. \tag{4}$$

The linear invariant (4) will be identified with the vector

$$H = [h_1, \cdots, h_n]^t \in F^n$$

so that the condition (3) is equivalent to

$$H^t X = H^t A X$$

holding true for any $X \in F^n$. Then, basic linear algebra leads to a simple description of the invariants of degree one. Namely, H is an invariant of degree one if and only if $H^t = H^t A$, if and only if

$$A^t H = H. \tag{5}$$

Therefore, H is an eigenvector of A^t corresponding to the eigenvalue 1. Consequently, the number of linearly independent linear invariants for a linear recurrence is the dimension of the kernel of $A^t - I_n \in M_{n \times n}(F)$. Notice that it is possible to have no linear invariants except the zero form.

Now suppose ϕ is homogeneous of degree $k > 1$.

With $A = [a_{ij}]$, we have the following (linear) relationship between the sets of variables $\{x_1, \cdots, x_n\}$ and $\{y_1, \cdots, y_n\}$:

$$y_1 = a_{11}x_1 + a_{12}x_2 + \cdots + a_{1n}x_n$$

$$y_2 = a_{21}x_1 + a_{22}x_2 + \cdots + a_{2n}x_n$$

$$\cdots\cdots\cdots\cdots\cdots\cdots\cdots\cdots\cdots\cdots$$

$$y_n = a_{n1}x_1 + a_{n2}x_2 + \cdots + a_{nn}x_n.$$

We remark first that the above relations imply the existence of a linear relationship between the set of $\binom{n+k-1}{k}$ monomials

$$x_1^{\alpha_1} x_2^{\alpha_2} \cdots x_n^{\alpha_n} \tag{6}$$

of degree k (i.e., $\alpha_1 + \cdots + \alpha_n = k$) which we agree to arrange in a natural lexicographic order (so that x_1^k comes first while x_n^k comes last) and the set of $\binom{n+k-1}{k}$ monomials

$$y_1^{\alpha_1} y_2^{\alpha_2} \cdots y_n^{\alpha_n} \tag{7}$$

of degree k arranged in lexicographic order. For example, if $n = k = 2$ one has

$$y_1^2 = a_{11}^2 x_1^2 + 2a_{11}a_{12}x_1 x_2 + a_{12}^2 x_2^2$$

$$y_1 y_2 = a_{11}a_{21}x_1^2 + (a_{11}a_{22} + a_{12}a_{21})x_1 x_2 + a_{12}a_{22}x_2^2$$

$$y_2^2 = a_{21}^2 x_1^2 + 2a_{21}a_{22}x_1 x_2 + a_{22}^2 x_2^2.$$

In the general case the matrix which makes the transition from (6) to (7) is the k-th symmetric power $S^k(A)$ of the matrix A, which is a square matrix of order $\binom{n+k-1}{k}$. For basic properties of the symmetric powers, see [2], p. 94-97. We notice that $S^k(A)$ is the symmetric analogue of the Kronecker power $A^{\times k}$ (which is a square matrix of order n^k). Indeed, assuming the indeterminates $x_1, \cdots, x_n$ are 'noncommutative entities' (which commute with the elements from F though), and that $y_1, \cdots, y_n$ are such that

$$y_i = \sum_{j=0}^{n} a_{ij} x_j, \quad i = 1, \cdots, n, \quad a_{ij} \in F$$

then the transition from the system of n^k (noncommutative) monomials of the form $x_{i_1} x_{i_2} \cdots x_{i_k}$, $1 \le i_j \le n$ (arranged in lexicographic order) to the system consisting of the products $y_{i_1} y_{i_2} \cdots y_{i_k}$, $1 \le i_j \le n$ is performed by the Kronecker product $A^{\times k}$.

Therefore, the linear transformation (1) induces in a natural way a linear transformation of the system of monomials (6), viewed as a point in the space $F^{\binom{n+k-1}{k}}$. The transition matrix is, in this later case, the symmetric power $S^k(A)$.

Any linear invariant for the dynamics on $F^{\binom{n+k-1}{k}}$ induced by $S^k(A)$ actually defines the coefficients of a degree k invariant form corresponding to the matrix A. Since the field F is infinite (in which case the equality of polynomial functions implies the equality between the underlying formal polynomials) we get a bijection between the invariant k-forms for A and the invariant 1-forms for $S^k(A)$. More explicitly, let the invariant k-form ϕ be given by

$$\phi(x_1, \cdots, x_n) = \sum_{\alpha \in T} C_\alpha x_1^{\alpha_1} \cdots x_n^{\alpha_n}$$

where the above sum is over the set T consisting of the $\binom{n+k-1}{k}$ sequences of nonnegative integers $\alpha = (\alpha_1, \cdots, \alpha_n)$ subject to $\alpha_1 + \cdots + \alpha_n = k$, ordered lexicographically. Then (3) together with the definition of the symmetric product leads to the relation

$$\phi(y_1, \cdots, y_n) = \sum_{\beta \in T} C_\beta y_1^{\beta_1} \cdots y_n^{\beta_n} = \sum_{\beta \in T} C_\beta \left(\sum_{\alpha \in T} S^k(A)_{\beta\alpha} x_1^{\alpha_1} \cdots x_n^{\alpha_n} \right)$$

valid for any $x_1, \cdots, x_n \in F$. Since F is infinite, the equality of polynomial functions is equivalent to the equality between the underlying formal polynomials:

$$\sum_{\beta \in T} C_\beta S^k(A)_{\beta\alpha} = C_\alpha$$

that is, the vector $(C_\alpha)_{\alpha \in T}$ is an eigenvector of $[S^k(A)]^t$ corresponding to the eigenvalue 1. This concludes the proof.

Examples: The Fibonacci recurrence

$$F_{n+2} = F_{n+1} + F_n, \quad n \ge 0 \tag{8}$$

corresponds to the matrix

$$A = \begin{pmatrix} 1 & 1 \\ 1 & 0 \end{pmatrix}.$$

We easily find that

$$[S^2(A)]^t - I_3 = \begin{pmatrix} 0 & 1 & 1 \\ 2 & 0 & 0 \\ 1 & 0 & -1 \end{pmatrix}$$

has trivial kernel, provided the characteristic of the (infinite) field F is odd, so there are no nonzero invariant 2-forms in this case. However, in the even characteristic case, any sequence satisfying the recurrence relation (8) is of the form

$$r, s, r + s, r, s, r + s, \cdots$$

and an invariant 2-form is $\phi(x,y) = x^2 + xy + y^2$. Notice that in this later case $R = 2$ so that the space of invariant 2-forms is one-dimensional.

In the general case of a second order linear recurrence

$$x_{n+2} = a x_{n+1} + b x_n, \quad n \geq 0$$

with $a, b \in F$, we have

$$A = \begin{pmatrix} a & b \\ 1 & 0 \end{pmatrix}$$

so that the necessary and sufficient condition for the existence of an invariant 2-form is that the determinant of

$$[S^2(A)]^t - I_3 = \begin{pmatrix} a^2 - 1 & a & 1 \\ 2ab & b - 1 & 0 \\ b^2 & 0 & -1 \end{pmatrix}$$

is zero. The case $a = 3$, $b = -1$ was considered in the introductory section.

3. LINEAR INVARIANTS OVER FINITE FIELDS: THE MACWILLIAMS IDENTITIES REVISITED

The study of the invariant linear forms for linear recurrences over a finite field $F = \mathbb{F}_q$ reveals a surprising 'dynamical' interpretation of the MacWilliams identities.

The fixed points of the linear transformation (1) induced by $A \in M_{n \times n}(\mathbb{F}_q)$ form subspace of $\mathbb{F}_q^n$. Notice that $\mathbb{F}_q^n$ is endowed with the (Hamming) distance

$$d(X,Y) = wt(X - Y)$$

that is, $d(X,Y)$ is the number of nonzero components (the weight) of the difference vector $X - Y$. The following extension of the concept of fixed point appears then natural:

Definition 2: Let $l \in \{0, \cdots, n\}$. A point of level l for the linear transformation induced by $A \in M_{n \times n}(\mathbb{F}_q)$ is an element $X \in \mathbb{F}_q^n$ such that $d(AX, X) = l$.

For example a point of level 0 is a fixed point, while if the level l is small with respect to n the concept of 'point of order l' might be seen as a reasonable approximation of the concept of fixed point. By using the notion of level as introduced above we give a natural extension of the concept of fixed point (which plays an important role in the classical theory of dynamical systems) within the framework of finite fields.

Notation: We will denote by $Fix_l(A)$ the number of points of level l of the linear transformation of $\mathbb{F}_q^n$ induced by $A \in M_{n \times n}(\mathbb{F}_q)$. In order to put the numbers $Fix_l(A)$, $l \in \{0, \cdots, n\}$, together we will use the generating function

$$Fix(A; t) = \sum_{l=0}^{n} Fix_l(A) t^l .$$

Now we will split the notion of linear invariant according to the Hamming weight.

Definition 3: We say that a linear invariant H of A is of level $l \in \{0, \cdots, n\}$ if the Hamming weight of $H \in \mathbb{F}_q^n$ is l. Also, we denote by $Inv_l(A)$ the number of linear invariants of level l of $A \in M_{n \times n}(\mathbb{F}_q)$. Thus, the generating function for the numbers $Inv_l(A)$, $l \in \{0, \cdots, n\}$ will be

$$Inv(A; t) = \sum_{l=0}^{n} Inv_l(A) t^l .$$

A linear invariant H is a solution of (5). One can, then, look at the matrix $A^t - I_n$ as a parity-check matrix defining a linear code (subspace of $\mathbb{F}_q^n$). If the $A^t - I_n$ has rank r then the space of linear invariants for the discrete dynamical system defined by A has dimension $n - r$. The dual of this $[n, n-r]$-code is the $[n, r]$-code spanned by any r linearly independent rows of $A^t - I_n$ which can be identified with the $[n, r]$-code spanned by any r linearly independent columns of $A - I_n$. An element $Y \in \mathbb{F}_q^n$ belongs to the column space of $A - I_n$ if and only if it can be written as

$$Y = (A - I_n)X$$

for some $X \in \mathbb{F}_q^n$.

The above considerations allow us to establish a connection with coding theory so that a sort of 'dynamical' perspective on the MacWilliams identities will emerge.

Thus, the space C of linear invariants for the discrete dynamics in $\mathbb{F}_q^n$ induced by A is the coding-theoretic dual of $C^{\perp}$, the image of the linear transformation $A - I_n$. Clearly, the polynomials

$$Inv(A; t)$$

and

$$\frac{1}{q^{n-r}} Fix(A; t)$$

belonging to $\mathbb{Z}[t]$ represent nothing else than the weight enumerators for the linear codes C and $C^{\perp}$, respectively. The MacWilliams identities ([1], p. 24, Theorem 5.2) applied to our case provide a nice connection (which might be though as a sort of 'reciprocity law') between $Inv(A; t)$ (describing the weight structure of the linear invariants of the discrete dynamics induced by A) and $Fix(A; t)$ (counting the fixed points, i.e., those of level zero, of that dynamics, together with the points of higher level):

$$q^r Inv(A; t) = [1 + (q-1)t]^n \left[\frac{1}{q^{n-r}} Fix\left(A; \frac{1-t}{1+(q-1)t} \right) \right]$$

or, equivalently,

$$q^n Inv(A; t) = [1(q-1)t]^n Fix\left(A; \frac{1-t}{1+(q-1)t} \right).$$

REFERENCES

[1]   Blake, Ian F. and Mullin, Ronald C. An Introduction to Algebraic and Combinatorial
      coding Theory. Academic Press, 1976.

[2]   Brown, William C. A Second Course in Linear Algebra. Wiley, 1988.

AMS Classification Numbers: 11B37, 11B39

BASE 10 RATS CYCLES AND ARBITRARILY LONG BASE 10 RATS CYCLES

Curtis Cooper and Robert E. Kennedy

1. INTRODUCTION

John Conway invented a digital game called RATS [1], an acronym for Reverse, Add, Then Sort. A game of RATS produces a sequence of positive integers. The decimal digits of each positive integer in the sequence are all nonzero and in nondecreasing order. The first number in a RATS sequence is a positive integer whose decimal digits are all nonzero and in nondecreasing order. To produce the $(i+1)$st number in a RATS sequence from the i number, for $i \geq 1$, reverse the digits of the ith number, add this number to the ith number, delete the zero digits in the sum, and then sort the remaining digits of the sum in nondecreasing order. The resulting sequence of positive integers is the RATS sequence for the first number.

For example, if we begin a game of RATS with 3, then the RATS sequence is

$$3, \quad 6, \quad 12, \quad 33, \quad 66, \quad 123, \quad 444, \quad 888, \quad 1677, \quad 3489, \quad 12333, \quad 44556,$$
$$111, \quad 222, \quad 444, \quad 888, \quad 1677, \quad 3489, \quad 12333, \quad 44556,$$
$$111, \quad 222, \quad 444, \quad 888, \quad 1677, \quad 3489, \quad 12333, \quad 44556,$$
$$111, \quad 222, \quad 444, \quad 888, \quad 1677, \quad 3489, \quad 12333, \quad 44556,$$
$$\vdots$$

Note that 6 follows 3 in the game since $3+3=6$, 123 follow 66, since $66+66=132$ and then sorting the digits 132 results in the number 123, and 111 follows 44556 in the sequence, since $44556 + 65544 = 110100$ and deleting the 0's leaves 111. It is clear that the last 8 numbers cycle. This is called a RATS cycle. The length of the cycle is 8 and the smallest number in the cycle is 111.

This paper is in final form and no version of it will be submitted for publication elsewhere.

Another interesting game starts with 1. The resulting RATS sequence is

1, 2, 4, 8, 16, 77, 145, 668, 1345, 6677, 13444, 55778, 133345,
666677, 1333444, 5567777, 12333445, 66666677, 133333444, 556667777,
12333334444, 55666667777,

123333334444, 556666667777,

1233333334444, 5566666667777,

12333333334444, 55666666667777,

123333333334444, 55666666667777,

1233333333334444, 556666666667777,

12333333333334444, 5566666666667777,

123333333333334444, 55666666666667777,

1233333333333334444, 556666666666667777,

12333333333333334444, 5566666666666667777,

123333333333333334444, 55666666666666667777,

1233333333333333334444, 556666666666666667777,

12333333333333333334444, 5566666666666666667777,

123333333333333333334444, 55666666666666666667777,

1233333333333333333334444, 5566666666666666666667777,

⋮

This recurring pattern is known as Conway's divergent sequence.

2. BASE 10 RATS CYCLES.

In [3], Curt McMullen gave a list of the base 10 RATS cycles he had discovered. We set out to search heuristically for some more base 10 RATS cycles. In [2], we compiled a list of the base 10 RATS cycles which included McMullen's and some new ones we found. Here, we will present some additional base 10 RATS cycles.

Due to the size and repetitive nature of the positive integers in our base 10 RATS games, we will use superscripts to denote repeated digits in a number. For example, the number 55666666666667777 will be represented by

$$5^2 6^{11} 7^4.$$

We had noticed that a number of the known base 10 RATS cycles contained positive integers consisting of 1's and 2's. Therefore, we decided to look for base 10 RATS cycles in this region. That is, we searched for base 10 RATS cycles by starting RATS games with numbers of the form

$$1^x2^y.$$

To automate our search as much as possible and because of the number of digits in some of our base 10 RATS games, we wrote several C programs to play our base 10 RATS games and search for base 10 RATS cycles. All of these programs can be found on the World-Wide Web at

http://www.math-cs.cmsu.edu/~curtisc/rats/.

Here are some highlights of what we found. We found base 10 RATS cycles of length 13, 15, 16, 17, 19, 20, 21, 22, 25, 26, 29, 30, 32, 34, 36, 40, 45, and 69. McMullen had not found any base 10 RATS cycles of these lengths. One of the new ones of length 13 is

$$1^{812}2^{1448}, \quad 3^{1624}4^{636}, \quad 6^{988}7^{1272},$$
$$134^{1975}5^{284}, \quad 6^{2}8^{1695}9^{564}, \quad 156^{3}7^{1133}8^{1124},$$
$$1345^{20}6^{2238}9, \quad 1^{4}2^{40}3^{2216}, \quad 4^{8}5^{80}6^{2172},$$
$$116_{2}160_{3}2084, \quad 4^{32}5^{320}6^{1908}, \quad 1^{64}2^{640}3^{1556}, \quad 4^{128}5^{1280}6^{852}.$$

We also found additional RATS cycles of lengths 4, 5, 6, 7, 8, 9, 10, 11, 12, 14, 18, and 24. One of the new ones of length 8 is

$$1^{6}5^{2}7^{8}, \quad 23^{3}8^{11}9, \quad 1^{2}2^{7}7^{8}, \quad 48^{4}9^{12},$$
$$1348^{8}9^{7}, \quad 1^{2}3^{2}4^{2}7^{4}8^{8}, \quad 1^{2}2^{3}3^{4}5^{2}0^{4}9^{2}, \quad 1^{3}08^{9}9^{3}.$$

Some of the base 10 RATS cycles we found of length 36, 45, and 69 begin with

$$1^{6}2^{281}3^{472}6^{8}2,$$
$$1^{6}354_{2}4140,$$
$$\text{and } 1^{6}780_{2}6814_{3}2_{4}4,$$

respectively. In particular, the base 10 RATS cycle of length 36 is

$1^6 2^{281} 3^{4726} 8^2$, $4^8 5^{562} 6^{4441} 9^4$, $1^9 2^{1124} 3^{3876} 4^7$,

$4^4 5^{2262} 6^{2750}$, $1^8 2^{4524} 3^{484}$, $4^{4064} 5^{952}$,

8^{3112}, 9^{1904}, $17^{1209} 8^{3807}$, $156^{2417} 7^{2597} 9$,

$1^2 3^2 4^{4834} 5^{179}$, $6^4 8^{4663} 9^{350}$, $156^7 7^{4317} 8^{692}$,

$1345^{3648} 6^{1366} 9$, $1^{2288} 2^{2728}$, $3^{4576} 4^{440}$,

$6^{4136} 7^{880}$, $13^{3257} 4^{1759}$, $5^2 6^{1499} 7^{3516}$,

$123^3 4^{2998} 5^{2015}$, $6^2 7^2 8^{994} 9^{4020}$, $156^3 7^4 8^{1988} 9^{3022}$,

$1^2 5^2 6^{678} 8^{3976} 9^{1025}$, $1^4 5^4 6^{12} 7^{2985} 8^{2014}$, $134^7 5^{1015} 6^{3988} 9^4$,

$1^{12} 2^{2030} 3^{2970} 4^4$, $4^{16} 5^{4068} 6^{932}$, $1^{3184} 2^{1832}$,

$2^{1352} 3^{3664}$, $5^{2704} 6^{2312}$, $1^{394} 2^{4623}$,

$3^{788} 4^{4229}$, $7^{1576} 8^{3441}$, $156^{3151} 7^{1865}$,

$23^{1289} 4^{3726} 89$, $1^3 2^2 7^{2575} 8^{2439}$, $1^4 5^{141} 6^{4868} 9^3$.

We found 3 base 10 RATS cycles of length 32. These RATS cycles begin with

$$1^{6235} 2^{106},$$
$$1^{2110} 2^{878} 3^2 4^4,$$
$$\text{and } 1^{2731} 2^{2838} 3^2 4^2.$$

We found families of base 10 RATS cycles of length 11, 13, 14, 16, and 18. One family of RATS cycles of length 14 starts at

$$1^{2957n + 73} 2^{2504n + 38} 3^2 4^4.$$

The proof that $1^{2957n + 73} 2^{2504n + 38} 3^2 4^4$ is the least member of a family of RATS cycle of length 14 follows from the RATS cycle

$$1^{2957n} + 73_2 504n + 38_3 2_4,$$
$$2^{453n} + 29_3 5008n + 76_4 5_8,$$
$$5^{906n} + 34_6 4555n + 67_7 16,$$
$$12^{1812n} + 37_3 3649n + 80,$$
$$4_2 5^{3624n} + 74_6 1837n + 42,$$
$$1^{1787n} + 38_2 3674n + 80,$$
$$3^{3574n} + 76_4 1887n + 42,$$
$$6^{1687n} + 34_7 3774n + 84,$$
$$134^{3374n} + 67_5 2087n + 50,$$
$$6_2 8^{1287n} + 21_9 4174n + 96,$$
$$156_3 8_2 2574n + 42_9 2887n + 73,$$
$$1_2 5_2 6_6 8^{5148n} + 84_9 313n + 26,$$
$$1_4 5_4 6^{1274835n} + 68_8 626n + 32,$$
$$134_7 5^{4209n} + 80_6 1252n + 24_9 4.$$

We also found some new smaller families of RATS cycles. One of length 8 starts with $1^{164n} + 110_7 91n + 61$. The proof that this is a family of RATS cycles of length 8 follows from the RATS cycle

$$1^{164n} + 110_7 91n + 61, \quad 2^{73n} + 49_8 182n + 122,$$
$$1^{146n} + 98_7 109n + 73, \quad 2^{37n} + 25_8 218n + 146,$$
$$1^{74n} + 50_7 181n + 121, \quad 45^{107n} + 70_8 148n + 99_9,$$
$$134^{214n} + 141_7 41n + 29, \quad 1_2 2^{82n} + 53_8 173n + 114_9 2.$$

We found

$$1^{128n} + 133_2 128n + 119_3 4_6 16n + 15_7$$

to be an interesting family of base 10 RATS cycles of length 4. The proof that this is a family of base 10 RATS cycles of length 4 follows from the RATS cycle

$$1^{128n} + 133_2 128n + 119_3 4_6 16n + 15_7,$$
$$3^{224n} + 226_4 16n + 14_7 32n + 30_8 2,$$
$$1^{64n} + 61_2 3_6 176n + 180_7 32n + 27_8,$$
$$23^{144n} + 143_7 71n + 59_8 64n + 59_9 3.$$

The entire list of 225 base 10 RATS cycles and 85 base 10 RATS families we know to date can be found on the WWW at

http://www.math-cs.cmsu.edu/~curtisc/rats/ratscycles.txt.

An excerpt of the beginning of the table follows. To save space in the table and to represent superscripts in computer output, we will use the notation $x\hat{\ }y$ or $x\hat{\ }\{y\}$ to denote x^y.

Base 10 RATS Cycles

Length	Least Member
2	$1\hat{\ }5\ 2\hat{\ }3\ 67$
2	$1\hat{\ }3\ 27$
2	$1\hat{\ }2\ 6\hat{\ }2\ 7$
2	$1\hat{\ }\{2n\}\ 7\hat{\ }n$
2	78
3	$1\hat{\ }\{6n\}\ 3\hat{\ }n$
4	$1\hat{\ }\{128n+133\}\ 2\hat{\ }\{128n+119\}\ 3\hat{\ }4\ 6\hat{\ }\{16n+15\}\ 7$
4	$1\hat{\ }\{17\}\ 2\hat{\ }\{15\}\ 67$
4	$1\hat{\ }9\ 2\hat{\ }3\ 678$
4	$1\hat{\ }9\ 2\hat{\ }7\ 7$
4	$1\hat{\ }\{21\}\ 2\hat{\ }3\ 7\hat{\ }3\ 8\hat{\ }3$
4	$1\hat{\ }3\ 4\hat{\ }6\ 5\hat{\ }3\ 6$
4	$1\hat{\ }8\ 6\hat{\ }8\ 7\hat{\ }4$
4	$1\hat{\ }\{4n+6\}\ 7\hat{\ }\{11n+15\}$
5	$1\hat{\ }2\ 2\hat{\ }6\ 3\hat{\ }\{58\}\ 4\hat{\ }\{20n+200\}\ 5\hat{\ }\{11n+265\}$
5	$1\hat{\ }\{24n\}\ 3\hat{\ }\{7n\}$.

3. ARBITRARILY LONG BASE 10 RATS CYCLES

In [4], Cooper and Kennedy asked if there are arbitrarily long base 10 RATS cycles. Our next goal is to answer that question in the affirmative.

Lemma 1: Let $t \geq 3$ be an odd integer. Then

$$1^{6\cdot2^{t-3}}3^{2\cdot2^{t-3}-1},$$

$$2^{4\cdot2^{t-3}}+1 4^{4\cdot2^{t-3}}-2,$$

$$4^3 6^{8\cdot2^{t-3}-4},$$

$$1^{6\cdot2^0}3^{8\cdot2^{t-3}-6\cdot2^0-1}$$

$$4^{6\cdot2^1}6^{8\cdot2^{t-3}-6\cdot2^1-1},$$

$$1^{6\cdot2^2}3^{8\cdot2^{t-3}-6\cdot2^2-1},$$

$$4^{6\cdot2^3}6^{8\cdot2^{t-3}-6\cdot2^3-1},$$

$$1^{6\cdot2^4}3^{8\cdot2^{t-3}-6\cdot2^4-1},$$

$$4^{6\cdot2^5}6^{8\cdot2^{t-3}-6\cdot2^5-1},$$

$$\vdots$$

$$1^{6\cdot2^{t-5}}3^{8\cdot2^{t-3}-6\cdot2^{t-5}-1},$$

$$4^{6\cdot2^{t-4}}6^{8\cdot2^{t-3}-6\cdot2^{t-4}-1},$$

is a base 10 RATS cycle of length t.

Lemma 2: Let $t \geq 8$ be an even integer. Then

$$1^{216 \cdot 2^{t-8}} 3^{40 \cdot 2^{t-8} - 1},$$

$$2^{176 \cdot 2^{t-8} + 1} 4^{80 \cdot 2^{t-8} - 2},$$

$$4^{96 \cdot 2^{t-8} + 3} 6^{160 \cdot 2^{t-8} - 4},$$

$$1^{192 \cdot 2^{t-8} + 6} 3^{64 \cdot 2^{t-8} - 7},$$

$$2^{128 \cdot 2^{t-8} + 13} 4^{128 \cdot 2^{t-8} - 14},$$

$$4^{27} 6^{256 \cdot 2^{t-8} - 28},$$

$$1^{54} 3^{256 \cdot 2^{t-8} - 55},$$

$$4^{108} 6^{256 \cdot 2^{t-8} - 109},$$

$$1^{216 \cdot 2^0} 3^{256 \cdot 2^{t-8} - 216 \cdot 2^0 - 1},$$

$$4^{216 \cdot 2^1} 6^{256 \cdot 2^{t-8} - 216 \cdot 2^1 - 1},$$

$$1^{216 \cdot 2^2} 3^{256 \cdot 2^{t-8} - 216 \cdot 2^2 - 1},$$

$$4^{216 \cdot 2^3} 6^{256 \cdot 2^{t-8} - 216 \cdot 2^3 - 1},$$

$$1^{216 \cdot 2^4} 3^{256 \cdot 2^{t-8} - 216 \cdot 2^4 - 1},$$

$$4^{216 \cdot 2^5} 6^{256 \cdot 2^{t-8} - 216 \cdot 2^5 - 1},$$

$$\vdots$$

$$1^{216 \cdot 2^{t-10}} 3^{256 \cdot 2^{t-8} - 216 \cdot 2^{t-10} - 1},$$

$$4^{216 \cdot 2^{t-9}} 6^{256 \cdot 2^{t-8} - 216 \cdot 2^{t-9} - 1}$$

is a base 10 RATS cycle of length t.

In addition, let n be a positive integer. Then, for any fixed odd integer $t \geq 3$, there is a base 10 RATS family of length t with smallest element

$$1^{(6 \cdot 2^{t-3})n} 3^{(2 \cdot 2^{t-3} - 1)n}$$

and for any fixed even integer $t \geq 8$, there is a base 10 RATS family of length t with smallest element

$$1^{(216 \cdot 2^{t-8})n} 3^{(40 \cdot 2^{t-8} - 1)n}.$$

We now have the following theorem.

Theorem: Let $t \geq 2$ be a positive integer. Then there exists a base 10 RATS cycle of length t.

Proof: The cases where $t \geq 3$ is an odd integer and $t \geq 8$ is an even integer are shown by Lemmas 1 and 2. The remaining cases can be handled by the following table.

Cycle of Length 2   117, 288.

Cycle of Length 4 $1^6 7^{15}$, $45^8 8^{11} 9$, $134^{17} 7^3$, $1^2 28^{16} 9^2$.

Cycle of Length 6 $1^4 2^6 5^2$, $3^4 4^4 6^4$, $8^4 9^8$, $178^7 9^4$, $1^2 7^7 8^4$, $15^6 6^3 9^2$.

This completes the proof.

We also found two other interesting collections of RATS cycles. They can be stated as follows.

Lemma 3: Let t be a positive integer. Then

$$1^{2^{2t-1}} + 1_2 2^{2t-1} - 1_7,$$

$$23 4^t - 2 8^2,$$

$$1^{2^1} + 1_2 2^1 - 1_6 4^t - 4^1 7,$$

$$23 4^t - 2 \cdot 4^1 7^{4^1} 8^{4^1} - 1_9,$$

$$1^{2^3} + 1_2 2^3 - 1_6 4^t - 4^2 7,$$

$$23 4^t - 2 \cdot 4^2 7^{4^2} 8^{4^2} - 1_9,$$

$$1^{2^5} + 1_2 2^5 - 1_6 4^t - 4^3 7,$$

$$23 4^t - 2 \cdot 4^3 7^{4^3} 8^{4^3} - 1_9,$$

$$\vdots$$

$$1^{2^{2t-3}} + 1_2 2^{2t-3} - 1_6 4^t - 4^{t-1} 7,$$

$$23 4^t - 2 \cdot 4^{t-1} 7^{4^{t-1}} 8^{4^{t-1}} - 1_9$$

is a base 10 RATS cycle of length $2t$.

Lemma 4: Let t be a positive integer. Then

$$1^{4^t} + 1_2 4^t - 1_6 7,$$

$$3^2 \cdot 4^t - 2_7 2_8 2,$$

$$1^{4^1} + 1_2 4^1 - 1_6 2 \cdot 4^t - 2 \cdot 4^1 + 1_7,$$

$$23^2 \cdot 4^t - 4^2 + 1_7 2 \cdot 4^1 8 2 \cdot 4^1 - 1_9,$$

$$1^{4^2} + 1_2 4^2 - 1_6 2 \cdot 4^t - 2 \cdot 4^2 + 1_7,$$

$$23^2 \cdot 4^t - 4^3 + 1_7 2 \cdot 4^2 8 2 \cdot 4^2 - 1_9,$$

$$1^{4^3} + 1_2 4^3 - 1_6 2 \cdot 4^t - 2 \cdot 4^3 + 1_7,$$

$$23^2 \cdot 4^t - 4^4 + 1_7 2 \cdot 4^3 8 2 \cdot 4^3 - 1_9,$$

$$\vdots$$

$$1^{4^{t-1}} + 1_2 4^{t-1} - 1_6 7,$$

$$23^2 \cdot 4^t - 4^t + 1_7 2 \cdot 4^{t-1} 8 2 \cdot 4^{t-1} - 1_9$$

is a base 10 RATS cycle of length $2t$.

4. QUESTIONS

We conclude this article with several open questions. One problem would be to find other base 10 RATS cycles and base 10 families of RATS cycles. In addition, we could study the game of RATS in other bases and find RATS cycles and families of RATS cycles in these bases. Finally, John Conway has a simple sounding, yet tremendously hard conjecture based on his base 10 RATS game. So far, every number with nonzero digits in nondecreasing order (up to 15 digits) which starts a base 10 RATS game either gets into a cycle or enters the divergent sequence. Conway's RATS conjecture is that this is true for every number with nonzero digits in nondecreasing order.

REFERENCES

[1]   Conway, J. "Play it again $\cdots$ and again $\cdots$." *Quantum* (November/December 1990): pp. 30-31 and 63.

[2]   Cooper, C. and Kennedy, R.E. "On Conway's RATS." *Mathematics in College*, (1998): pp. 28-35.

[3]   Guy, R.K. "Conway's RATS and Other Reversals." *The American Mathematical Monthly*, Vol. *96* (1989): pp. 425-428.

[4]   Guy, R.K. and Nowakowski, R.J. "Monthly Unsolved Problems, 1969-1997." *The American Mathematical Monthly*, Vol. *104* (1997): pp. 967-973.

AMS Classification Numbers: 11A63

QUINTICS $x^5 - 5x - k$, THE GOLDEN SECTION, AND SQUARE LUCAS NUMBERS

Michele Elia and Piero Filipponi

1. INTRODUCTION

Here we consider a special case of the problem of characterizing the real numbers that, once multiplied by the integer a, have the same fractional parts as those of their n^{th} powers ($n \geq 2$ an integer). These numbers, which will be referred to as numbers possessing the property $\mathcal{P}(n, a)$, are clearly given (cf (1.1) of [3]) by the real roots of the equations

$$x^n - ax = k, \tag{1.1}$$

where the integer k can be assumed to be nonnegative without loss of generality. It is quite obvious that all the integers possess $\mathcal{P}(n, a)$ for all n and a: they emerge from (1.1) when $k = m^n - am$ (m an integer). The closed-form expressions for the above mentioned numbers can be readily found for $2 \leq n \leq 4$ by using (1.1) and the well-known formulas for the solution of second-, third- and fourth-degree equations. Finding them for $n \geq 5$ is a much more difficult task. The expressions for the numbers possessing $\mathcal{P}(n, 1)$ were worked out in [3] (see also [4]) for $n \leq 5$. The special case $n = 5$ shows an interesting connection with Fibonacci numbers.

In this paper we extend this result by finding all the numbers possessing $\mathcal{P}(5, 5)$ that can be expressed by radicals. More precisely, we let $(n, a) = (5, 5)$ in (1.1), and prove that the only values of k for which the roots of the quintics of the Bring-Jerrard form

$$q(x) = x^5 - 5x - k \quad (k \text{ a nonnegative integer}) \tag{1.2}$$

can be expressed by radicals are either the integers of the form $m^5 - 5m$, or 12, or one of the integers 3, 28 and 396 which are Lucas number products. For the sake of completeness, the corresponding expressions, some of which involve the golden section, are also given.

This paper is in final form and no version of it will be submitted for publication elsewhere.

2. SOLVING $q(x)$

The quintic $q(x)$ above may be either irreducible or reducible over the rational field $\mathbb{Q}$. If it is reducible over $\mathbb{Q}$, then it is reducible over the integer ring $\mathbb{Z}$ as well [9, Thm. 23, p. 24], and the real root may entail either a linear factor or a cubic irreducible factor. Necessary and sufficient conditions for its decomposition are given in Subsection 2.2. We consider first the irreducible case.

2.1. The irreducible case

We shall prove that, if $q(x)$ is irreducible, then it can be solved by radicals iff $k = 12$. To this aim, we need the following theorem of Dummit [2] that we quote in a form specialized to our Bring-Jerrard quintic.

Theorem 1 (Dummit). The irreducible quintic $q(x) = x^5 - 5x - k$ is solvable by radicals iff the polynomial

$$s(x) = x^6 - 40x^5 + 1000x^4 - 20000x^3 + 250000x^2$$
$$- 3125(512 + k^4)x + 15625(256 + 3k^4) \qquad (2.1)$$

has a rational root.

Now, we are in a position to state our main theorem.

Theorem 2. If $q(x) = x^5 = 5x - k$ is irreducible over $\mathbb{Q}$, then it can be solved by radicals iff $k = 12$.

Proof. By Theorem 1, it is sufficient to prove that the monic sextic $s(x)$ has a rational root u' iff $k = \pm 12$. After observing that, by virtue of the Rational Root Theorem (see [5. p. 253] or [6. p. 38]), a rational root of a monic polynomial is an integer, we may assume that u' is an integer. Since all (but the first) coefficients of $s(x)$ are divisible by 5, the integral root u' is as well. Consequently, it is convenient to make the replacement $x = 5y$ in (2.1), thus getting

$$s(y) = 5^6[y^6 - 8y^5 + 40y^4 - 160y^3 + 40y^2 - (512 + k^4)y + 256 + 3k^4]. \qquad (2.2)$$

Let $u = u'/5$ be the integral root of $s(y)$. Replace y by u in the r.h.s. of (2.2), equate this polynomial to zero, and solve for k^4, thus getting the equality

$$k^4 = (u - 2)^4 \, \frac{u^2 + 16}{u - 3} \qquad (2.3)$$

which can be rewritten in the form

$$u^2 + 16 = (u - 3)\left(\frac{k}{u - 2}\right)^4 \qquad (2.4)$$

whence one can see that $v := k/(u-2)$ must be an integer because g.c.d. $(u-3, u-2) = 1$. Therefore, from (2.4), the integer u can be viewed as a root of the quadratic equation

$$u^2 - v^4 u + 3v^4 + 16 = 0 \qquad (2.5)$$

whose discriminant $v^8 - 12v^4 - 64$ must be a perfect square (say, w^2). Consequently, v must be a root of the quadratic equation in v^4

$$v^8 - 12v^4 - 64 - w^2 = 0 \qquad (2.6)$$

whose discriminant $4(w^2 + 100)$ must, in turn, be a perfect square (say, $4t^2$). In other words, w and t must satisfy the diophantine equation

$$w^2 + 100 = t^2 \qquad (2.7)$$

whose solutions (see (3.5) of [3]) are $(w, t) = (0, 10)$ and $(24, 26)$. Letting $w = 24$ and 0 in (2.6) yields the couple of equations

$$v^8 - 12v^4 - 640 = 0, \qquad (2.8)$$

$$v^8 - 12v^4 - 64 = 0, \qquad (2.9)$$

respectively. The positive root of (2.8) is $v^4 = 32$ which is not a fourth power and has to be disregarded, whereas the positive root of (2.9) is $v^4 = 16$. Replace this value in (2.5) to obtain the equation

$$u^2 - 16u + 64 = 0 \qquad (2.10)$$

which has the double root $u = 8$.

Finally, replace u by 8 in (2.3) thus getting $k = 12$.   Q.E.D.

By means of Theorems 1 and 2, we have proved that the only irreducible quintic of the form (1.2) that can be solved by radicals is $x^5 - 5x - 12$. Its solution is given by Dummit in [2], and is reported here in a more compact form. The single real root y_0 of $x^5 - 5x - 12$ is

$$y_0 = 5^{-2/5} \left(\sqrt[5]{R_1} + \sqrt[5]{R_2} + \sqrt[5]{R_3} + \sqrt[5]{R_4} \right) \qquad (2.11)$$

where

$$\begin{cases} R_1 = 20\alpha + 15 + 6\sqrt{10(\alpha+2)} + 3\sqrt{10(\beta+2)} \\[2mm] R_2 = 20\alpha + 15 - 6\sqrt{10(\alpha+2)} - 3\sqrt{10(\beta+2)} \\[2mm] R_3 = 20\beta + 15 - 3\sqrt{10(\alpha+2)} + 6\sqrt{10(\beta+2)} \\[2mm] R_4 = 20\beta + 15 + 3\sqrt{10(\alpha+2)} - 6\sqrt{10(\beta+2)} \end{cases} \qquad (2.12)$$

and $\alpha = (1+\sqrt{5})/2 = 1 - \beta$ is the golden section. The numbers $\pm y_0$ clearly possess $\mathcal{P}(5,5)$.

2.2 The reducible case

Rabinowitz [8] considered the factorization of the quintic $x^5 - x - k$ and showed that it factors over $\mathbf{Q}$ iff k is either an integer of the form $m^5 - m$, or has some special values depending on square Fibonacci numbers. Hereafter, following [3], we discuss in full the factorization of the quintic $q(x)$ given by (1.2).

First, if $k = m^5 - 5m$, then

$$x^5 - 5x - (m^5 - 5m) = (x - m)(x^4 + mx^3 + m^2x^2 + m^3x + m^4 - 5) \tag{2.13}$$

whence it is plain that all the integers (say, m) possess $\mathcal{P}(5,5)$. The quartic factor on the r.h.s. of (2.13) has real roots iff $m = 0$ or $m = \pm 1$. To prove this fact, it is sufficient to replace x by mx thus reducing the question to seeing that $x^4 + x^3 + x^2 + x = 5/m^4 - 1$ has no real roots for $|m| \geq 2$, which is easy to confirm. More precisely, for $m = 0$, we have the roots

$$z_0 = \pm\sqrt[4]{5} \tag{2.14}$$

whereas, for $m = \pm 1$, we have the roots ± 1 and

$$z_1 = \pm\left(\sqrt[3]{\frac{35}{27}+\sqrt{\frac{50}{27}}} + \sqrt[3]{\frac{35}{27}-\sqrt{\frac{50}{27}}}+\frac{2}{3}\right). \tag{2.15}$$

Of course, both z_0 and z_1 possess $\mathcal{P}(5,5)$.

Then, we state the following.

Theorem 3. If $k \neq m^5 - 5m$, then $q(x)$ is reducible over $\mathbf{Q}$ iff

$$k = \begin{cases} L_0L_5L_6 & = 396 \\ L_3L_4 & = 28 \\ L_2 & = 3. \end{cases} \tag{2.16}$$

Proof. If $k \neq m^5 - 5m$, then the only possible factorization of $q(x)$ has the form

$$x^5 - 5x - k = (x^2 + ax + b)(x^3 - ax^2 + cx + d) \tag{2.17}$$

whence one gets readily the system

$$\begin{cases} b + c - a^2 = 0 \\ a(b - c) - d = 0 \\ ad + bc = -5 \\ bd = -k. \end{cases} \tag{2.18}$$

By using the first two equation of (2.18) to eliminate a and d, one obtains the couple of equations

$$\begin{cases} bc + b^2 - c^2 = -5 \\ b^2(b-c)^2(b+c) = k^2 \end{cases} \tag{2.19}$$

the first of which implies that the couple (b,c) must belong to the set of the couples that represent 5 by means of the quadratic form $Q(b,c) = c^2 - bc - b^2$. Since $(-1,2)$ is a solution to $Q(b,c) = 5$, Gauss' theory of quadratic forms (e.g., see [7] and [10]) tells us that all the solutions are given by

$$(b,c) = \pm(L_{2n-1}, L_{2n}) \quad (n \in \mathbb{Z}) \tag{2.20}$$

where L_n is the n-th Lucas number. We recall that $L_{-n} = (-1)^n L_n$.

From (2.20) and the second equation of (2.19), we see that

$$k^2 = L_{2n-1}^2 (L_{2n-1} - L_{2n})^2 (\pm L_{2n-1} \pm L_{2n}) = L_{2n-1}^2 L_{2n-2}^2 (\pm L_{2n+1}) \tag{2.21}$$

where the minus sign in the last factor occurs when $n < 0$. From (2.21), it is patent that, for k to be an integer, L_{2n+1} must be a square. Since it is a well-known fact [1] that the only Lucas numbers that are a square are $L_1 = 1$ and $L_3 = 4$, four possibilities are available to us. Namely, we have

$$k^2 = \begin{cases} L_{-5}^2 L_{-6}^2(-L_{-3}) & \text{for } n = -2 \text{ in (2.21)} \\ L_{-3}^2 L_{-4}^2(-L_{-1}) & \text{for } n = -1 \text{ in (2.21)} \\ L_{-1}^2 L_{-2}^2 L_1 & \text{for } n = 0 \text{ in (2.21)} \\ L_1^2 L_0^2 L_3 & \text{for } n = 1 \text{ in (2.21).} \end{cases} \tag{2.22}$$

After observing that the constraint $k \neq m^5 - 5m$ removes the last equality in (2.22) which gives $k = L_1 L_0 \sqrt{L_3} = 4 = (-1)^5 - 5(-1)$, we get readily

$$k = \begin{cases} (-L_{-5})L_{-6}\sqrt{-L_{-3}} = L_5 L_6 L_0 = 396 \\ (-L_{-3})L_{-4}\sqrt{-L_{-1}} = L_3 L_4 L_1 = 28 \\ (-L_{-1})L_{-2}\sqrt{L_1} = L_1 L_2 L_1 = 3, \end{cases}$$

as desired. Q.E.D.

The decomposition of $q(x)$ for the above values of k are

$$x^5 - 5x - 3 = (x^2 - x - 1)(x^3 + x^2 + 2x + 3),$$ (2.23)

$$x^5 - 5x - 28 = (x^2 - x + 4)(x^3 + x^2 - 3x - 7),$$ (2.24)

$$x^5 - 5x - 396 = (x^2 - 2x + 11)(x^3 + 2x^2 - 7x - 36).$$ (2.25)

The real roots of the quadratic and cubic factor in (2.23) are

$$x_1 = \alpha,$$ (2.26)

$$x_2 = \beta,$$ (2.27)

$$x_3 = \sqrt[3]{\frac{-65}{54} + \sqrt{\frac{175}{108}}} + \sqrt[3]{\frac{-65}{54} - \sqrt{\frac{175}{108}}} - \frac{1}{3},$$ (2.28)

respectively, whereas the real roots of (2.24) and (2.25) are

$$x_4 = \sqrt[3]{\frac{80}{27} + \sqrt{\frac{200}{27}}} + \sqrt[3]{\frac{80}{27} - \sqrt{\frac{200}{27}}} - \frac{1}{3},$$ (2.29)

$$x_5 = \sqrt[3]{\frac{415}{27} + \sqrt{\frac{5800}{27}}} + \sqrt[3]{\frac{415}{27} - \sqrt{\frac{5800}{27}}} - \frac{2}{3},$$ (2.30)

respectively. Obviously, the numbers $\pm x_j (1 \le j \le 5)$ possess $\mathcal{P}(5,5)$.

ACKNOWLEDGEMENTS

The contribution of the first author has been financially supported by CNR (Italian Council for National Research), whereas the contribution of the second author has been given within the framework of an agreement between the Italian PT Administration (Istituto Superiore PT) and the Fondazione Ugo Bordoni.

The authors thank the anonymous referee whose valuable suggestions led to a real improvement of this paper.

REFERENCES

[1]   Cohn, J.H.E. "Square Fibonacci Numbers, etc." *The Fibonacci Quarterly*, Vol. *2.2* (1964): pp. 109-113.

[2]   Dummit, D.S. "Solving Solvable Quintics." *Mathematics of Computation*, Vol. *57.195* (1991): pp. 387-401.

[3]   Elia, M. and Filipponi, P. "Equations of the Bring-Jerrard Form, the Golden Section, and Square Fibonacci Numbers." *The Fibonacci Quarterly*, Vol. *36.3* (1998): pp. 282-286.

[4]   Filipponi, P. "A Curious Property of the Golden Section." *Int. J. Math. Educ. Sci. Technol.*, Vol. *23.5* (1992): pp. 805-808.

[5]   Gallian, J.A. <u>Contemporary</u> <u>Abstract</u> <u>Algebra</u>. Lexington: D.C. Heath & Co., 1990.

[6]   Hadlock, C.R. <u>Field</u> <u>Theory</u> <u>and</u> <u>Its</u> <u>Classical</u> <u>Problems</u>. Washington: The
        Mathematical Association of America, 1978.

[7]   Jones, J.P. "Diophantine Representation of the Fibonacci Numbers." *The Fibonacci*
        Quarterly, Vol. *13.1* (1975): pp. 84-88.

[8]   Rabinowitz, S. "The Factorization of $x^5 \pm x + n$." *Mathematics Magazine*, Vol. *61.3*
        (1988): pp. 191-193.

[9]   Rotman, J. <u>Galois</u> <u>Theory</u>. New York: Springer-Verlag, 1990.

[10]   Venkov, B.A. <u>Elementary</u> <u>Number</u> <u>Theory</u>. Groningen: Wolters-Noordhoff, 1970.

AMS Classification Numbers: 12D05, 11B39, 14H52

THE PASCAL-DE MOIVRE MOMENTS AND THEIR GENERATING FUNCTIONS

Larry Ericksen

1. INTRODUCTION

The moment generating functions for polynomials in two variables are presented here in closed and exponential formats. Each equivalent format offers an unique perspective on the generating function structure. For example, the closed formulas for the generating functions are ratios of polynomials which can have a hyperbolic representation, whereas the exponential sum formulas are composed of power series of special number sequences.

Moment generating equations for the polynomial terms and their coefficients are generalized to any arbitrary moment configuration. Particular attention is given to polynomials with coefficients of the generalized Pascal triangle [2]. And generating functions for normalized distributions are discussed with some examples given for binomial and trinomial distributions.

After deriving a moment generating strategy for the general polynomial, we apply the moment generating approach to Pascal-DeMoivre polynomials. With special characteristics like coefficient symmetry, these polynomials create some special generating function features.

2. GENERAL POLYNOMIAL MOMENTS

We define a homogeneous polynomial $P(z, y)$ in two variables, whose polynomial terms $P_h(z, y)$ of coefficients g_h are summed over $0 \leq h \leq n$. Taking this polynomial to a power N yields a polynomial of coefficients C_h given by

---

This paper is in final form and no version of it will be submitted for publication elsewhere.

$$P^N(z,y) \equiv (P(z,y))^N \equiv \left(\sum_{h=0}^{n} g_h z^h y^{n-h} \right)^N = \sum_{h=0}^{Nn} C_h z^h y^{Nn-h}, \tag{1}$$

which is also homogenous since all $P_h^N(z,y)$ terms have the same degree Nn. The single variable case of equation (1) is a polynomial $P(z,1)$ of order n, and the power series expansion of $P^N(z,1)$ is a polynomial of order Nn.

For all polynomial terms $C_h z^h y^{Nn-h}$ in (1), we obtain the moments about an arbitrary point k (any real number) as a polynomial moment equation:

$$\mathbf{pm}_R(z,y) = \sum_{h=0}^{Nn} (h-k)^R C_h z^h y^{Nn-h}. \tag{2}$$

Each Rth polynomial moment in (2) is found as the coefficient of $t^R/R!$ in the expansion of its exponential generating function:

$$\mathbf{pm}_{(egf)}(z,y) = \sum_{R=0}^{\infty} \mathbf{pm}_R(z,y) \frac{t^R}{R!} = \sum_{R=0}^{\infty} \sum_{h=0}^{Nn} (h-k)^R C_h z^h y^{Nn-h} \frac{t^R}{R!}. \tag{3}$$

This moment generating function can be written as (4), in terms of the original polynomial $P(z,y)$, as derived by

$$\mathbf{pm}_{(egf)}(z,y) = \sum_{h=0}^{Nn} C_h z^h y^{Nn-h} \sum_{R=0}^{\infty} (h-k)^R \frac{t^R}{R!}$$

$$= \sum_{h=0}^{Nn} C_h z^h y^{Nn-h} e^{(h-k)t}$$

$$= e^{-kt} P^N(ze^t, y). \tag{4}$$

Equation (4) shows that polynomial moments of (2) can be obtained from the moment generating function with the variable (z) being replaced by (ze^t) in the Nth power of the original $P(z,y)$ polynomial. And the calculation of these moments about any point k from the origin affects the generating function simply by a factor e^{-kt}.

By setting $k = \omega - \alpha$ in equations (3,4), we can state the general theorem for generating polynomial moments, measured at an arbitrary distance α from any desired reference point ω, as

Theorem: For any polynomial $P^N(z,y) \equiv \sum_{h=0}^{Nn} C_h z^h y^{Nn-h}$, the polynomial moment generating function

$$\mathbf{pm}_{(egf)}(z,y) \equiv \sum_{R=0}^{\infty} \sum_{h=0}^{Nn} (h-\omega+\alpha)^R C_h z^h y^{Nn-h} \frac{t^R}{R!} = e^{(\alpha-\omega)t} P^N(ze^t, y). \tag{5}$$

The transformation strategy (5) from polynomial to moment generating function forms the basis for all subsequent formulas. This paper applies this principle to the generating functions for the polynomial moments, the coefficient moments, and their normalized distribution moments.

Moment generating function theorem (5) applies also to polynomials $f(z)$ in one variable from polynomials $P^N(z,y)$ in (1) such as, if $g_h = (-1)^h$, for

- a general polynomial $f(z) \equiv P(z,1)$ of order n, for $N = 1$ and $y = 1$.

- a polynomial $f(z) \equiv P^N(z,y_0)$ of order N and with root y_0 of multiplicity N, when $n = 1$ and $y = y_0 = $ a constant.

- a polynomial combination $f(z) \equiv \prod_j P_j^{N_j}(z,y_j)$ of order $N = \sum_j N_j$, with each root y_j of multiplicity N_j, for all $n_j = 1$ and $y_j \in \mathbb{Z}$.

- a polynomial $f(z) \equiv (z+1)P(z,1)$ of order $n+1$ with $n+1$ roots of unity, for $N = 1$ and $y = 1$.

For symmetrical distributions, we may choose ω as the first moment ν_1 about the origin, so that the Rth moments about ν_1 will be zero valued for $R = $ odd > 1. By taking $\omega = \nu_1 = \rho N n$, equation (5) can be rewritten for moment generating functions (6) by the polynomial transformations of (7).

$$\text{pm}_{(egf)}(z,y) = \sum_{R=0}^{\infty} \sum_{h=0}^{Nn} (h - \rho N n + \alpha)^R C_h z^h y^{Nn-h} \frac{t^R}{R!} \tag{6}$$

$$= e^{(\alpha - \rho N n)t} y^{Nn} \sum_{h=0}^{Nn} C_h \left(\frac{z}{y}\right)^h e^{ht}$$

$$= e^{\alpha t} P^N\left(z e^{(1-\rho)t}, y e^{-\rho t}\right). \tag{7}$$

For normalized symmetrical distributions discussed later, we may want to change variables (z,y) to $(x\theta, x(1-\theta))$ with a real common multiplier x, in order to simplify the calculation of the moment generating functions.

3. PASCAL-DE MOIVRE COEFFICIENTS

The Pascal-DeMoivre coefficients $C_h(N,J)$ will represent the coefficients C_h in the polynomial expansion (1) of $P^N(z,y)$. For any positive integer combination of $N\&J$, these Pascal-DeMoivre coefficients $C_h(N,J)$ are generated from the expansion of $(z^{J-1} + z^{J-2}y + \cdots + y^{J-1})^N$ with variable $n = J - 1$ and coefficients $g_h = 1$ of a symmetrical $P(z,y)$ polynomial. The Pascal-DeMoivre polynomial (9) is thus a special case of the general polynomial equation (1), as shown by its generating function (8) and its homogeneous polynomial of symmetrical $C_h(N,J)$ coefficients (9).

$$P^N(z,y) = \left(\sum_{h=0}^{J-1} z^h y^{J-1-h} \right)^N = \left(\frac{z^J - y^J}{z - y} \right)^N \tag{8}$$

$$= \sum_{h=0}^{N(J-1)} C_h(N,J) z^h y^{N(J-1)-h}. \tag{9}$$

An explicit formula to calculate the multinomial $C_h(N,J)$ coefficients was given by DeMoivre in 1756 as:

$$C_h(N,J) = \sum_{a=0}^{A} (-1)^a \binom{h - aJ + N - 1}{N-1} \binom{N}{a},$$

where $A = \lfloor h/J \rfloor$ is the largest integer not exceeding the bracketed value.

N	$h=0$	1	2	3	4	5	6	7	8
1	1	1	1						
2	1	2	3	2	1				
3	1	3	6	7	6	3	1		
4	1	4	10	16	19	16	10	4	1

Table 1: Pascal-DeMoivre Triangle $C_h(N,3)$

An unique Pascal-DeMoivre triangle is created from the $C_h(N,J)$ coefficients for each positive integer value J. In the Table 1 with $J = 3$, the $C_h(N,J)$ terms of the Pascal-DeMoivre triangle are shown for row numbers $1 \leq N \leq 4$. Each $C_h(N,J)$ sequence of $N(J-1)+1$ nonzero terms comes from the entries in the Nth row of the Pascal-DeMoivre triangle.

With z replaced by $-z$ in (8,9), the alternating signed counterpart to the Pascal-DeMoivre polynomial is the expansion $(y^{J-1} - zy^{J-2} + \cdots + (-z)^{J-1})^N$, shown with its generating function (10) and $(-1)^h C_h(N,J)$ coefficients (11):

$$P^N(-z,y) = \left(\sum_{h=0}^{J-1} (-1)^h z^h y^{J-1-h} \right)^N = \left(\frac{y^J - (-z)^J}{y + z} \right)^N \tag{10}$$

$$= \sum_{h=0}^{N(J-1)} (-1)^h C_h(N,J) z^h y^{N(J-1)-h}. \tag{11}$$

The polynomial generating function at the right of equation (10) has two distinct numerator forms: $(y^J + z^J)^N$ and $(y^J - z^J)^N$, depending on whether J is an odd or even integer.

4. PASCAL-DE MOIVRE POLYNOMIAL MOMENTS

We will identify the Pascal-DeMoivre moments with $C_h(N,J)$ coefficients by **cm**, to differentiate them from the **pm** notation of the more general polynomials with C_h coefficients.

The structures of the Pascal-DeMoivre moment generating functions $cm_{(egf)}$ follow the variable substitution method from (5,7) for the $P^N(\pm z, y)$ polynomial forms (8,10).

Using the polynomial generator of (8) with the variable transformations from (5,7), we obtain the Pascal-DeMoivre moment generating formulas:

$$cm_{(egf)}(z,y) = \sum_{R=0}^{\infty} cm_R(z,y) \frac{t^R}{R!} = P^N(ze^t, y)e^{(\alpha - \omega)t} \tag{12}$$

$$= \left(\frac{z^J e^{Jt} - y^J}{ze^t - y}\right)^N e^{(\alpha - \omega)t} \tag{13}$$

$$= \left(\frac{z^J e^{(1-\rho)Jt} - y^J e^{-\rho Jt}}{ze^{(1-\rho)t} - ye^{-\rho t}}\right)^N e^{\alpha t}. \tag{14}$$

From the alternating signed polynomial generator in (10) with the polynomial variables (z,y) replaced according to (5,7), we derive the corresponding Pascal-DeMoivre formulas:

$$cm_{(egf)}(-z,y) = \sum_{R=0}^{\infty} cm_R(-z,y) \frac{t^R}{R!} = P^N(-ze^t, y)e^{(\alpha - \omega)t} \tag{15}$$

$$= \left(\frac{y^J - (-z)^J e^{Jt}}{y + ze^t}\right)^N e^{(\alpha - \omega)t} \tag{16}$$

$$= \left(\frac{y^J e^{-\rho Jt} - (-z)^J e^{(1-\rho)Jt}}{ye^{-\rho t} + ze^{(1-\rho)t}}\right)^N e^{\alpha t}. \tag{17}$$

Although we could simply use (12,13,14) with $z = -z$, we choose to keep equations (15,16,17) as a distinct reference for some unique features of alternating signed generating functions, such as their dependence on the J value.

Optimal values for ω in (13,16), as the first moment ν_1 about the origin will be derived later in the discussion of moments as exponential infinite sums. For example, the optimal ω values, and their related ρ values in (14,17), may be selected to make zero the odd moments about ν_1 for symmetrical distributions, like the Pascal-DeMoivre coefficients.

5. PASCAL-DE MOIVRE COEFFICIENT MOMENTS

The moment generating functions for the Pascal-DeMoivre coefficients are obtained from the polynomial $P(z,y)$ with $(z,y) = (1,1)$. For these symmetrical distributions, the arithmetic mean M is calculated over $0 \le h \le N(J-1)$. If we take $\omega = M = N(J-1)/2$ and $\rho = 1/2$, the Pascal-DeMoivre coefficient formulas (13,16) will have a simple closed exponential form and formulas (14,17) will have a hyperbolic structure.

For polynomial $P(z,y)$ at $(z,y) = (1,1)$, the closed exponential form of (13) becomes the $cm_{(egf)}$ equation (18). The hyperbolic form of (14) is given by the half angle equation in (19). Using the half-argument formulas from [7], we rewrite equation (19) as the full angle hyperbolic equation of (20).

$$cm_{(egf)}(1,1) = \left(\frac{e^{Jt}-1}{e^t-1}\right)^N \frac{e^{\alpha t}}{e^{N(J-1)t/2}} \tag{18}$$

$$= \left(\frac{\sinh{(Jt/2)}}{\sinh{(t/2)}}\right)^N e^{\alpha t} \tag{19}$$

$$= \left(\frac{\cosh{(Jt)}-1}{\cosh{(t)}-1}\right)^{N/2} e^{\alpha t}. \tag{20}$$

For polynomial $P(-z,y)$ at $(z,y) = (1,1)$ with odd J values, the exponential formula from (16) gives $cm_{(egf)}$ equation (21). The hyperbolic equation (17) yields the half angle and full angle formulas for $cm_{(egf)}$ in (22,23).

$$\left.\begin{array}{r} cm_{(egf)}(-1,1) \\ J = \text{odd} \end{array}\right\} = \left(\frac{1+e^{Jt}}{1+e^t}\right)^N \frac{e^{\alpha t}}{e^{N(J-1)t/2}} \tag{21}$$

$$= \left(\frac{\cosh{(Jt/2)}}{\cosh{(t/2)}}\right)^N e^{\alpha t} \tag{22}$$

$$= \left(\frac{\cosh{(Jt)}+1}{\cosh{(t)}+1}\right)^{N/2} e^{\alpha t}. \tag{23}$$

For polynomial $P(-z,y)$ at $(z,y) = (1,1)$ with even J values, the exponential equation in (16) yields the $cm_{(egf)}$ equation (24). The hyperbolic formula in (17) also becomes the half angle and full angle equations in (25,26).

$$\left.\begin{array}{r} cm_{(egf)}(-1,1) \\ J = \text{even} \end{array}\right\} = \left(\frac{1-e^{Jt}}{1+e^t}\right)^N \frac{e^{\alpha t}}{e^{N(J-1)t/2}} \tag{24}$$

$$= \left(\frac{-\sinh{(Jt/2)}}{\cosh{(t/2)}}\right)^N e^{\alpha t} \tag{25}$$

$$= \left(\frac{1-\cosh{(Jt)}}{1+\cosh{(t)}}\right)^{N/2} e^{\alpha t}. \tag{26}$$

The hyperbolic formulas (20,23,26) with $(z,y) = (1,1)$ can also be created from $P^N(z,y)$ in (8,10) by replacing z^J by cosh (Jt) and N by $N/2$.

All exponential forms in (18,21,24) simplify to the generating functions $\mathbf{cm}_{(egf)}$ for moments about the origin, when $\alpha = M$ and $e^{(\alpha - M)t} = 1$. The hyperbolic versions in (19,20,22,23,25,26) become the generating functions $\mathbf{cm}_{(egf)}$ for moments about the mean M, when $\alpha = 0$ and $e^{\alpha t} = 1$.

And whenever the polynomial variables (z,y) are equal in (13,14,16,17), these simple exponential and hyperbolic forms are maintained as

$$\mathbf{cm}_{(egf)}(\pm z, z) = z^M \mathbf{cm}_{(egf)}(\pm 1, 1). \tag{27}$$

6. THE DISTRIBUTION MOMENTS

Distribution moments are defined as the moments of the distribution's probability densities. The Pascal-DeMoivre distribution moments m_R are coefficients in their generating function expansion (28), with the coefficient moments reduced by a normalization factor cn. The normalization is taken so that the first finite nonzero moment $m_R = \mathbf{cm}_R/cn$ shall equal one.

$$m_{(egf)} = \sum_{R=0}^{\infty} m_R \frac{t^R}{R!} = \frac{1}{cn} \sum_{R=0}^{\infty} \mathbf{cm}_R \frac{t^R}{R!} = \frac{\mathbf{cm}_{(egf)}}{cn}. \tag{28}$$

The normalization factor cn in (28) is the first nonzero coefficient moment $\mathbf{cm}_{kN}$ where k is minimal, or equivalently when the least k^{th} derivative of $P(ze^t, y)$ in (12) is nonzero:

$$cn = \mathbf{cm}_{kN} = \left(\frac{d^k}{dt^k} P(ze^t, y) \Big|_{t=0} \frac{t^k}{k!} \right)^N, \tag{29}$$

where $cn = \mathbf{cm}_0 = P^N(z,y)$ if $P(z,y) \neq 0$, or

$= \mathbf{cm}_N = (P'(ze^t, y)t)^N$ if $P(z,y) = 0$ and $P'(ze^t, y) \neq 0$, etc.

And we shall see that only those derivatives at $k = \{0,1\}$ are required to obtain nonzero Pascal-DeMoivre coefficient moments.

For positive coefficient moments obtained from (12), the first nonzero moment $\mathbf{cm}_R$ occurs at $R = 0$ since $P(z,y) \neq 0$, except for $z = -y$ when $J =$ even. And for that exception, the first nonzero moment $\mathbf{cm}_R$ occurs at $R = N$ where $P'(ze^t, y)$ computed at $t = 0$ is $\frac{J}{2}z^{(J-1)}$. Thus for the generating function (28) of these distribution moments, we have the normalization factors $cn = \mathbf{cm}_0$ or $\mathbf{cm}_N$, with values obtained from (13,29) and the $P(z,y)$ formulas of (8,9):

$$cn = cn(z,y) = \begin{cases} \left(\dfrac{x^J - y^J}{x - y}\right)^N & \text{for } z \neq \pm y \\[2ex] z^{N(J-1)}J^N & \text{for } z = +y \\[2ex] z^{N(J-1)} & \text{for } z = -y \text{ and } J = \text{odd} \\[2ex] z^{N(J-1)}\left(\dfrac{Jt}{2}\right)^N & \text{for } z = -y \text{ and } J = \text{even} \end{cases} \tag{30}$$

For alternating signed coefficient moments from (15), the first nonzero R^{th} moment cm_R generally occurs at $R = 0$ because $P(-z,y) \neq 0$. With z replaced by $-z$ in (29), the corresponding normalization factor cn equals $cm_0 = P(-z,y)$, except for $z = y$ when $J = $ even. In that case, the first nonzero coefficient moment occurs at $R = N$ with $P'(-ze^t,y)$ computed in (29) at $t = 0$ to be $-\frac{J}{2}z^{(J-1)}$. The normalization factors cn for the generating function (28) of the alternating signed distribution moments can thus be gotten from (16,29) and the $P(-z,y)$ formulas from (10,11):

$$cn = cn(-z,y) = \begin{cases} \left(\dfrac{y^J + z^J}{y + z}\right)^N & \text{for } z \neq \pm y \text{ and } J = \text{odd} \\[2ex] \left(\dfrac{y^J - z^J}{y + z}\right)^N & \text{for } z \neq \pm y \text{ and } J = \text{even} \\[2ex] (-z)^{N(J-1)}J^N & \text{for } z = -y \\[2ex] z^{N(J-1)} & \text{for } z = +y \text{ and } J = \text{odd} \\[2ex] z^{N(J-1)}\left(-\dfrac{Jt}{2}\right)^N & \text{for } z = +y \text{ and } J = \text{even} \end{cases} \tag{31}$$

For the distribution moment calculations, we may decompose variables (z,y) by a real common multiplier x, thus rewriting the original polynomial as $P(xp, xq)$. Using this variable substitution in equations (9,11,29) allows the distribution moment equations to take the reduced form given by

$$m_{(egf)}(xp, xq) = \frac{P^N(xpe^t, xq)e^{(\alpha - \omega)t}}{cn(xp, xq)} = \frac{P^N(pe^t, q)}{cn(p,q)}e^{(\alpha - \omega)t}, \tag{32}$$

which is independent of the common factor x for this variable selection in the Pascal-DeMoivre polynomials. We may choose that the variables be related by $q = f(p)$, say for example $p + q = 1$ which would give

$$m_{(egf)}(xp, x(1-p)) = \frac{P^N(pe^t, 1-p)}{cn(p, 1-p)} e^{(\alpha - \omega)t}, \tag{33}$$

reducing the distribution moment's dependence to just a single variable p.

7. SPECIAL MOMENT CASES

Using the notation in [1], the generating functions (28) for the distribution moments in two special moment cases can be uniquely identified by:

$$m_{(egf)} = \begin{cases} \nu_{(egf)} & \text{for moments about the origin} \\ \\ \mu_{(egf)} & \text{for moments about the mean } (M), \end{cases} \tag{34}$$

with $\alpha = \omega$ for $\nu_{(egf)}$ and $(\alpha, \omega) = (0, M)$ for $\mu_{(egf)}$, and applied in (12,15,28).

7.1 The Binomial Distribution

For positive distribution moments obtained from (13,28,30), the generating function in the binomial case for $J = 2$, in all cases when $z \neq -y$, becomes

$$m_{(egf)}(z, y) = \frac{1}{P^N(z, y)} \left(\frac{z^2 e^{2t} - y^2}{ze^t - y} \right)^N e^{(\alpha - \omega)t}$$

$$= P^{-N}(z, y) \left(ze^t + y \right)^N e^{(\alpha - \omega)t}, \tag{35}$$

where we may select ω as the mean $M = N(J-1)/2 = N/2$ for $J = 2$.

If we choose special values for variables (z, y) as $(x\theta, x(1-\theta))$, we get $P(z, y) = x \sum_{h=0}^{1} \theta^h (1-\theta)^{1-h} = x$. So for the generating function of the polynomial moments about the origin, with $\alpha = M$, the binomial equation (35) becomes the known moment generating function in [1,7] as

$$m_{(egf)}(z, y) = \nu_{(egf)}(x\theta, x(1-\theta)) = \frac{(x\theta e^t + x(1-\theta))^N}{x^N} = (\theta e^t + (1-\theta))^N. \tag{36}$$

Looking at the symmetrical binomial coefficients themselves, we choose variables $(z, y) = (1, 1)$. Since the Nth row sum of all Pascal-DeMoivre coefficients is J^N, the binomial coefficient sum of any Pascal triangle row equals $P^N(z, y) = 2^N$, so that the moment generating function (35) is written as

$$m_{(egf)}(1, 1) = \left(\frac{1}{2}\right)^N (e^t + 1)^N e^{(\alpha - M)t}. \tag{37}$$

In the case $\alpha = M = N/2$, the binomial moment equation (37) gives the moment generating function $\nu_{(egf)}$ in (38) for the symmetrical $C_h(N, J)$ coefficients. This $\nu_{(egf)}$ form

(38) is the same as equation (36) with $\theta = \frac{1}{2}$ as the symmetry condition. And when $\alpha = 0$, the moment equation in (37) becomes equation (39) for $\mu_{(egf)}$ in exponential and hyperbolic formats.

$$\nu_{(egf)}(1,1) = 2^{-N}(e^t + 1)^N, \tag{38}$$

$$\mu_{(egf)}(1,1) = 2^{-N}(e^t + 1)^N(e^{-Nt/2}) = (\cosh(t/2))^N. \tag{39}$$

The coefficient moment generators for the binomial coefficients are just the equations (38,39) times the normalizing factor $cn = 2^N$, giving:

$$c\nu_{(egf)}(1,1) = (e^t + 1)^N, \tag{40}$$

$$c\mu_{(egf)}(1,1) = (2\cosh(t/2))^N = (1 + \cosh t)^{N/2}. \tag{41}$$

7.2 Binomial Coefficient Moments - Stirling Numbers

In the alternating signed coefficient case for polynomial $P(-z, y)$, the coefficient moment generating function (16) at $J = 2$ reduces to

$$cm_{(egf)}(-z, y) = \left(\frac{y^2 - z^2 e^{2t}}{y + ze^t}\right)^N \frac{e^{\alpha t}}{e^{\omega t}} = (y - ze^t)^N e^{(\alpha - \omega)t}. \tag{42}$$

By taking variables $(z, y) = (1, 1)$ and choosing $\omega = M = N/2$, the generating function (42) for the binomial coefficients becomes:

$$cm_{(egf)}(-1, 1) = (1 - e^t)^N e^{(\alpha - N/2)t} \tag{43}$$

$$= (-\sinh(t/2))^N e^{\alpha t}.$$

The referee observed that equation (43) gives the generating function for weighted Stirling numbers $S(R, N, x)$ of the second kind, so we can write

$$cm_{(egf)}(-1, 1) = (-1)^N N! \sum_{R=N}^{\infty} S\left(R, N, \alpha - \frac{N}{2}\right) \frac{t^R}{R!}. \tag{44}$$

7.3 The Trinomial Distribution

The distribution moments for positive trinomial coefficients $C_h(N, J)$ at $J = 3$ are obtained from equation (18), reduced by the normalization factor cn from (30) with $cn = P^N(z, y) = 3^N$ at $z = y = 1$, as

$$m_{(egf)}(1,1) = \left(\frac{1}{3}\right)^N \left(\frac{e^{3t}-1}{e^t-1}\right)^N e^{(\alpha-N)t}$$

$$= 3^{-N}(1+2\cosh t)^N e^{\alpha t}, \tag{45}$$

since we take $\omega = M = N(J-1)/2 = N$ for $J = 3$.

Trinomial moments for alternating signed Pascal-DeMoivre coefficients $(-1)^h C_h(N,J)$ at $J = 3$ are obtained from their generating function (21), normalized per (31) by $cn = P^N(-z,y) = 1$ at $z = y = 1$, as

$$m_{(egf)}(-1,1) = \mathbf{cm}_{(egf)}(-1,1) = \left(\frac{1+e^{3t}}{1+e^t}\right)^N e^{(\alpha-N)t}$$

$$= ((2\cosh t)-1)^N e^{\alpha t}. \tag{46}$$

7.4 Spaced Coefficient Moments

A version of the Pascal-DeMoivre polynomial generating function (12,13) evaluates the Rth coefficient moments with an initial displacement value k and a spacing variable d, as defined by the spaced coefficient moment formula:

$$\mathbf{scm}_{(egf)}(z,y) = \sum_{R=0}^{\infty} \sum_{h=0}^{N(J-1)} (dh-k)^R C_h(N,J) z^h y^{N(J-1)-h} \frac{t^R}{R!}$$

$$= \left(\frac{z^J e^{Jdt} - y^J}{z e^{dt} - y}\right)^N e^{-kt}, \tag{47}$$

with variable k in (3) replaced by k/d and t replaced by dt. With the distance k scaled by the spacing variable d, these spaced coefficient moments correspond to coefficient moments about a point k/d from the origin.

The polynomial generating function (15,16) for alternating signed Pascal-DeMoivre coefficients becomes the spaced coefficient moment as

$$\mathbf{scm}_{(egf)}(-z,y) = \sum_{R=0}^{\infty} \sum_{h=0}^{N(J-1)} (-1)^h (dh-k)^R C_h(N,J) z^h y^{N(J-1)-h} \frac{t^R}{R!}$$

$$= \left(\frac{y^J - (-z)^J e^{Jdt}}{y + z e^{dt}}\right)^N e^{-kt}. \tag{48}$$

If we choose $(N,J) = (1, n+1)$ and $z = y = 1$, generating functions (47,48) become the positive and alternating signed sum of powers equations in [5]. Using the general formulas of (47,48), explicit formulas for the Rth spaced coefficient moments $\mathbf{scm}_R$ are presented in [3].

8. THE EXPONENTIAL INFINITE SUMS

The Pascal-DeMoivre coefficient moments $\mathbf{cm}_R(z,y)$ can be described as Sheffer sequences, which allows their exponential generating functions to be expressible in exponential form, according to the Sheffer criterion that

$$\mathbf{cm}_{(egf)}(z,y) = \sum_{R=0}^{\infty} \mathbf{cm}_R(z,y) \frac{t^R}{R!} = \mathcal{A}(t)\, \mathrm{Exp}\{N\mathcal{B}(t)\}, \tag{49}$$

where $\mathcal{A}(t) = \sum_{r=0}^{\infty} a_r t^r$ and $\mathcal{B}(t) = \sum_{r=1}^{\infty} b_r t^r$. From the section on distribution moments, we see that $\mathcal{A}(t)$ is the normalization factor cn from (30,31), which either has a constant value or is a constant times t^N.

The exponential representation for the distribution moment generating function $m_{(egf)}$ as $\mathrm{Exp}\{N\mathcal{B}(t)\}$ in (49) was made possible by the normalization of the coefficient moment generating function $\mathbf{cm}_{(egf)}$ to have an initial constant term of one. In generating function theory, only those series with constant term of 1 can formally be expressed as exponentials of power series.

We will derive the $\mathcal{B}(t)$ power series coefficients, corresponding to generating functions $m_{(egf)}$, the normalized forms of polynomial moments (13,16) and coefficient moments (18,21,24). Whereas these formulas are ratios of polynomials, like $(e^{Jt}-1)^n/(e^t-1)^N$ in (18), the exponential series (49) is a simple coefficient difference in a unique number system. The exponential infinite sum format exposes the fundamental number sequence $\mathcal{N}_r$ for each Pascal-DeMoivre moment generating function as $\mathcal{B}(t) = \sum_{r=1}^{\infty} \beta_r \mathcal{N}_r t^r$.

First we show that the Euler and Bernoulli number sequences are the fundamental number sequences $\mathcal{N}_r$ associated with the coefficient moments (18,21,24) whose variables $(z,y) = (\pm 1, 1)$. Later we present the Eulerian numbers as the number sequences $\mathcal{N}_r$ for polynomial moments (13,16) in the general $(\pm z, y)$ format.

8.1 Coefficient Moments

To examine the moment generating function (18) for positive coefficient moments, we look to the Bernoulli number sequences. The Bernoulli numbers B_r are the coefficients in the expansion (50) of their exponential generating function [7]. The first few Bernoulli numbers are $\left\{1, -\frac{1}{2}, \frac{1}{6}, 0, -\frac{1}{30}, \cdots\right\}$ for $r \geq 0$.

$$B_{(egf)} = \frac{t}{e^t - 1} = \sum_{r=0}^{\infty} B_r \frac{t^r}{r!}. \tag{50}$$

If we take equation (50), replace t by $-t$, subtract 1, divide both sides by t, and integrate both sides from 0 to T, we obtain

$$\log\left(\frac{e^T - 1}{T}\right) = \sum_{r=1}^{\infty} \frac{(-1)^r B_r T^r}{r \cdot r!}. \tag{51}$$

We exponentiate both sides of (51), then raise it to the Nth power and replace T by Jt, to get the Bernoulli representation for

$$\left(\frac{e^{Jt} - 1}{Jt}\right)^N = \text{Exp}\left\{ N \sum_{r=1}^{\infty} \frac{(-1)^r B_r J^r}{r} \frac{t^r}{r!}\right\}. \tag{52}$$

We note that equation (52) is the reciprocal of a generating function for the generalized Bernoulli numbers $B_r^{(N)}$.

By repeated application of $(e^{Jt} - 1)^N$ from (52), we get the exponential equivalent of the moment generating function for the Pascal-DeMoivre coefficients from (18) as:

$$\text{cm}_{(egf)}(1,1) = \left(\frac{e^{Jt} - 1}{e^t - 1}\right)^N \frac{e^{\alpha t}}{e^{N(J-1)t/2}}$$

$$= J^N \text{Exp}\left\{ (\alpha - M)t + N \sum_{r=1}^{\infty} \frac{(-1)^r B_r (J^r - 1)}{r} \frac{t^r}{r!}\right\} \tag{53}$$

$$= J^N \text{Exp}\left\{ \alpha t + N \sum_{r=2}^{\infty} \frac{(-1)^r B_r (J^r - 1)}{r} \frac{t^r}{r!}\right\}, \tag{54}$$

since the mean $M = N(J-1)/2$ and $B_1 = -1/2$.

Next we examine the moment generating function (21) for alternating signed coefficient moments by identifying a number sequence having Euler polynomial values. Euler values E_r are defined by equation (55) as coefficients in the expansion of the generating function [7]. Not to be confused with Euler numbers or Euler constants which have different mathematical definitions, the Euler values in this paper are the constant terms (as coefficients of x^0) in Euler polynomials $E_r(x)$ with the first few values of $\left\{1, -\frac{1}{2}, 0, \frac{1}{4}, 0, -\frac{1}{2}, \cdots\right\}$.

$$E_{(egf)} = \frac{2}{e^t + 1} = \sum_{r=0}^{\infty} E_r \frac{t^r}{r!}. \tag{55}$$

The relationship between Euler values and Bernoulli numbers can be expressed in the known identity:

$$E_{(r-1)} = 2(1 - 2^r)\frac{B_r}{r}. \tag{56}$$

Taking the Euler value generating function (55), we replace t by $-t$, divide by 2, and integrate from 0 to T, and get the identity:

$$\log\left(\frac{e^T+1}{2}\right) = \sum_{r=1}^{\infty} \frac{(-1)^r E_{(r-1)} T^r}{(-2)\cdot r!}. \tag{57}$$

If we exponentiate equation (57) and take both sides to the Nth power, the Euler value representation of the distribution moments is given in the form of equation (58) with T replaced by Jt.

$$\left(\frac{e^{Jt}+1}{2}\right)^N = \mathrm{Exp}\left\{N \sum_{r=1}^{\infty} \frac{(-1)^r E_{(r-1)} J^r}{(-2)} \frac{t^r}{r!}\right\}. \tag{58}$$

We could describe equation (58) as the reciprocal of the generating function for generalized Euler values $E_r^{(N)}$.

Repeated application of $(e^{Jt}+1)^N$ from (58) gives the exponential equivalent of the moment generating function for the alternating signed coefficients from (21) as:

$$\begin{aligned}
\left.\begin{array}{c} \mathbf{cm}_{(egf)}(-1,1) \\ J = \text{odd} \end{array}\right\} &= \left(\frac{1+e^{Jt}}{1+e^t}\right)^N \frac{e^{\alpha t}}{e^{N(J-1)t/2}} \\[2mm]
&= \mathrm{Exp}\left\{(\alpha - M)t + N \sum_{r=1}^{\infty} \frac{(-1)^r E_{(r-1)}(J^r-1)}{(-2)} \frac{t^r}{r!}\right\} \tag{59} \\[2mm]
&= \mathrm{Exp}\left\{\alpha t + N \sum_{r=2}^{\infty} \frac{(-1)^r E_{(r-1)}(J^r-1)}{(-2)} \frac{t^r}{r!}\right\}. \tag{60}
\end{aligned}$$

Applying $(e^{Jt}-1)^N$ from (52) and $(e^t+1)^N$ from (58), the exponential equivalent of the moment generating function for the alternating signed coefficients from (24) becomes equation (61). Substitution of the Bernoulli number identity (56) for the Euler values in (61) gives equation (62).

$$\begin{aligned}
\left.\begin{array}{c} \mathbf{cm}_{(egf)}(-1,1) \\ J = \text{even} \end{array}\right\} &= \left(\frac{1-e^{Jt}}{1+e^t}\right)^N \frac{e^{\alpha t}}{e^{N(J-1)t/2}} \\[2mm]
&= \left(\frac{-Jt}{2}\right)^N \mathrm{Exp}\left\{(\alpha - M)t + N \sum_{r=1}^{\infty} (-1)^r\left(\frac{B_r J^r}{r} + \frac{E_{(r-1)}}{2}\right)\frac{t^r}{r!}\right\} \tag{61} \\[2mm]
&= \left(\frac{-Jt}{2}\right)^N \mathrm{Exp}\left\{\alpha t + N \sum_{r=2}^{\infty} \frac{(-1)^r B_r(J^r - 2^r + 1)}{r} \frac{t^r}{r!}\right\}. \tag{62}
\end{aligned}$$

We note that the derivation of equations (54,60,62) proves and generalizes the conjecture in [2] that the moment generating equations, for $\nu_{(egf)}(\pm 1, 1)$ and $\mu_{(egf)}(\pm 1, 1)$, can be expressed by these exponential infinite sums with Bernoulli number and Euler value coefficients.   This result can also be proven by using a more elaborate algorithmic transformation from [4].

8.2 Polynomial Moments

The moment generating functions (13,16) for the Pascal-DeMoivre polynomial moments will use the properties of the Eulerian number sequence, in order to accommodate both polynomial variables (z, y). An explicit formula for the Eulerin number $\left\langle {r \atop h} \right\rangle$ is given by

$$\left\langle {r \atop h} \right\rangle = \sum_{i=0}^{h} (-1)^i \binom{r+1}{i} (h-i)^r,$$

which will create the entries for the Euler number triangle in Table 2.

r	$h = 0$	1	2	3	4
0	1				
1	0	1			
2	0	1	1		
3	0	1	4	1	
4	0	1	11	11	1

Table 2: Eulerian Number Triangle $\left\langle {r \atop h} \right\rangle$

The polynomial moment generating function (13) of positive terms can be written in exponential format (49) with the aid of these Eulerian numbers. Based on an exponential generating function given by Euler [6] in 1755, we write the Eulerian numbers $\left\langle {r \atop h} \right\rangle$ as coefficients in the expansion

$$\frac{1-\lambda}{e^t - \lambda} = \sum_{r=0}^{\infty} \sum_{h=0}^{r} \left\langle {r \atop h} \right\rangle \lambda^{r-h} \frac{t^r}{(\lambda-1)^r \cdot r!}. \tag{63}$$

If we take equation (63), subtract $1 - 1/\lambda$, then divide by $-(1 - 1/\lambda)$, and integrate from 0 to T, we get

$$\log\left(\frac{e^T - \lambda}{1 - \lambda}\right) = \frac{T}{1-\lambda} - \sum_{r=2}^{\infty} \sum_{h=0}^{r-1} \left\langle {r-1 \atop h} \right\rangle \lambda^{r-h} \frac{T^r}{(\lambda-1)^r \cdot r!}. \tag{64}$$

Exponentiating equation (64), replacing T by Jt, and taking it to the power N, we obtain

$$\left(\frac{e^{Jt}-\lambda}{1-\lambda}\right)^N = \mathrm{Exp}\left\{\frac{NJt}{1-\lambda} - N\sum_{r=2}^{\infty}\sum_{h=0}^{r-1}\left\langle {r-1 \atop h} \right\rangle \lambda^{r-h}\frac{(Jt)^r}{(\lambda-1)^r \cdot r!}\right\}. \tag{65}$$

By repeatedly applying $(e^{Jt}-\lambda)^N$ from equation (65) and taking λ as the rational value y^J/z^J, we can write our generating function (13) for Pascal-DeMoivre moments as

$$\mathrm{cm}_{(egf)}(z,y) = \left(\frac{z^J e^{Jt}-y^J}{ze^t - y}\right)^N e^{(\alpha-\omega)t} =$$

$$\left(\frac{z^J - y^J}{z-y}\right)^N \mathrm{Exp}\left\{(\alpha-\omega+\nu_1)t - N\sum_{r=2}^{\infty}\sum_{h=0}^{r-1}\left\langle {r-1 \atop h}\right\rangle(\Phi_J^+ - \Phi_1^+)\frac{t^r}{r!}\right\}, \tag{66}$$

$$\text{where } \Phi_J^+ = J^r(z^J)^h(y^J)^{r-h}/(y^J - z^J)^r, \tag{67}$$

with $\nu_1 = NJz^J/(z^J - y^J) - Nz/(z-y)$.

The generating function (16) for alternating signed polynomial moments will also use Eulerian numbers to create the needed exponential coefficients. Replacing λ by $-\lambda$ in the positive signed formula (65), we create the corresponding alternating signed formula as

$$\left(\frac{e^{Jt}+\lambda}{1+\lambda}\right)^N = \mathrm{Exp}\left\{\frac{NJt}{1+\lambda} - N\sum_{r=2}^{\infty}\sum_{h=0}^{r-1}(-1)^h\left\langle {r-1 \atop h}\right\rangle \lambda^{r-h}\frac{(Jt)^r}{(1+\lambda)^r \cdot r!}\right\}. \tag{68}$$

By repeatedly applying $(e^{Jt}+\lambda)^N$ from equation (68) and taking λ as the rational value y^J/z^J, we can write our generating function (16) for Pascal-DeMoivre moments with $J = $ odd as

$$\mathrm{cm}_{(egf)}(-z,y) = \left(\frac{z^J e^{Jt}+y^J}{ze^t + y}\right)^N e^{(\alpha-\omega)t} =$$

$$\left(\frac{z^J + y^J}{z+y}\right)^N \mathrm{Exp}\left\{(\alpha-\omega+\nu_1)t - N\sum_{r=2}^{\infty}\sum_{h=0}^{r-1}(-1)^h\left\langle {r-1 \atop h}\right\rangle(\Phi_J^- - \Phi_1^-)\frac{t^r}{r!}\right\}, \tag{69}$$

$$\text{where } \Phi_J^- = J^r(z^J)^h(y^J)^{r-h}/(z^J + y^J)^r, \tag{70}$$

with $\nu_1 = NJz^J/(z^J + y^J) - Nz/(z+y)$.

When $J = $ even, the other generating function (16) for alternating signed polynomial moments uses the ratio of $(e^{Jt}-\lambda)^N$ from (65) and $(e^t + \lambda)^N$ from (68) with $\lambda = y^J/z^J$ to give

$$\mathbf{cm}_{(egf)}(-z,y) = \left(\frac{y^J - z^J e^{Jt}}{y + z e^t}\right)^N e^{(\alpha - \omega)t} =$$

$$\left(\frac{y^J - z^J}{y + z}\right)^N \operatorname{Exp}\left\{(\alpha - \omega + \nu_1)t - N \sum_{r=2}^{\infty} \sum_{h=0}^{r-1} \binom{r-1}{h}(\Phi_J^+ - (-1)^h \Phi_1^-)\frac{t^r}{r!}\right\}, \qquad (71)$$

with $\nu_1 = NJz^J/(z^J - y^J) - Nz/(z+y)$, and where Φ_J^+ comes from (67) and Φ_1^- comes from (70).

An identity found in [6] for the relationship between Eulerian numbers, Euler values and Bernoulli numbers can be derived from (56,60,69) by setting $(z,y) = (1,1)$ and equating coefficients to get

$$\frac{1}{2^r} \sum_{h=0}^{r-1} (-1)^{h+1} \binom{r-1}{h} = \frac{(-1)^r E_{(r-1)}}{(-2)} = (-1)^r (2^r - 1)\frac{B_r}{r}. \qquad (72)$$

As the number sequences are linked to the creation of moment generating functions, we hope further analysis of Pascal-DeMoivre polynomial moments will allow us to appreciation the structure of these special number sequences.

ACKNOWLEDGMENT

The author wishes to thank the anonymous referee for the stamina needed to review this document. The referee's insightful suggestions made the proofs more succinct and inspired the author to generalize the paper.

REFERENCES

[1]   Beyer, W.H. CRC Handbook of Tables For Probability And Statistics, Cleveland: Chemical Rubber Company, 1966: pp. 3, 16.

[2]   Ericksen, L. "The Pascal-DeMoivre Triangles." *The Fibonacci Quarterly*, Vol. *36.1* (1998): pp. 20-33.

[3]   Ericksen, L. "Sum of Powers From Uniform distribution Moments." *Journal of Statistical Planning and Inference*, submitted (1998).

[4]   Fielder, D.C. and Alford, C.O. "More Applications of a Partition Driven Symmetric Table." Applications of Fibonacci Numbers, Volume 6. Edited by G.E. Bergum, A.F. Horadam and A.N. Philippou, Dordrecht, The Netherlands, 1996: pp. 93-103.

[5]   Howard, F.T. "Sum of Powers of Integers Via Generating Functions." *The Fibonacci Quarterly*, Vol. *34.3* (1996): pp. 244-256.

L. ERICKSEN

[6]   Lehmer, D.H. "Generalized Eulerian Numbers." *Journal of Combinatorial Theory,*
        Series A, Vol. *32* (1982): pp. 195-215.

[7]   Zwillinger, D. CRC Standard Mathematical Tables and Formulae 30th ed. Boca Raton
        : CRC Press, 1996: pp. 17-19, 476-481, 581.

AMS Classification Numbers: 11B68, 33B10, 62E15

INVESTIGATING SPECIAL BINARY SEQUENCES WITH SOME COMPUTER HELP

Daniel C. Fielder and Cecil O. Alford

INTRODUCTION

In this note, we study exclusively n-length binary sequences where at least one 1 is adjacent to (or touching) another 1. For brevity, we refer to them as "binary sequences with some touching 1's." They are not to be confused with n-length binary sequences where every 1 is adjacent to at least one other 1 [1]. We let C_n denote the collection with unrestricted content, and we let D_n denote the subset of C_n with even content. (Content is defined as the number of 1's in a binary sequence.)

Vadja [5, p. 13] shows that the number of n-length binary sequences with no touching 1's equals the Fibonacci number F_{n+2}. Accordingly, the number of binary sequences with some touching 1's is $|C_n| = 2^n - F_{n+2}$.

Based on observations of enclosure patterns of 1's and 0's in tables of binary numbers, we develop a combinatorial, second expression for $|C_n|$. As a bonus, equating the two count expressions for $|C_n|$ leads to interesting Fibonacci identities.

We continue with the count, $|D_n|$, of binary sequences with some touching 1's which are restricted to even content. A novel feature of its count is the emergence of an oscillatory sequence which adjusts counts by regularly adding and subtracting values from Fibonacci sequences [2].

For completeness, each count sequence is accompanied by its closed generating function, difference equation, and general term. Because the difference equations are all linear with constant coefficients, we make use of the features of z-Transforms [3]. Since the z-Transform is

---

This paper is in final form and no version of it will be submitted for publication elsewhere.

a sequence in powers of $\frac{1}{z}$, z may be replaced by $\frac{1}{x}$ to obtain the more conventional expansions in x.

The computer language *Mathematica* [6] is used to calculate numerical and literal expansions and to compute entries for tables of representative values.

COUNTING WITH ENCLOSURE PATTERNS

Consider binary tables arranged in the conventional manner where the rows of binary sequences are ordered from 0 through 2^{n-1}. The columns are labeled in ascending order from right to left starting with column 0. *Enclosure patterns* are those binary arrangements with touching 1's shown in boxes, open on the right and having adjacent 1's on the left. Examples of binary tables with enclosure patterns for $n = 4$ and $n = 5$ are shown in Table 1.

N	$n=4$				N	$n=5$					N	$n=5$				
0	0 0 0 0				0	0 0 0 0 0					16	1 0 0 0 0				
1	0 0 0 1				1	0 0 0 0 1					17	1 0 0 0 1				
2	0 0 1 0				2	0 0 0 1 0					18	1 0 0 1 0				
3	0 0 1 1				3	0 0 0 1 1					19	1 0 0 1 1				
4	0 1 0 0				4	0 0 1 0 0					20	1 0 1 0 0				
5	0 1 0 1				5	0 0 1 0 1					21	1 0 1 0 1				
6	0 1 1 0				6	0 0 1 1 0					22	1 0 1 1 0				
7	0 1 1 1				7	0 0 1 1 1					23	1 0 1 1 1				
8	1 0 0 0				8	0 1 0 0 0					24	1 1 0 0 0				
9	1 0 0 1				9	0 1 0 0 1					25	1 1 0 0 1				
10	1 0 1 0				10	0 1 0 1 0					26	1 1 0 1 0				
11	1 0 1 1				11	0 1 0 1 1					27	1 1 0 1 1				
12	1 1 0 0				12	0 1 1 0 0					28	1 1 1 0 0				
13	1 1 0 1				13	0 1 1 0 1					29	1 1 1 0 1				
14	1 1 1 0				14	0 1 1 1 0					30	1 1 1 1 0				
15	1 1 1 1				15	0 1 1 1 1					31	1 1 1 1 1				

Table 1. Binary sequences and enclosure patterns for $n = 4$ and $n = 5$

The arrangement of binary sequences induces natural classifications for counting sequences with touching 1's based on the position of leftmost pairs of 1's regardless of any contiguous condition other than there be no touching 1's to the left of the chosen pair of 1's. This concept becomes clear from an illustration for $n = 5$.

y	y	0	1	1	$pattern(0)$
y	0	1	1	x	$pattern(1)$
y	1	1	x	x	$pattern(2)$
1	1	x	x	x	$pattern(3)$

The y's can be any mix of 1's and 0's except for no touching 1's. A 0 appears where necessary at the immediate left of the pair of 1's as a "non-1" buffer. The x's can be any mix of 1's and 0's. The above is general and all touching 1 possibilities are included.

We designate an enclosure as $pattern(\omega)$ according to the column number ω of the rightmost 1 of the pair of 1's. For $pattern(\omega)$ there are 2^ω ways of arranging the x's. This is because the ω free binary positions to the right can be filled with 2^ω rows of ω length binary sequences. There are $(n - \omega - 3)$ binary positions to the left of the buffer 0 for which the y's cannot have touching 1's. Thus, according to the previous section, the number of ways of arranging the y's in rows is $F_{n-\omega-1}$. From the product rule of combinatorics, $pattern(\omega)$ contributes $(2^\omega F_{n-\omega-1})$ sequences with some touching 1's. By summing over all patterns, we find a new way of counting sequences with some touching 1's. The new and old counting expressions are equated in (1) as

$$|C_n| = \sum_{k=1}^{(n-1)} 2^{n-k-1} F_k = 2^n - F_{n+2}, \quad n \geq 0 \tag{1}$$

(By convention, if the upper index is less than the lower index in a summation, the summation is zero.)

Rearrangement of (1) leads to other interesting identities. Here are two.

$$F_{n+2} + \sum_{k=2}^{(n-1)} 2^{n-k-1} F_k = 2^n\left(\frac{3}{4}\right), \quad n \geq 2 \tag{2}$$

$$F_n = 2^{n-2}\left(1 - \sum_{k=1}^{(n-3)} \frac{F_k}{2^{k+1}}\right), \quad n \geq 2 \tag{3}$$

In particular, we see that the table for $n = 5$ has 1 pattern of size $2^3 = 8$, 1 pattern of size $2^2 = 4$, 2 patterns of size $2^1 = 2$, and 3 patterns of size $2^0 = 1$. Thus, the number of sequences with some touching 1's can be counted as $1 \times 2^3 + 1 \times 2^2 + 2 \times 2^1 + 3 \times 2^0 = 19$. Note the obvious presence of the consecutive Fibonacci numbers, 1, 1, 2, 3, as coefficients within the sum. The count is also, of course, $|C_5| = 2^5 - F_7 = 19$.

DIFFERENCE EQUATION, GENERATING FUNCTIONS, GENERAL TERM

Generally, the difference equation of a sum of functions is not the simple sum of the difference equations for each function. To combine difference equations, we use a simple but effective approach. It is assumed that the initial conditions of all difference equations are known.

Instead of a simple sum or difference of difference equations, we found what we call an *adjusted* sum. The term "adjusted" is used because each difference equation is adjusted to have the same recursion appearance as every other difference equation in the sum, except for function designation. To explain further, suppose $f_1 + f_2 = f_3$. For the sum of the similarly appearing recursion portions we would interpret $f_1(n+2) + f_2(n+2)$ as $\{f_1 + f_2\}(n+2) = f_3(n+2)$, etc. All the extra terms accumulated in the adjusting process are totaled to determine the inhomogeneous part of the resultant difference equation.

The method is illustrated in finding the difference equation for $f_3(n) = 2^n - F_{n+2}$, where $f_1(n) = 2^n$ and $f_2(n) = F_{n+2}$.

The difference equation for 2^n is $f_1(n+1) - 2f_1(n) = 0$ and that for F_{n+2} is $f_2(n+2) - f_2(n+1) - f_2(n) = 0$. The equations (with the adjusted inhomogeneous part for each in **bold** type) become

$$f_1(n+2) - f_1(n+1) - f_1(n) - \boldsymbol{f_1(n+1)} + \boldsymbol{f_1(n)} = 0 \tag{4}$$

$$f_2(n+2) - f_2(n+1) - f_2(n) = 0 \tag{5}$$

$$f_3(n+2) - f_3(n+1) - f_3(n) - \boldsymbol{f_1(n+1)} + \boldsymbol{f_1(n)} = 0 \tag{6}$$

We have $(4) - (5) = (6)$. If we display (6) in letter sequence terms and evaluate the inhomogeneous part as 2^n, we have the difference equation for the count of n-length binary sequences with some touching 1's as (7). The initial conditions are $c_0 = 0$ and $c_1 = 0$.

$$c_{n+2} - c_{n+1} - c_n = 2^n \tag{7}$$

Through z-Transform methods [3], the closed form generating functions in z is found and the corresponding function in x, where $x = \frac{1}{z}$, follows.

$$f_3(z) = \frac{z}{z^3 - 3z^2 + z + 2}, f_3\left(\frac{1}{x}\right) = g_3(x) = \frac{x^2}{1 - 3x + x^2 + 2x^3} \tag{8}$$

By expanding $g_3(x)$ from (8), we obtain the conventional series including the general term, $(2^n - F_{n+2})x^n$, as

$$g_3(x) = 0 + 0x + x^2 + 3x^3 + 8x^4 + 19x^5 + 43x^6 + 94x^7 + 201x^8 + 423x^9$$

$$+ 880x^{10} + 1815x^{11} + \cdots + (2^n - F_{n+2})x^n + \cdots \tag{9}$$

SEQUENCES WITH SOME TOUCHING ONES AND EVEN CONTENT

In studying restricted distribution of 1's and 0's in binary sequences, the counts are often symmetrically divided as even-odd, 1's-0's, etc., so that knowing one type of count immediately gives the other. As a case in point, Liu [4, p.7] includes a discussion of equality of sequence counts among various odd and even 1's and 0's distributions. Preliminary *Mathematica* work, however indicated that having <u>close</u> to (but not exactly) half as many sequences with even content as with full content occurs frequently. The "exactly half" condition occurs regularly but less frequently.

Our earlier approach with patterns is useful here. All $pattern(\omega)$'s for $0 < \omega \le n - 2$ contain an even number of binary sequences with some touching 1's since that count is 2^{ω}. The value of the first sequence in these patterns is always even, and thus has a 0 in the 0 position. The next value is odd, necessitating the same 1's elsewhere but with a 1 in the 0 position. The regular binary table arrangement induces consecutive pairs of binary sequences with even-odd content. We see that exactly half of the $pattern(\omega)$'s for $0 < \omega \le n - 2$ have an even number of 1's.

Each $pattern(0)$, however, has but one sequence because $2^0 = 1$. That sequence has exactly two touching 1's, and they are in the rightmost positions. Since there are F_{n-1} pattern(0)'s, whenever F_{n-1} is odd, it is impossible for half of them to have an even number of 1's. The $pattern(0)$ count is responsible for all count asymmetries in D_n.

$Pattern(0)$ Contribution to $|D_n|$. Since each $pattern(0)$ sequence in D_n always has exactly two touching 1's and a 0 buffer immediately to their left, it is sufficient to find the count of $(n-3)$-length non-touching binary sequences with even content. To simplify the calculations, we let $m = n - 3$ and convert back to n as convenient later.

Let G_m be the set of m-length binary sequences with no touching 1's and even content, and let H_m be the comparable set with odd content. Sequences from G_m and H_m each consist of a 0 or a 1 followed by $m-1$ binary digits with no touching 1's. The part of G_m starting with 0 must be G_{m-1} to maintain even content. The part of G_m starting with 1 must be followed by part of H_{m-1} to maintain even content. However, it cannot be that part of H_{m-1} which starts with a 1 because of the non-touching 1 condition. That part of H_{m-1} which starts with a 0 is H_{m-2}. Thus, $G_m = 0G_{m-1}+1, 0H_{m-2}$. Through similar reasoning, $H_m = 0H_{m-1}+1, 0G_{m-2}$. When index-adjusted for z-Transform application [3], there results the simultaneous difference equations, $g_{m+2} = g_{m+1} + h_m$ and $h_{m+2} = h_{m+1} + g_m$. By subtracting g_{m+3} from g_{m+4}, replacing $h_{m+2} - h_{m+1}$ by g_m and transposing across the equals sign, the following index-adjusted, 4th order homogeneous difference equation with initial conditions $g_0 = g_1 = g_2 = 1$ and $g_3 = 2$ is obtained.

$$g_{m+4} - 2g_{m+3} + g_{m+2} - g_m = 0 \tag{10}$$

The closed z-Transform of (10) subject to the initial conditions becomes

$$\frac{z^4 - z^3 + z}{z^4 - 2z^3 + z^2 - 1} = \frac{z^4 - z^3 + z}{(z^2 - z - 1)(z^2 - z + 1)} = \frac{1}{2}\left[\frac{z^2 + z}{(z^2 - z - 1)} + \frac{z^2 - z}{(z^2 - z + 1)}\right] \tag{11}$$

Upon application of the inverse z-Transform and restorative replacement of m by $n-3$, we obtain from (11)

$$|\text{even content } pattern(0)\text{'s}| = \frac{1}{2}\left[F_{n-1} + \frac{2}{\sqrt{3}}\sin\frac{(n-1)\pi}{3}\right] \tag{12}$$

When (12) is combined with counts of the other $pattern(k)$'s with even content, the expression for $|D_n|$ becomes

$$|D_n| = \frac{1}{2}\left[\sum_{k=1}^{(n-1)} 2^{n-k-1}F_k\right] + \frac{1}{\sqrt{3}}\sin\frac{(n-1)\pi}{3} = \frac{1}{2}[2^n - F_{n+2}] + \frac{1}{\sqrt{3}}\sin\frac{(n-1)\pi}{3}, \quad n > 0 \tag{13}$$

CALCULATIONS

Calculated values appear in Table 2. In the last column of Table 2, the periodic oscillation of the computed even count about one-half the full count is obvious from the ceiling ($\lceil\ \rceil$) and floor ($\lfloor\ \rfloor$) functions.

n	Even 1's count	Full 1's count	Twice even count	Even count from full count
0	0	0	0	$\lfloor\frac{0}{2}\rfloor$
1	0	0	0	$\frac{0}{2}$
2	1	1	2	$\lceil\frac{1}{2}\rceil$
3	2	3	4	$\lceil\frac{3}{2}\rceil$
4	4	8	8	$\frac{8}{2}$
5	9	19	18	$\lfloor\frac{19}{2}\rfloor$
6	21	43	42	$\lfloor\frac{43}{2}\rfloor$
7	47	94	94	$\frac{94}{2}$
8	101	201	202	$\lceil\frac{201}{2}\rceil$
9	212	423	424	$\lceil\frac{423}{2}\rceil$
10	440	880	880	$\frac{880}{2}$
11	907	1815	1814	$\lfloor\frac{1815}{2}\rfloor$

Table 2. *Mathematica* counts and computed even counts

FUNCTIONS FOR D_n

By using *additive modulation* [2] in the form of a modulation factor $\dfrac{2}{\sqrt{3}}\sin\dfrac{(n-1)\pi}{3}$ to add $+1$, -1, or 0 at selected discrete n, we produce coefficients for <u>twice</u> the number of n-length binary sequences with some touching 1's and even content. Division by 2 returns the correct count.

The z-Transform of $2^n - F_{n+2}$ is given in (8) as $f_3(z) = \dfrac{z}{z^3 - 3z^2 + z + 2}$. In anticipation of canceling an unwanted -1 added at the origin by the modulation factor, a $+1$ added to $f_3(z)$ changes it to

$$f_4(z) = \frac{z^3 - 3z^2 + 2z + 2}{z^3 - 3z^2 + z + 2} \tag{14}$$

When the z-Transform of $\dfrac{2}{\sqrt{3}}\sin\dfrac{(n-1)\pi}{3}$ is added to (14), and the sum divided by 2, we have

$$f_5(z) = \frac{1}{2}\left[\frac{-z^2 + z}{z^2 - z + 1} + \frac{z^3 - 3z^2 + 2z + 2}{z^3 - 3z^2 + z + 2}\right] = \frac{z^3 - 2z^2 + z + 1}{z^5 - 4z^4 + 5z^3 - 2z^2 - z + 2} \tag{15}$$

If z is replaced by $\dfrac{1}{x}$ in (15), we have the corrected generating function in x for n-length binary sequences having even numbers of touching 1's as

$$g_5(x) = \frac{x^2 - 2x^3 + x^4 + x^5}{1 - 4x + 5x^2 - 2x^3 - x^4 + 2x^5} \tag{16}$$

When expanded, the first 12 terms of (16) become sequence values for counting n-length binary numbers with some touching 1's and even content.

$$0 + 0x + x^2 + 2x^3 + 4x^4 + 9x^5 + 21x^6 + 47x^7 + 101x^8 + 212x^9 + 440x^{10} + 907x^{11}\cdots + \tag{17}$$

Through z-Transform methods a recursion expression which satisfies (16) and (17) with $a_0 = 0$, $a_1 = 0$, $a_2 = 1$, $a_3 = 2$, and $a_4 = 4$ is found as

$$a_{n+5} - 4a_{n+4} + 5a_{n+3} - 2a_{n+2} - a_{n+1} + 2a_n = \int_{-\epsilon}^{+\epsilon} \delta(n)dn \tag{18}$$

where $\delta(n)$ is the Dirac delta function and ϵ is arbitrarily small. $\displaystyle\int_{-\epsilon}^{+\epsilon}\delta(n)dn = 1$ at $n = 0$ and 0 elsewhere. The general counting term for all even content is

$$\frac{1}{2}\left[2^n - F_{n+2} + \frac{2}{\sqrt{3}}\sin\frac{(n-1)\pi}{3} + \int_{-\epsilon}^{+\epsilon}\delta(n)dn\right] \tag{19}$$

SUMMARY

With combinatorial help from the structure of binary tables, we developed an expression for counting the number of n-length binary sequences with some touching 1's and general content. We used that expressions with other known equalities to generate Fibonacci identities. We extended the count to sequences with even content. We included closed generating functions, difference equations, as well as finite sequences for both counting cases.

ACKNOWLEDGEMENTS

We thank our colleagues for their comments, particularly Dr. Dick Kenan for convincing us to rethink our conditions for the Dirac integral. We are more than grateful for the generous help and encouragement from Dr. Marjorie Johnson. To our anonymous referee, we are in real debt since he furnished suggestions for completing a key part of the analytic proof.

REFERENCES

[1]   Austin, Richard B. and Guy, Richard K. "Binary Sequences Without Isolated Ones." *The Fibonacci Quarterly*, Vol. *16.1* (1978): pp. 84-86.

[2]   Fielder, Daniel C. and Alford, Cecil O. "Sinusoidally Modulated Fibonacci Sequences." Presented at the 77th annual meeting of the Southeastern Section of the MAA, The College of Charleston and The Citadel, Charleston, SC, March 13-14, 1998.

[3]   Jury, Eliuh I. Theory and Application of the z-Transform Method. New York: John Wiley & Sons, 1964.

[4]   Liu, C. L. Introduction to Combinatorial Mathematics. New York: McGraw-Hill Book company, 1968.

[5]   Vajda, S. Fibonacci & Lucas Numbers, and the Golden Section. Chichester, UK: Ellis Horwood Limited, 1989.

[6]   Wolfram, Stephan Mathematica, A System for Doing Mathematics by Computer. (Second Edition) Reading, MA: Addison-Wessley, 1991.

AMS Classification Numbers: 05A15, 11B39, 11Y56

INTEGRATION SEQUENCES OF JACOBSTHAL AND JACOBSTHAL-LUCAS POLYNOMIALS

Piero Filipponi and Alwyn F. Horadam

1. AIM OF THE PAPER

Here we are concerned with the *Jacobsthal polynomials* $J_n(x)$ and the *Jacobsthal-Lucas polynomials* $j_n(x)$ (e.g., see [4] and [5]) which are a natural extension of the *Jacobsthal numbers* J_n and the *Jacobsthal-Lucas numbers* j_n which, in turn, have been investigated in [3]. These polynomials are defined by the second-order recurrence relations

$$J_{n+2}(x) = J_{n+1}(x) + 2xJ_n(x), \quad [J_0(x) = 0, \ J_1(x) = 1] \tag{1.1}$$

and

$$j_{n+2}(x) = j_{n+1}(x) + 2xj_n(x), \quad [j_0(x) = 2, \ j_1(x) = 1] \tag{1.2}$$

respectively, where x is an indeterminate. Since throughout this paper we shall make use of the notation and the formulas found in [4], the reader is assumed to be aware of its contents.

The same idea that led to writing [2] as a companion paper of [1] now leads to writing this paper as a companion of [5].

Definitions: Following [2], let us define the polynomials $\Gamma_n(x)$ and $\Theta_n(x)$ as

$$\Gamma_n(x) = \int_0^x J_n(s)ds \tag{1.3}$$

and

$$\Theta_n(x) = \int_0^x j_n(s)ds. \tag{1.4}$$

The aim of this paper is to study the basic properties of the above sequences. As done in [1], [2] and [5], we shall confine ourselves to considering the case $x = 1$. Since, letting $x = 1$ in (1.1) and (1.2) will yield the Jacobsthal numbers and the Jacobsthal-Lucas numbers {cf. (2.3)

This paper is in final form and no version of it will be submitted for publication elsewhere.

and (2.4) of [3]}

$$J_n = \frac{2^n - (-1)^n}{3} \quad \text{and} \quad j_n = 2^n + (-1)^n, \tag{1.5}$$

the sequences $\{\Gamma_n(1)\}$ and $\{\Theta_n(1)\}$ will be referred to as *Jacobsthal and Jacobsthal-Lucas integration sequences*. For notational convenience, their terms $\Gamma_n(1)$ and $\Theta_n(1)$ will be denoted by Γ_n and Θ_n, respectively.

This paper is set out as follows. After establishing closed-form expressions for the numbers Γ_n and Θ_n (Section 2), several identities involving them are exhibited (Section 3); a discussion on a rather interesting aspect of these numbers (namely, conditions for them to be integers) concludes our study (Section 4).

2. THE NUMBERS Γ_n AND Θ_n

Closed-form expressions for $\Gamma_n(x)$ and $\Theta_n(x)$ are, quite obviously, useful tools for discovering properties of these polynomials.

Theorem 1: $\Gamma_n(x) = \dfrac{j_{n+1}(x) - 1}{2(n+1)}.$ \hfill (2.1)

Theorem 2: $\Theta_n(x) = \dfrac{j_{n+2}(x)}{n+2} - \dfrac{j_{n+1}(x)}{2(n+1)} - \dfrac{n}{2(n+1)(n+2)}.$ \hfill (2.2)

Proof of Theorem 1: From (1.3), and (3.3) of [4], we have

$$\Gamma_n(x) = \int_0^x \frac{\alpha^n(s) - \beta^n(s)}{\Delta(s)} ds$$

whence, by letting $\Delta(s) = y$ and by taking (1.4)-(1.5) of [4] into account,

$$\Gamma_n(x) = \frac{1}{4} \int_1^{\Delta(x)} \left[\left(\frac{1+y}{2} \right)^n - \left(\frac{1-y}{2} \right)^n \right] dy$$

$$= \frac{1}{2^{n+2}} \left\{ \left[\frac{(1+y)^{n+1}}{n+1} \right]_1^{\Delta(x)} + \left[\frac{(1-y)^{n+1}}{n+1} \right]_1^{\Delta(x)} \right\}$$

$$= \frac{1}{2(n+1)} \left[\alpha^{n+1}(x) + \beta^{n+1}(x) - 1 \right] = \frac{j_{n+1}(x) - 1}{2(n+1)} \quad \text{(from (3.4) of [4]).}$$

Proof of Theorem 2: Analogously, from (1.4), and (3.4) of [4], we have

$$\Theta_n(x) = \frac{1}{2^{n+2}} \int_1^{\Delta(x)} \left[y(1+y)^n + y(1-y)^n \right] dy$$

$$= \frac{1}{2^{n+2}} \left[\int_1^{\Delta(x)} (y+1-1)(1+y)^n dy - \int_1^{\Delta(x)} (1-y-1)(1-y)^n dy \right]$$

$$= \frac{1}{2^{n+2}} \left\{ \left[\frac{(1+y)^{n+2}}{n+2} - \frac{(1+y)^{n+1}}{n+1} \right]_1^{\Delta(x)} + \left[\frac{(1-y)^{n+2}}{n+2} - \frac{(1-y)^{n+1}}{n+1} \right]_1^{\Delta(x)} \right\}$$

$$= \frac{\alpha^{n+2}(x)}{n+2} - \frac{\alpha^{n+1}(x)}{2(n+1)} + \frac{\beta^{n+2}(x)}{n+2} - \frac{\beta^{n+1}}{2(n+1)} - \frac{n}{2(n+1)(n+2)}$$

$$= \frac{j_{n+2}(x)}{n+2} - \frac{j_{n+1}(x)}{2(n+1)} - \frac{n}{2(n+1)(n+2)} \quad \text{(from (3.4) of [4])}. \qquad \square$$

As stated earlier, the numbers Γ_n and Θ_n are the object of our study. Their closed-form expressions can be immediately obtained by putting $x = 1$ in (2.1) and (2.2). Namely, we get

$$\Gamma_n = \frac{j_{n+1} - 1}{2(n+1)} \tag{2.3}$$

$$= \frac{2^{n+1} - (-1)^n - 1}{2(n+1)} \quad \text{[from (1.5)]}, \tag{2.3'}$$

and

$$\Theta_n = \frac{j_{n+2}}{n+2} - \frac{j_{n+1}}{2(n+1)} - \frac{n}{2(n+1)(n+2)} \tag{2.4}$$

$$= \frac{n(j_{n+2} - 1) + 2(n+2)j_n}{2(n+1)(n+2)} \quad \text{[by letting } x = 1 \text{ in (1.2)]} \tag{2.4'}$$

$$= \frac{n\left[3 \cdot 2^{n+1} + 3(-1)^n - 1\right] + 2^{n+2} + 4(-1)^n}{2(n+1)(n+2)} \quad \text{[from (1.5)]}. \tag{2.4''}$$

As an illustration, the numbers Γ_n and Θ_n are shown in Table 1 for the first few nonnegative values of n.

n	0	1	2	3	4	5	6	7	8	9	10
Γ_n	0	1	1	2	3	$\frac{16}{3}$	9	16	$\frac{85}{3}$	$\frac{256}{5}$	93
Θ_n	2	1	3	4	$\frac{23}{3}$	$\frac{38}{3}$	23	$\frac{122}{3}$	$\frac{1111}{15}$	$\frac{674}{5}$	$\frac{745}{3}$

Table 1. The numbers Γ_n and Θ_n for $0 \leq n \leq 10$.

3. BASIC PROPERTIES OF Γ_n AND Θ_n

A list of basic relations involving Γ_n and Θ_n is exhibited in Subsection 3.1. It includes generating functions, some identities, and the evaluation of several finite sums. To save space, the proofs are either given in detail or sketched in Subsection 3.2 only for a few of these relations.

3.1. Relations

Generating functions

$$\sum_{n=0}^{\infty} \Gamma_n y^n = \frac{1}{2y} \ln \frac{1-y}{(1-2y)(1+y)} \overset{\text{def}}{=} G_\Gamma(y), \tag{3.1}$$

$$\sum_{n=0}^{\infty} \Theta_n y^n = \frac{2-y}{y} G_\Gamma(y). \tag{3.2}$$

Identities

$$\Theta_n = (n\Gamma_{n+1} + j_n)/(n+1) \qquad (n \geq 0) \quad \text{[from (2.4') and (2.3)]}, \tag{3.3}$$

$$\Gamma_n = (n\Theta_{n-1} - j_{n-1})/(n-1) \quad (n > 1) \quad \text{[from (3.3)]}, \tag{3.4}$$

$$\Gamma_n + \Theta_n = 2\Gamma_{n+1}, \tag{3.5}$$

$$\Gamma_{n+1} + 2\Gamma_{n-1} = [(n+1)\Theta_n - 1]/n, \tag{3.6}$$

$$\Gamma_n + 2\Gamma_{n-1} = \frac{n+2}{n+1}\Gamma_{n+1} + \frac{j_n - n - 1}{n(n+1)}, \tag{3.7}$$

$$\Theta_n + 2\Theta_{n-1} = \frac{n(n+3)}{(n+1)^2}\Theta_{n+1} + \frac{9nJ_n - j_n - n^2 + 1}{n(n+1)^2}, \tag{3.8}$$

Simson formula analogs

Letting (for any given sequence of numbers $\{A_n\}$) $\sigma[A_n] = A_{n-1}A_{n+1} - A_n^2$, we get

$$\sigma[\Gamma_n] = \frac{X_n}{n(n+1)^2(n+2)} + (-1)^n \frac{2}{n+1}\Gamma_n \tag{3.9}$$

where

$$X_n = \begin{cases} 4^n + n(n+2) & (n \text{ even}) \\ (2^{n-1} - 1)[2^{n+1} - n(n+2) - 1] & (n \text{ odd}), \end{cases} \tag{3.9'}$$

and

$$\sigma[\Theta_n] = \sigma[\Gamma_n] + 4\sigma[\Gamma_{n+1}] - \frac{Y_n}{n(n+1)(n+2)(n+3)} \tag{3.10}$$

where

$$Y_n = \begin{cases} 2^n[2^{n+3} + 3n(n+3) - 2] & (n \text{ even}) \\ 2^{n+1}[2^{n+2} - 3n(n+3) - 8] & (n \text{ odd}). \end{cases} \tag{3.10'}$$

Finite sums

Letting

$$H_k^{(y)} \overset{\text{def}}{=} \sum_{n=1}^{k} y^n/n = H_k^{(1)} + \sum_{n=1}^{k} \binom{k}{n}\frac{(y-1)^n}{n} \quad \text{(from [6, 13, p. 77])} \tag{3.11}$$

be the *generalized harmonic numbers* (see [6, Section 1.2.7., p. 73]), we have

$$\sum_{n=0}^{N} \Gamma_n \overset{\text{def}}{=} S_\Gamma(N) = \frac{1}{2}\left(H_{N+1}^{(2)} + H_{N+1}^{(-1)} - H_{N+1}^{(1)} \right) \tag{3.12}$$

$$= \frac{1}{2}H_{N+1}^{(2)} - \sum_{n=0}^{\lfloor N/2 \rfloor} \frac{1}{2n+1} \tag{3.12'}$$

$$= \frac{1}{2}\left(H_{N+1}^{(2)} + H_{\lfloor N/2 \rfloor}^{(1)} \right) - H_{2\lfloor N/2 \rfloor+1}^{(1)} \quad \text{(from [6, 16, p. 490])}, \tag{3.12''}$$

$$\sum_{n=1}^{N} \Gamma_n/n = \frac{1}{2}H_N^{(2)} - H_N^{(-1)} - \Gamma_N, \tag{3.13}$$

$$\sum_{n=0}^{N} n\Gamma_n = \frac{N+3}{2}(\Gamma_{N+2}-1) - S_\Gamma(N), \tag{3.14}$$

$$\sum_{n=0}^{N} \Theta_n = 2\Gamma_{N+1} + S_\Gamma(N), \tag{3.15}$$

$$\sum_{n=1}^{N} \Theta_n/n = H_N^{(2)} + H_N^{(-1)} - \Gamma_{N+1} + 1, \tag{3.16}$$

$$\sum_{n=0}^{N} n\Theta_n = \frac{N+3}{2}(\Gamma_{N+2}-1) + 2N\Gamma_{N+1} - 3S_\Gamma(N), \tag{3.17}$$

$$\sum_{n=0}^{N} \binom{N}{n}\Gamma_n = \frac{3^{N+1}-2-(-1)^N}{2(N+1)} - \Gamma_N, \tag{3.18}$$

$$\sum_{n=0}^{N} \binom{N}{n}\Theta_n = \Theta_N + \frac{N[3^{N+2}-2^{N+3}+2-3(-1)^N]+4[1-(-1)^N]}{2(N+1)(N+2)}, \tag{3.19}$$

and

$$\sum_{n=1}^{N} 1/\Gamma_{2n-1} = \frac{4^{N+1}-3N-4}{9 \cdot 4^{N-1}} \quad \text{from (2.3') and (3.1) of [1]} \tag{3.20}$$

whence

$$\sum_{n=1}^{\infty} 1/\Gamma_{2n-1} = 16/9. \tag{3.21}$$

Remarks:

(i)   Identity (3.5) is an important feature of Γ_n and Θ_n, being analogous to $J_n + j_n = 2J_{n+1}$ (see (2.20) of [3]) for Jacobsthal and Jacobsthal-Lucas numbers.

(ii)   We do not exclude the possibility that more compact forms can be found for some of the above relations and, in particular, for the Simson formula analogs (3.9) and (3.10).

3.2 Some proofs

Proof of (3.5): From (2.3) and (2.4), the left-hand side of (3.5) can be written as

$$\frac{j_{n+2}}{n+2} - \left[\frac{1}{2(n+1)} + \frac{n}{2(n+1)(n+2)}\right] = \frac{j_{n+2}}{n+2} - \frac{1}{n+2}$$

$$= \frac{j_{n+2}-1}{n+2} = 2\Gamma_{n+1} \quad \text{[from (2.3)]}.$$

Proof od (3.9) a sketch): First, from (2.3′) write

$$\Gamma_{n-1}\Gamma_{n+1} = \frac{2^{2n+1} - (5\cdot 2^{n-1}-1)[1-(-1)^n]}{2n(n+2)} \tag{3.22}$$

and

$$\Gamma_n^2 = \frac{2^{2n+1} - (2^{n+1}-1)[1+(-1)^n]}{2(n+1)^2}. \tag{3.23}$$

Then, use (3.22) and (3.23) to get, after a good deal of calculation,

$$\sigma[\Gamma_n] = \begin{cases} \dfrac{2^{n+1}-1}{(n+1)^2} + \dfrac{4^n}{n(n+1)^2(n+2)} & (n \text{ even}) \\[2em] -\dfrac{5\cdot 2^{n-1}-1}{(n+1)^2} + \dfrac{4^n - 5\cdot 2^{n-1}+1}{n(n+1)^2(n+2)} & (n \text{ odd}). \end{cases} \tag{3.24}$$

Finally, derive the right-hand side of (3.9) from (3.24), after some simple manipulation involving the use of (2.3′).

Proof of (3.13): By using (2.3′) and the identity $1/[n(n+1] = 1/n - 1/(n+1)$, the left-hand side of (3.13) can be written as

$$\frac{1}{2}\left[\sum_{n=1}^{N} \frac{2^{n+1}-(-1)^n-1}{n} - \sum_{n=1}^{N} \frac{2^{n+1}-(-1)^n-1}{n+1}\right]$$

$$= \frac{1}{2}\left[\sum_{n=1}^{N} \frac{2^n}{n} - \frac{2^{N+1}}{N+1} + \frac{1}{N+1} - 2\sum_{n=1}^{N} \frac{(-1)^n}{n} - \frac{(-1)^{N+1}}{N+1}\right]$$

$$= \frac{1}{2}\left[H_N^{(2)} - 2H_N^{(-1)} - 2\Gamma_N\right] \quad \text{[from (3.11) and (2.3′)]}.$$

Proof of (3.18): From (2.3'), the left-hand side of (3.18) can be written as

$$\sum_{n=0}^{N}\binom{N}{n}\frac{2^n}{n+1}-\frac{1}{2}\sum_{n=0}^{N}\binom{N}{n}\frac{(-1)^n}{n+1}-\frac{1}{2}\sum_{n=0}^{N}\binom{N}{n}\frac{1}{n+1}$$

$$= A_N - \tfrac{1}{2}(B_N + C_N). \tag{3.25}$$

Let us manipulate the quantities A_N, B_N and C_N, separately.

(i) $$A_N = \sum_{n=1}^{N+1}\binom{N}{n-1}\frac{2^{n-1}}{n}$$

$$= \frac{1}{2}\left[\sum_{n=1}^{N+1}\binom{N+1}{n}\frac{2^n}{n}-\sum_{n=1}^{N}\binom{N}{n}\frac{2^n}{n}\right] \text{ (from [8, pp. 1-2])}$$

$$= \frac{1}{2}\left[H^{(3)}_{N+1}-H^{(1)}_{N+1}-H^{(3)}_{N}+H^{(1)}_{N}\right]=\frac{3^{N+1}-1}{2(N+1)} \text{ [from (3.11)]}. \tag{3.26}$$

(ii) Analogously, we get

$$B_N = 1/(N+1) \text{ and } C_N = (2^{N+1}-1)/(N+1). \tag{3.27}$$

From (3.25)-(3.27), we see that the left-hand side of (3.18) is given by

$$\frac{3^{N+1}-2^{N+1}-1}{2(N+1)}=\frac{3^{N+1}-2-(-1)^N}{2(N+1)}-\Gamma_N \text{ [from (2.3')]}.$$

4. ON THE INTEGER VALUES OF Γ_n AND O_n

The problem of finding the set of all n for which Γ_n and Θ_n are integers arises quite naturally after the inspection of Table 1. Its solution is given in Theorem 3 and (partially) in Theorems 4 and 5 below. Throughout this section, the notation 2-psp (e.g., see [7, p. 86]) stands for the *Fermat pseudoprimes* in base 2 (or *Poulet numbers*), that is for those odd composites m for which $2^{m-1} \equiv 1$ (mod m) ($341 = 11 \cdot 31$ being the smallest among them).

Theorem 3: Γ_n is an integer iff $n+1$ is either a power of 2, or a prime, or a 2-psp.

Theorem 4: If $n = 2^k - 1$ with $k = 2^h$ ($h = 0, 1, 2, \cdots$), then Θ_n is an integer.

Theorem 5: If $n = 2^q - 2$ with q either a prime or a 2-psp, then Θ_n is an integer.

Proof of Theorem 3: From (2.3'), the theorem is trivially true for $n = 0$.

Case 1 ($n \geq 2$, even): From (2.3'), we have $\Gamma_n = (2^n - 1)/(n+1)$ whence, for Γ_n to be an integer, the congruence $2^n \equiv 1$ (mod $n+1$) must necessarily be satisfied. This happens clearly iff $n+1$ is either a prime (Fermat's little theorem) or 2-psp.

Case 2 (n odd): From (2.3'), we have $\Gamma_n = 2^n/(n+1)$ whence, for Γ_n to be an integer, $n+1$ must necessarily be a power of 2.

Proof of Theorem 4: From (2.4''), after some manipulation, we have

$$\Theta_n = 2\,\frac{2^n - 1 + n2^{n-k-1}}{n+2} \quad \text{(for } n = 2^k - 1 \text{ and } k \geq 1). \tag{4.1}$$

Letting $k = 1$ (i.e., $n = 1$) in (4.1) yields $\Theta_1 = 1$, as expected. For $k \geq 2$, since g.c.d.$(2, n+2) = 1$, it is sufficient to prove the equivalent congruences

$$2^n + n2^{n-k-1} \equiv 1 \;(\text{mod } n+2) \quad (n = 2^k - 1, \; k \geq 2),$$

$$2^{2^k - k} + 1 \equiv 0 \;(\text{mod } 2^k + 1) \quad (k \geq 2). \tag{4.2}$$

Since $k = 2^h$ by hypothesis, it is readily seen that $(2^k - k)/k$ is an odd integer, so that we can invoke the property (e.g., see [9, p. 248])

$$a^s + b^s \equiv 0 \;(\text{mod } a^t + b^t) \quad (\text{if } s/t \text{ is an odd integer}) \tag{4.3}$$

to ascertain the validity of (4.2).

Proof of Theorem 5: First, observe that, if $q = 2$ (i.e., $n = 2$), then (2.4'') yields $\Theta_n = 3$ (cf. Table 1). Then, for $q \geq 3$ odd (i.e., for n even), use (2.4'') again to obtain

$$\Theta_n = \frac{(n+1)(2^{n+1} + 1) + n2^n + 1}{(n+1)(n+2)} \overset{\text{def}}{=} \frac{M}{(n+1)(n+2)}. \tag{4.4}$$

Since g.c.d.$(n+1, n+2) = 1$, it is sufficient to prove that: (i) $M \equiv 0 \;(\text{mod } n+1)$ and (ii) $M \equiv 0 \;(\text{mod } n+2)$.

Proof of (i): Write $M = (n+1)(2^{n+1} + 1) + (n+1)2^n - (2^n - 1)$, whence it remains clearly to prove that (since $n = 2^q - 2$, given)

$$2^n - 1 = 2^{2(2^{q-1})} - 1 \equiv 0 \;(\text{mod } 2^q - 1). \tag{4.5}$$

If q is either a prime or a 2-psp, then $2^{q-1} - 1 \equiv 0 \;(\text{mod } q)$, so that we can invoke the property (e.g., see [9, p. 248])

$$a^s - b^s \equiv 0 \;(\text{mod } a^t - b^t) \quad [\text{if } s \equiv 0 \;(\text{mod } t)] \tag{4.6}$$

to ascertain the validity of (4.5).

Proof of (ii): Write $M = (n+2)(3 \cdot 2^n + 1) - 2^{n+2}$, whence it remains clearly to prove the congruence $2^{n+2} \equiv 0 \;(\text{mod } n+2)$ which is clearly true if $n = 2^q - 2$.

4.1. On the primality of the integer values of Γ_n and Θ_n

Whenever a new sequence of integers comes on the scene, one is led to ask himself about their divisibility properties and, in particular, about their primality. For the integers under study, we established the following.

(i) Γ_n is a prime only for $n = 3$ and 4. In fact, if $n = 2^k - 1$, then $\Gamma_n = 2^{2^k - k - 1}$ which is a prime (namely, 2) only for $k = 2$ (i.e., $n = 3$) whereas, for $n = q - 1$, we have $\Gamma_{q-1} = (2^{q-1} - 1)/q = (2^{(q-1)/2} - 1)(2^{(q-1)/2} + 1)/q$ which is clearly a prime (namely, 3) only for $q = 5$ (i.e., $n = 4$).

(ii) $\Theta_0 = 2$ is a prime. For n as in Theorem 4, Θ_n is either 1 (for $n = 1$) or an even integer greater than 2 (for $n \geq 3$). A proof of the statement is given in the Appendix. For n as in Theorem 5, we have

$$\Theta(q) \overset{\text{def}}{=} \Theta_{2^q - 2} = \left[2^{2^q - q}\left(3 \cdot 2^{q-2} - 1\right) + 1\right] / (2^q - 1)$$

whence $\Theta(2) = 3$ and $\Theta(3) = 23$ are primes, while $\Theta(5) = 99580895$ is composite; the 37-digit number $\Theta(7)$ is composite as well as it is divisible by 26993. Finding all n for which Θ_n is prime remains an open problem.

5. CONCLUSIONS

The Jacobsthal and Jacobsthal-Lucas integration sequences have been investigated and some of their properties have been established. A natural extension of this work is the study of the sequences $\{\Gamma_n(x)\}$ and $\{\Theta_n(x)\}$, defined by (2.1) and (2.2), for $x \neq 1$. This will be the aim of a future paper.

Just to taste the flavour of the interrelationships existing among these generalized sequences, we leave to the perseverance of the interested reader the detailed proof of the following proposition.

Proposition: If t_k is the k-th triangular number, and

$$U_N(n) \overset{\text{def}}{=} \sum_{k=1}^{N} (-1)^{N-k} \Gamma_n(t_k), \tag{5.1}$$

then

$$\sum_{n=1}^{\infty} \frac{1}{U_{2N-1}(2n-1)} = \left[\frac{4N}{(2N-1)(2N+1)}\right]^2. \tag{5.2}$$

Proof (a hint): After recalling that $t_k = (k^2 + k)/2$, use (3.4) and (1.4) of [4] to find $j_n(t_k)$, and Theorem 1 to find $\Gamma_n(t_k)$. After some manipulations, one gets $U_{2N-1}(2n-1) = (2N)^{2n}/(4n)$. Replace this quantity in the left-hand side of (5.2) and evaluate the sum.

It can be immediately observed that, for $N = 1$, (5.2) reduces to (3.21).

APPENDIX

Here we give the detailed proof of the statement at point (ii) of Subsection 4.1.

Statement: If $n = 2^{2^h} - 1$ $(h = 0, 1, 2, \cdots)$, then Θ_n is either 1 (for $h = 0$), or an even integer greater than 2 (for $h \geq 1$).

Proof: Put $2^h = k$ and use (2.4″) to write

$$\Theta_n = \Theta_{2^k - 1} = \frac{(2^k - 1)\left[3 \cdot 2^{2^k} + 3(-1)^{2^k - 1} - 1\right] + 2^{2^k + 1} + 4(-1)^{2^k - 1}}{2(2^k)(2^k + 1)}. \tag{A1}$$

For $h = 0$ (i.e., $k = n = 1$), (A1) reduces to

$$\Theta_1 = \frac{1 \cdot [3 \cdot 4 - 3 - 1] + 8 - 4}{2 \cdot 2 \cdot 3} = 1 \tag{A2}$$

as expected (see Table 1).

For $h \geq 1$ (i.e., $k \geq 2$, $n \geq 3$), (A1) becomes

$$\begin{aligned}
\Theta_{2^k - 1} &= \frac{(2^k - 1)\left(3 \cdot 2^{2^k} - 4\right) + 2^{2^k + 1} - 4}{2^{k+1}\left(2^k + 1\right)} \\[2mm]
&= \frac{3 \cdot 2^{2^k + k} - 2^{k+2} - 3 \cdot 2^{2^k} + 2^{2^k + 1}}{2^{k+1}\left(2^k + 1\right)} \\[2mm]
&= \frac{3 \cdot 2^{2^k - 1} - 2 - 3 \cdot 2^{2^k - k - 1} + 2^{2^k - k}}{2^k + 1}.
\end{aligned} \tag{A3}$$

If $k \geq 2$ (as assumed), then the numerator of (A3) is even, whereas its denominator is odd. It follows that the integer (see Thm. 4) $\Theta_{2^k - 1} \geq 4$ must necessarily be even.

ACKNOWLEDGEMENTS

The contribution of the first author (P.F.) has been given within the framework of an agreement between the Italian PT Administration and the Fondazione Ugo Bordoni.

The authors wish to thank the anonymous referee for correcting some inaccuracies in the proof of Theorem 4.

REFERENCES

[1]   Filipponi, P. and Horadam, A.F. "Derivative Sequences of Fibonacci and Lucas Polynomials." Applications of Fibonacci Numbers, Volume 4. Edited by G.E. Bergum, A.F. Horadam and A.N. Philippou, Kluwer Academic Publishers, Dordrecht, The Netherlands, 1991: pp. 99-108.

[2]   Horadam, A.F. and Filipponi, P. "Integration Sequences of Fibonacci and Lucas Polynomials." Applications of Fibonacci Numbers, Volume 5. Edited by G.E. Bergum, A.F. Horadam and A.N. Philippou, Kluwer Academic Publishers, Dordrecht, The Netherlands, 1993: pp. 317-330.

[3]   Horadam, A.F. "Jacobsthal Representation Numbers." The Fibonacci Quarterly, Vol. 34.1 (1996): pp. 40-54.

[4]   Horadam, A.F. "Jacobsthal Representation Polynomials." The Fibonacci Quarterly, Vol. 35.2 (1997): pp. 137-148.

[5]   Horadam, A.F. and Filipponi, P. "Derivative Sequences of Jacobsthal and Jacobsthal-Lucas Polynomials." The Fibonacci Quarterly, Vol. 35.4 (1997): pp. 352-357.

[6]   Knuth, D.E. The Art of Computer Programming, Volume 1, 2nd ed. Reading (Mass.): Addison-Wesley, 1973.

[7]   Ribenboim, P. The Book of Prime Number Records. New York: Springer, 1988.

[8]   Riordan, J. Combinatorial Identities. New York: Wiley, 1968.

[9]   Sierpinski, W. Elementary Theory of Numbers. Warsaw: Panstwowe Wydawnictwo Naukowe, 1964.

AMS Classification Numbers: 11B37, 11B83, 26A06

RECURSIVE SEQUENCES

Herta T. Freitag, Marjorie Bicknell-Johnson and George M. Phillips

1. INTRODUCTION

In this paper we show that a sequence $\{A_n\}$ defined by the second order recurrence relation

$$A_{n+2} = u_1 A_{n+1} + u_2 A_n,\tag{1}$$

satisfies the congruence relation

$$A_{n+2k} \equiv u_1 A_{n+k} + u_2 A_n \pmod{10}\tag{2}$$

for all choices of integers u_1 and u_2 and all initial values A_1 and A_2 if and only if $k = 1$ or 5 (mod 24). In the rest of this section we summarize the relevant earlier work in this area.

At the First International Conference on Fibonacci Numbers and Their Applications the first author [1] showed that the generalized Fibonacci sequence $\{H_n\}$ defined by the recurrence relation

$$H_{n+2} = H_{n+1} + H_n\tag{3}$$

(where H_1 and H_2 are arbitrary positive integers) satisfies the congruence relation

$$H_{n+2k} \equiv H_{n+k} + H_n \pmod{10},\tag{4}$$

for all initial values H_1 and H_2 if and only if $k \equiv 1$ or 5 (mod 12). For such $k \geq 5$, let us arrange the members of the sequence $\{H_n\}$ as the elements of a matrix with k columns, where the jth column has elements $H_{j+(i-1)k}$, $i = 1, 2, \cdots$. Then any three consecutive elements in any row of the matrix satisfy (3) and any three consecutive elements in any column of this matrix satisfy a congruence relation of the form (4).

This paper is in final form and no version of it will be submitted for publication elsewhere.

The particular case where $k = 5$ was generalized in [2] as follows. If the sequence $\{A_n\}$ satisfies (1), where the u_i and A_i, $i = 1, 2$, are any integers, then for any prime $p \geq 5$

$$A_{n+2p} \equiv u_1 A_{n+p} + u_2 A_n \pmod{2p}. \tag{5}$$

The latter result was extended to recurrence relations of any order in [3], in a slightly weaker form, to show that if, for any $m \geq 2$,

$$A_{n+m} = \sum_{i=1}^{m} u_i A_{n+m-i}, \tag{6}$$

where the u_i and A_i, $i = 1, 2, \cdots, m$, are any integers and the characteristic polynomial of the recurrence relation has distinct zeros, then for any prime p

$$A_{n+mp} \equiv \sum_{i=1}^{m} u_i A_{n+(m-i)p} \pmod{p}. \tag{7}$$

In [4] some further congruence results are given for the sequence satisfying the mth order recurrence relation (6). It is shown that it is not necessary for the roots of the characteristic polynomial to be distinct and that (7) can be strengthened to give

$$A_{n+mp^b} \equiv \sum_{i=1}^{m} u_i A_{n+(m-i)p^b} \pmod{p}. \tag{8}$$

for all nonnegative integers b and any prime p. It is also shown in [4] that, given any fixed positive integer c such that $(c, u_m) = 1$, there exists an integer g such that

$$A_{n+mk} \equiv \sum_{i=1}^{m} u_i A_{n+(m-i)k} \pmod{c}, \tag{9}$$

provided that $k \equiv 1 \pmod{g}$.

Although the material in [2], [3] and [4] is of greater generality and, as noted above, the latter two papers are concerned with higher order recurrence relations, none of these papers explicitly generalizes all the results obtained in the parent paper [1]. This is what motivates the present work.

2. A GENERALIZATION

The sequence $\{A_n\}$ generated by the general second order recurrence relation (1) is completely determined by its starting values, say $A_1 = a$, $A_2 = b$. We will use $\{F_n\}$ to denote the special case where $a = 1$ and $b = u_1$, since the classical Fibonacci sequence is recovered when $u_1 = u_2 = 1$. Thus we define

$$F_{n+2} = u_1 F_{n+1} + u_2 F_n, \tag{10}$$

where $F_1 = 1$, $F_2 = u_1$ and u_1, u_2 are any fixed integers. It is easily verified by induction that the general sequence $\{A_n\}$ is very simply expressed in terms of this sequence $\{F_n\}$ by the relation

$$A_n = bF_{n-1} + u_2 aF_{n-2}, \tag{11}$$

and this holds for $n \geq 1$ if we extend the sequence $\{F_n\}$ backwards by defining $F_0 = 0$ and $F_{-1} = 1/u_2$. (Thus F_{-1} will be an integer only if $u_2 = \pm 1$.)

It is clear from the recurrence relation (10) that each F_n is a polynomial in u_1 and u_2, and it may be verified by induction that

$$F_{n+1} = \sum_{r=0}^{[\frac{n}{2}]} \binom{n-r}{r} u_1^{n-2r} u_2^r. \tag{12}$$

We are interested in the nature of the relation (12) modulo 10 and we note that powers of u_1 and u_2 may be reduced modulo 10 using the fact that, for $r \geq 1$,

$$m^{4+r} \equiv m^r \pmod{10} \tag{13}$$

for all positive integer m. We may deduce from (12) and (13) that, for $n > 1$, F_n satisfies a congruence relation of the form

$$F_n \equiv u_1^{n-1} + a_n u_1^{n+1} u_2 + b_n u_1^{n-1} u_2^2 + c_n u_1^{n+1} u_2^3$$
$$+ d_n u_1^{n-1} u_2^4 + e_n u_2^{(n-1)/2} \pmod{10}, \tag{14}$$

where a_n, b_n, c_n and d_n belong to the set $\{0, 1, \cdots, 9\}$ and $e_n = 0$ or 1 when n is even or odd respectively. By (13), the exponents $n-1$, $n+1$ and $(n-1)/2$ in (14) may be reduced (mod 4) to members of the set $\{1, 2, 3, 4\}$. From (10) and (14) the coefficients a_n, b_n, c_n and d_n may be computed recursively, for $n \geq 2$, using the following algorithm.

ALGORITHM

$a_2 = a_3 = b_2 = b_3 = c_2 = c_3 = d_2 = d_3 = 0$

for $n = 4, 5, 6, \cdots$ do

$\quad a_n \equiv a_{n-1} + d_{n-2} + 1 \pmod{10}$

$\quad$ if $n \equiv 4 \pmod 8$ then $a_n \equiv a_n + 1 \pmod{10}$

$\quad b_n \equiv b_{n-1} + a_{n-2} \pmod{10}$

$\quad$ if $n \equiv 6 \pmod 8$ then $b_n \equiv b_n + 1 \pmod{10}$

$\quad c_n \equiv c_{n-1} + b_{n-2} \pmod{10}$

$\quad$ if $n \equiv 0 \pmod 8$ then $c_n \equiv c_n + 1 \pmod{10}$

$\quad d_n \equiv d_{n-1} + c_{n-2} \pmod{10}$

$\quad$ if $n \equiv 2 \pmod 8$ then $d_n \equiv d_n + 1 \pmod{10}$

next n

For what follows, we require the following definition.

<u>Definition:</u> We will say that a sequence $\{s_n\}$ has period q (mod m) if q is the smallest positive integer such that $s_{n+q} \equiv s_n$ (mod m) for all $n \geq n_0$ for some n_0.

For example, it is easily verified that the classical Fibonacci sequence has period 60 (mod 10). On applying the above Algorithm, which is readily translated into a program in any of the well-known symbolic programming languages, such as *Maple or Mathematica*, we find that $F_{121} \equiv u_1^4 + 9u_1^4 u_2^4 + u_2^4$, $F_{122} \equiv u_1$ and $F_{123} \equiv u_1^2 + u_2$ (mod 10). Clearly $F_{122} \equiv F_2$ and $F_{123} \equiv F_3$ (mod 10), and an induction argument using (10) shows that $F_{n+120} \equiv F_n$ (mod 10) for all $n \geq 2$, independent of the values of u_1 and u_2. (We observe that F_{121} is not always congruent to F_1 modulo 10.) Thus the period of the sequence $\{F_n\}$ is a divisor of 120 and we will see presently that the precise period for a given choice of u_1 and u_2 depends on the congruence classes modulo 10 of u_1 and u_2. Given this property of the sequence $\{F_n\}$, we can deduce the following result for the more general sequence $\{A_n\}$ directly from (11).

<u>Theorem 1:</u> *The period of the sequence $\{A_n\}$ which satisfies the general second order linear recurrence relation (1) is a divisor of 120.*

We next show how a congruence relation connecting A_{n+2k}, A_{n+k} and A_n for all $n \geq 0$ follows from two congruence relations which depend only on k and not on n.

<u>Theorem 2:</u> *If, for some positive integer k, the two congruences*

$$F_{2k} - u_1 F_k \equiv 0 \ (mod \ 10) \tag{15}$$

and

$$F_{2k+1} - u_1 F_{k+1} - u_2 \equiv 0 \ (mod \ 10) \tag{16}$$

both hold, then

$$A_{n+2k} \equiv u_1 A_{n+k} + u_2 A_n \ (mod \ 10)$$

for all $n \geq 0$.

<u>Proof:</u> From (15) and (16) and the recurrence relation (10) it follows immediately by induction on n that, for all $n \geq 0$,

$$F_{n+2k} \equiv u_1 F_{n+k} + u_2 F_n \ (mod \ 10) \tag{17}$$

and the proof is completed by using (11).

We now determine the set S of values of k for which both (15) and (16) hold for all integer values of u_1 and u_2. We know from the choice of $u_1 \equiv u_2 \equiv 1$ (mod 10) in [1] that S is at most the set of all k such that $k \equiv 1$ or 5 (mod 12). The following choice of u_1 and u_2 shows that, in fact, S is not the whole of the latter set: we choose $u_1 \equiv 0$ and $u_2 \equiv 2$ (mod 10). an elementary calculation shows that the resulting sequence $\{F_n\}$ is congruent (mod 10) to the sequence

$$\{1, 0, 2, 0, 4, 0, 8, 0, 6, 0, 2, 0, \cdots\},$$

which has period 8. In this case, (15) and (16) respectively become

$$F_{2k} \equiv 0 \ (\text{mod } 10) \tag{18}$$

and

$$F_{2k+1} - 2 \equiv 0 \ (\text{mod } 10). \tag{19}$$

For these sequences we see that (18) is always satisfied and that (19) is satisfied only for $k = 1$ or 5 (mod 8). Thus the set S must be at most the set of k for which $k \equiv 1$ or 5 (mod 24). Further investigation, which we will now summarize, shows that S is the whole of this set.

Since $F_{n+120} \equiv F_n$ (mod 10) for all $n \geq 2$, it suffices to verify that (15) and (16) hold for $k = 1, 5, 25, 29, 49, 53, 73, 77, 97$ and 101, and that (15) holds for $k = 121$. The latter result is easily settled, since

$$F_{242} - u_1 F_{121} \equiv F_2 - u_1 F_{121} \equiv u_1 - u_1(u_1^4 + 9u_1^4 u_2^4 + u_2^4)$$

and, using (13), we see that the last expression is indeed congruent to zero modulo 10. Having generated algebraically the values of F_n by the above algorithm, for $1 \leq n \leq 123$, we verify case by case that (15) and (16) hold for the ten values $k = 1, 5, 25, \cdots, 101$ cited above. As this is very elementary, we will give only one example, namely $k = 101$. For this we require the values of F_{101}, F_{102}, F_{202} and F_{203}, which are as follows:

$$F_{202} \equiv F_{82} \equiv u_1 + 2u_1^3 u_2 + 5u_1 u_2^2 + 4u_1^3 u_2^3 + 9u_1 u_2^4 \ (\text{mod } 10)$$

$$F_{203} \equiv F_{83} \equiv u_1^2 + u_1^4 u_2 + 7u_1^2 u_2^3 + 7u_1^2 u_2^4 + u_2 \ (\text{mod } 10)$$

$$F_{101} \equiv u_1^4 + 7u_1^2 u_2 + 4u_1^4 u_2^2 + 4u_1^2 u_2^3 + 4u_1^4 u_2^4 + u_2^2 \ (\text{mod } 10)$$

$$F_{102} \equiv u_1 + 6u_1^3 u_2 + 5u_1 u_2^2 + 2u_1^3 u_2^3 + 2u_1 u_2^4 \ (\text{mod } 10).$$

Then for (15) with $k = 101$ we obtain

$$F_{202} - u_1 F_{101} \equiv F_{82} - u_1 F_{101} \equiv 5u_1 u_2(-u_1^2 + u_2^3) \ (\text{mod } 10)$$

and, since the right side is always even, it is clearly congruent to zero modulo 10. For (16) with $k = 101$ we obtain

$$F_{203} - u_1 F_{102} - u_2 \equiv 5u_1^2 u_2(-u_1^2 - u_2 + u_1^2 u_2^2 + u_2^3)$$

which is also congruent to zero modulo 10.

Having checked that (15) and (16) both hold for the other nine values of k quoted above, then in view of Theorem 2 we have proved the main result of this paper:

Theorem 3: *The sequence $\{A_n\}$ which is generated by the general second order recurrence relation (1) satisfies the congruence equation*

$$A_{n+2k} \equiv u_1 A_{n+k} + u_2 A_n \pmod{10}$$

for all integer values of u_1 nd u_2 and all initial values A_1 and A_2 if and only if $k \equiv 1$ or 5 (mod 24).

3. FURTHER RESULTS

We saw above that the period of the sequence $\{a_n\}$ is a divisor of 120 and that if $u_1 \equiv u_2 \equiv 1$ (mod 10) the sequence $\{F_n\}$, defined by (10), is congruent to the classical Fibonacci sequence, which has period 60 (mod 10). Thus for $u_1 \equiv u_2 \equiv 1$ (mod 10) the period of the sequence $\{A_n\}$ is a divisor of 60. We also saw that if $u_1 \equiv 0$ and $u_2 \equiv 2$ (mod 10), the sequence $\{F_n\}$, defined by (10), has period 8. Since there are only 10 choices for each of u_1 and u_2 (mod 10), there are only 100 possible sequences (mod 10). We can find the periods of the 100 equivalence classes of sequences more easily by using congruences modulo 2 and 5.

Table 1 gives the periods for the sequences $\{F_n\}$, for all four possible values of u_1 and u_2 modulo 2. In Table 2 we give the periods of all 25 possible sequences $\{F_n\}$ modulo 5.

u_2 \ u_1	0	1
0	1	1
1	2	3

Table 1: The periods of $\{F_n\}$ for given values of u_1 and u_2 (mod 2)

u_2 \ u_1	0	1	2	3	4
0	1	1	4	4	2
1	2	20	12	12	20
2	8	4	24	24	4
3	8	24	4	4	24
4	4	6	5	10	3

Table 2: The periods of $\{F_n\}$ for given values of u_1 and u_2 (mod 5)

If $p_m(S)$ denotes the period of the sequence $\{S_n\}$ (mod m), then we see that $p_{10}(S)$ is the least common multiple of $p_2(S)$ and $p_5(S)$. Thus from Tables 1 and 2 we can construct Table 3 which gives the periods of the sequences $\{F_n\}$ for all possible values of u_1 and u_2 (mod 10), and we note that there is no sequence $\{F_n\}$, and thus no sequence $\{A_n\}$, with period 120. The only sequences with the largest period of 60 are those with $u_1 \equiv u_2 \equiv 1$, which are congruent to the classical Fibonacci numbers, and those with $u_1 \equiv 9$ and $u_2 \equiv 1$, which are congruent to the Fibonacci sequence with alternating signs, $1, -1, 2, -3, 5, -8, \cdots$. We also observe that all the proper divisors of 120, except for 40, occur in Table 3.

u_2 \ u_1	0	1	2	3	4	5	6	7	8	9
0	1	1	4	4	2	1	1	4	4	2
1	2	60	12	12	20	6	20	12	12	60
2	8	4	24	24	4	8	4	24	24	4
3	8	24	4	12	24	24	24	12	4	24
4	4	6	5	10	3	4	6	5	10	3
5	2	3	4	12	2	3	2	12	4	6
6	2	20	12	12	20	2	20	12	12	20
7	8	12	24	24	4	24	4	24	24	12
8	8	24	4	4	24	8	24	4	4	24
9	4	6	10	30	6	12	6	15	10	3

Table 3: The periods of $\{F_n\}$ for given values of u_i and u_2 (mod 10)

ACKNOWLEDGEMENTS

The second and third authors give thanks for the joy of working with Herta Taussig Freitag, and especially in this her ninetieth year, 1998.

REFERENCES

[1]   Freitag, H.T. "A Property of Unit Digits of Fibonacci Numbers." Applications of Fibonacci Numbers, Volume 2. Edited by G.E. Bergum, A.F. Horadam and A.N. Philippou. Kluwer Academic Publishers, Dordrecht, The Netherlands, 1986: pp. 39-42.

[2]   Freitag, H.T. and Phillips, G.M. "A congruence relation for certain recursive sequences." The Fibonacci Quarterly, Vol. 24.4 (1986): pp. 332-335.

[3]   Freitag, H.T. and Phillips, G.M. "A congruence relation for a linear recursive sequence of
        arbitrary order." Applications of Fibonacci Numbers, Volume 3. Edited by G.E.
        Bergum, A.F. Horadam and A.N. Philippou. Kluwer Academic Publishers, Dordrecht,
        The Netherlands, 1988: pp. 39-44.

[4]   Somer, Lawrence. "Congruence Relations for kth Order Linear Recurrences." *The
        Fibonacci Quarterly*, Vol. *27.1* (1989): pp. 25-30.

AMS Classification Numbers: 11B37, 11B39

ON GENERAL DIVISIBILITY OF SUMS OF INTEGRAL POWERS OF THE GOLDEN RATIO

Herta T. Freitag and Daniel C. Fielder

INTRODUCTION

The well-known *Golden Ratio*, $\alpha = (1 + \sqrt{5})/2$, is the limit as $n \to \infty$ of the ratio of the Fibonacci numbers F_n/F_{n-1} and the Lucas numbers L_n/L_{n-1}. Eq. (1) served to illustrate in [2,3] that unique integer solutions $b = 7$ and $c = 11$ exist.

$$\frac{1}{\alpha^b} \sum_{i=1}^{10} \alpha^i = \frac{\left(\alpha^1 + \alpha^2 + \alpha^3 + \alpha^4 + \alpha^5 + \alpha^6 + \alpha^7 + \alpha^8 + \alpha^9 + \alpha^{10}\right)}{\alpha^b} = c \tag{1}$$

In [2,3], (1) verified b and c for divisibility by α^b for a specific *span*, s, of sums of powers of α and a specific starting power for the span. However, a literature search revealed no discussion of conditions for a general s and general starting powers. We develop and present those conditions. We saw this as an opportunity to show that conjectures from properly oriented computer experiments can often lead directly to conditions for derivations of general analytic proofs. We used the computer algebra language *Mathematica* [6] to search for those conjectures.

CALCULATIONS, COMPUTER EXPERIMENTS, AND CONJECTURES

The Golden Ratio often appears as $\alpha = \left(L_1 + F_1\sqrt{5}\right)/2$. According to Hoggatt [1, p. 60], $\alpha^b = \left(L_b + F_b\sqrt{5}\right)/2$. This quadratic surd is separable and linear in L_b and F_b for positive integral, negative integral, or zero b. Thus, a way of stating the divisibility problem becomes

$$\frac{\sum_{i=p}^{q} \alpha^i}{\alpha^b} = \frac{\sum_{i=p}^{q}\left(L_i + F_i\sqrt{5}\right)}{\left(L_b + F_b\sqrt{5}\right)} = \frac{\left(L_{q+2} - L_{p+1}\right) + \left(F_{q+2} - F_{p+1}\right)\sqrt{5}}{\left(L_b + F_b\sqrt{5}\right)} = c. \tag{2}$$

This paper is in final form and no version of it will be submitted for publication elsewhere.

149

In (2), a valid solution exists only for integral or zero values for p (the starting power), q (the ending power), and b (the power of the denominator α). The values of c (the quotient) and $s = q - p + 1$ (the span) must be non-zero, positive integers.

In reducing (2), we applied the identity, $\sum_{i=p}^{q} G_i = G_{q+2} - G_{p+1}$, for consecutive G where G is either a Fibonacci or Lucas number (see Vajda [2, p. 39]), and we canceled a common factor, $\frac{1}{2}$.

Because of the separability features of the surd, a sufficient condition for divisibility becomes

$$\frac{\left(F_{q+2} - F_{p+1}\right)}{\left(L_{q+2} - L_{p+1}\right)} = \frac{F_b}{L_b}. \tag{3}$$

For those p, q, and b combinations for which divisibility occurs, the value of c can be found from either of the expressions derived from (3) as

$$\frac{L_{q+2} - L_{p+1}}{L_b} = \frac{F_{q+2} - F_{p+1}}{F_b} = c. \tag{4}$$

In our *Mathematica* computer experiments, we selected ranges for p, s, and b and compared the left fraction of (3) with the right fraction for matches. For a given p and s, the upper summation index, q, is $q = p + s - 1$. Specifically, we can choose any integer value for p and to avoid a trivial span of 1, any $s \geq 2$. To assure that we examine all potential b's, we let b range from p to $p + s$. If a combination of p, q and b satisfies (3), we have a match for integral b and c. In our program experiments, the values of p, q, b, c, and s are calculated and checked, but only successful sets survive to be recorded as output.

Division by 0 in *Mathematica* is to be avoided even though in some cases the calculations may continue correctly. A "division by zero" error statement could have occurred in our experiments (a) when $b = 0$, (b) when F_0 was used, and (c) when $L_{q+2} - L_{p+1} = 0$. For (a), we anticipated a valid division, calculated it as a special case and jumped over the calculation which would have given an error notice. For (b), we chose the form of (3) above rather than its reciprocal to avoid division by zero on the right side since L_b is never 0, while F_b might be. Whenever $L_{q+2} - L_{p+1} = 0$, there cannot be a valid division. For (c), then, the *Mathematica* computer experiment is programmed to recognize this and move on to the next calculation without using $L_{q+2} - L_{p+1} = 0$. These factors and others like them must be considered in any successful experiment.

Since the ratios compared can sometimes be rational fractions with very large numerators and denominators, conventional computer methods often yield approximate values-- sometimes very good approximate values, but approximate, nonetheless. A distinct, practical feature of *Mathematica* is the manipulation of <u>exact</u> rational fractions with the size of the integers involved limited only by a very efficiently managed available memory. Moreover, each rational fraction is automatically reduced to its lowest terms, making identification and comparison easy. Our *Mathematica* program, **test13.ma**, based on (3) was used for experimentation.

Program **test13.ma** features an adjustable lower limit, p, and an adjustable span, s. Since the upper limit, q, of the numerator summation is directly related to s, q is implicitly adjustable. Only when (3) is satisfied by integer values, are the values of p, q, b, c, and s shown as printed output. For those interested, details of **test13.ma** may be obtained from the second author (DCF).

test13.ma with p ranging from -3 to 2 and applicable span ranging from 2 to 10 produced the following output.

p	q	b	c	s		p	q	b	c	s	
-3	-2	-1	1	2		0	1	2	1	2	
-3	-1	-1	2	3		**0**	**2**	**2**	**2**	**3**	
-3	2	1	4	6		0	5	4	4	6	
-3	6	3	11	10		0	9	6	11	10	
-2	-1	0	1	2		1	2	3	1	2	
-2	0	0	2	3		**1**	**3**	**3**	**2**	**3**	(5)
-2	3	2	4	6		1	6	5	4	6	
-2	7	4	11	10		1	10	7	11	10	
-1	0	1	1	2		2	3	4	1	2	
-1	1	1	2	3		**2**	**4**	**4**	**2**	**3**	
-1	4	3	4	6		2	7	6	4	6	
-1	8	5	11	10		2	11	8	11	10	

Analysis of (5): Consider the entries of (5) for any p <u>exclusive</u> of entries with span $s = 3$. (Entries for $s = 3$ are shown in **bold** type above.) We note that s for a successful span is <u>exactly</u> twice a positive <u>odd</u> integer. The span $s = q - p + 1 = 4k - 2$ for $k = 1$, 2, 3, $\cdots$. This means that $q = p + 4k - 3$. The power, b, of the denominator α is $b = (p + q + 3)/2$. In terms of k, $b = p + 2k$. Note that the values of c in (5) for the selected entries are always odd-indexed Lucas numbers L_1, L_3, L_5, L_7, L_9, L_{11}, L_{13} and that the index is one-half the corresponding span. In general, the value of c is $L_{(q - p + 1)/2} = L_{(4k - 2)/2}$. Hence, for integer p in the range $-\infty < p < \infty$ and integer k in the range $k \geq 1$, based on (5) we conjecture

$$\frac{\sum_{i=p}^{p+4k-3} \alpha^i}{\alpha^{p+2k}} = L_{2k-1}. \tag{6}$$

We justify our conjecture with an analytic proof of (6) in a later section.

A Singular Case: Consider the entries of (5) in which $s = 3$ or $q = p + 2$. These are the entries excluded in the development of (6). With the exception of the trivial $s = 1$, $s = 3$ appears to be the <u>only</u> odd s. By setting the upper and lower span limits at 3 in **test13.ma**, the following typical values for the singular case appear as output.

p	q	b	c	s
-4	-2	-2	2	3
-3	-1	-1	2	3
-2	0	0	2	3
-1	1	1	2	3
0	2	2	2	3
1	3	3	2	3
2	4	4	2	3

(7)

It is obvious that the behavior of the singular case is not governed by (6). Here is an instance where computer experimentation aided greatly in finding and separating two cases. From (7) it appears that the power, b, of the denominator α is $b = q = p + 2$, and the value of c is L_0. Hence, for integer p in the range $-\infty < p < \infty$, we are able to conjecture

$$\frac{\sum_{i=p}^{p+2} \alpha^i}{\alpha^{p+2}} = L_0. \tag{8}$$

We justify the validity of the conjecture later.

RELATED WORK

When we first considered computer experimentation in this area, we wondered if anyone had generalized the problem and had reported proofs where our experiments had produced conjectures. Our search of the usual sources did not uncover justification of (6) and (8). Even if we had found proofs, we would have proceeded anyway but with an altered approach, since the problem would still demonstrate the benefits of computer experimentation.

Because of the separability of the surds involved, corresponding ratio identities between sums of L's (or F's) and a suitably indexed L (or F) can be conjectured directly from (6) and (8). After we had completed our work on (6) and (8), we could not dismiss the feeling that someone must have considered at least the separate ratios. We finally uncovered reference [6] in which S_n is defined as $S_n = \sum_{i=1}^{n} F_i$. Reference [6] noted "There is an unexpected relation between the F-sequence and the S-sequence, namely that S_{4r-2} is a multiple of $F_{2r+1}\cdots$ and the multiplier is a Lucas number $c_{2r-1}\cdots$." The proof in [6] is quite involved.

Fortunately, the objective of reference [6] was in no way similar to ours. Nevertheless, if the authors had multiplied their S_n by $\sqrt{5}$, found a similar summation for L_i, and added the results, they might have observed an interpretation of our (6) and (8).

In our terms, S_n becomes $S_{4r-2} = F_{2r+1}L_{2r-1}$. By restricting r to always be even, we arrive at the simpler form (9) which we prove later.

$$\sum_{i=1}^{2r-2} F_i = F_{r+1}L_{r-1} \qquad (9)$$

PROOFS AND SOME EXTRAS

Proof of conjecture, equation (6): Let $S = \sum_{i=p}^{p+4k-3} \alpha^i$. Then, by the Binet form (see [1, p. 11]) and the use of $\alpha - 1 = \alpha^{-1}$, we have $S = \alpha^{4k+p-1} - \alpha^{p+1}$. (An alternate equivalent would have been to use the identity, $\sum_{i=p}^{q} G_i = G_{q+2} - G_{p+1}$ [5] in much the same way as used in the development of (2).) Now we divide by α^{p+2k} to get

$$\frac{S}{\alpha^{p+2k}} = \alpha^{2k-1} - \alpha^{-(2k-1)}. \qquad (10)$$

Since $\alpha^n = F_n\alpha + F_{n-1}$, and $\alpha^{-n} = (-1)^{n-1}(F_n\alpha - F_{n+1})$, $\alpha^{2k-1} = F_{2k-1}\alpha + F_{2k-2}$ and $\alpha^{-(2k-1)} = F_{2k-1}\alpha - F_{2k}$. It follows that $\alpha^{2k-1} - \alpha^{-(2k-1)} = F_{2k-2} + F_{2k} = L_{2k-1}$. Hence, (6) is proved and appears as

$$\frac{\sum_{i=p}^{p+4k-3} \alpha^i}{\alpha^{p+2k}} = L_{2k-1}.$$

Proof of conjecture, equation (8): We have

$$\frac{\sum_{i=p}^{p-2} \alpha^i}{\alpha^{p-2}} = \alpha^{-2} + \alpha^{-1} + 1. \qquad (11)$$

But, since $\alpha^{-1} = \alpha - 1$ and $\alpha^{-2} = -\alpha + 2$, (11) always reduces to $L_0 = 2$.

Proof of equation (9): Since $\sum_{i=1}^{n} F_i = F_{n+2} - 1$, we have $\sum_{i=1}^{2(r-1)} F_i = F_{2r} - 1$. Next, through the Binet formula, $F_{r+1}L_{r-1} = \frac{1}{\alpha-\beta}\{(\alpha^{r+1} - \beta^{r+1})(\alpha^{r-1} + \beta^{r-1})\} =$

$\frac{1}{\alpha-\beta}\{(\alpha^{2r} - \beta^{2r}) + (\alpha^{r+1}\beta^{r-1} - \alpha^{r-1}\beta^{r+1})\} = F_{2r} + (-1)^{r-1}\frac{(\alpha^2 - \beta^2)}{\alpha - \beta} = F_{2r} - 1$, as r is even. It follows that for all _even_ r,

$$\sum_{i=1}^{2(r-1)} F_i = F_{r+1}L_{r-1}.$$

An interesting relationship: While making the proofs of (6), (8), and (9) we encountered several interesting identities. We share one with you.

Since $\alpha^n = F_n\alpha + F_{n-1}$ and $\alpha^{-n} = (-1)^{n-1}(F_n\alpha - F_{n+1})$,
$\alpha^n + (-1)^{n-1}\alpha^{-n} = 2F_n\alpha - F_n = (2\alpha - 1)F_n = (\alpha + \alpha^{-1})F_n.$ Thus,

$$\frac{\sum_{i=1}^{n}\left\{\alpha^i + (-1)^{i-1}\alpha^{-i}\right\}}{\alpha + \alpha^{-1}} = \sum_{i=1}^{n} F_i = F_{n+2} - 1. \tag{12}$$

A natural consequence: Because of the separability of F_k and L_k in $L_k + F_k\sqrt{5} = \alpha^k$, it follows that conjectures (6) and (8) and their subsequent proofs are valid when all α^k's are replaced by F_k's or L_k's.

SUMMARY

By breaking up powers of the Golden Ratio into separable Lucas and Fibonacci parts, we experimentally investigated divisibility conditions of sums of consecutive powers of α, and by implication, sums of consecutively indexed Fibonacci and Lucas numbers. Computer experimentation suggested that divisibility takes place for exactly one odd length sequence span and for even length sequence spans only when the length is twice an odd integer. We obtained our results as general computer-experimental conjectures and subsequently proved the conjectures to establish new divisibility conditions. Other Fibonacci relations were found and are included.

REFERENCES

[1]   Hoggatt, V.E., Jr. Fibonacci and Lucas Numbers. Boston: Houghton-Mifflin, 1969, republished Santa Clara, CA: The Fibonacci Association, 1979.

[2]   Problem B-690 (Proposal). *The Fibonacci Quarterly*, Vol. *29.2* (1991): p. 181.

[3]   Problem B-690 (Solution). *The Fibonacci Quarterly*, Vol. *30.2* (1992): p. 184.

[4]   Rumney, M. and Primrose, E.F.J. "Relations Between a Sequence of Fibonacci Type and the Sequence of Its Partial Sums." *The Fibonacci Quarterly*, Vol. *34.3* (1971): pp. 296-298.

[5]   Vajda, S. Fibonacci and Lucas Numbers, and the Golden Section. Chichester, UK: Ellis Horwood Ltd., 1989.

[6]   Wolfram Research, Inc. *Mathematica*, Ver. *2.2*. Champaign, IL: Wolfram Research Inc., 1991.

AMS Classification Numbers: 11B39, 05A15

SYLVESTER'S ALGORITHM AND FIBONACCI NUMBERS

Herta T. Freitag and George M. Phillips

1. SYLVESTER'S ALGORITHM

We begin by describing an elementary process for expressing a positive proper fraction as a sum of reciprocals of positive integers. Starting with m_1/n_1, where $m_1 < n_1$, we subtract the largest fraction with numerator unity to leave either zero or another positive fraction, say m_2/n_2. We then subtract the largest fraction with numerator unity from m_2/n_2 and repeat this process until we end up with zero. Eves [1] attributes this process to J. J. Sylvester (1814-1897), whose original account is given in [12]. In this section we will review Sylvester's algorithm and summarize his findings. In Section 2 we make some further observations on such sums of reciprocals and in Section 3 we extend the Sylvester process from positive proper fractions to all real numbers between 0 and 1. In Section 4 we give some related results concerning sums of reciprocals of Fibonacci numbers.

We therefore begin with positive integers m_1 and n_1, with $1 < m_1 < n_1$. Then, from the division algorithm (see Grimaldi [3]), there exist positive integers q_1 and r_1 such that

$$n_1 = q_1 m_1 + r_1 \tag{1}$$

with $q_1 \geq 1$ and $0 \leq r_1 < m_1$. If $r_1 = 0$ we terminate the process and write $m_1/n_1 = 1/q_1$. Otherwise, we have

$$\frac{m_1}{n_1} = \frac{1}{q_1 + 1} + \frac{m_2}{n_2}, \tag{2}$$

say, where

$$m_2 = m_1 - r_1, \ n_2 = n_1(q_1 + 1). \tag{3}$$

This paper is in final form and no version of it will be submitted for publication elsewhere.

155

Thus

$$n_2 > n_1 > m_1 > m_2.$$

If $m_2 = 1$ we terminate the process; otherwise we continue by applying the division algorithm to n_2 and m_2, to construct q_2, r_2, m_3 and n_3. Since the sequence of positive integers $m_1, m_2, m_3,$ $\cdots$ is strictly decreasing, then $r_k = 0$ for some k and we have

$$\frac{m_1}{n_1} = \frac{1}{q_1+1} + \frac{1}{q_2+1} + \cdots + \frac{1}{q_{k-1}+1} + \frac{1}{q_k}. \tag{4}$$

Also, it is clear that $k \leq m_1$. The representation of a proper fraction in this particular way has been named in many ways: Mays [6] calls it the Fibonacci-Sylvester expansion or the "greedy algorithm expansion" and quotes Salzer's [8] description of it as an R-expansion. A related algorithm for representing rational numbers as a sum of unit fractions, called the Engel expansion, has been studied by Knopfmacher and Mays [5] and Shallit [11]. In general, Engel expansions and Sylvester series are distinct because of the much weaker growth condition for the digits in the Engel expansion. With respect to Mays [6] we will drop the reference to Fibonacci (having need of this name later) and, swayed by chronology, will refer to (4) as simply the Sylvester series for the fraction m_1/n_1. From (1) and (2) and successive relations concerning m_j/m_j and m_{j+1}/n_{j+1} it follows that

$$\frac{1}{q_j+1} < \frac{m_j}{n_j} \leq \frac{1}{q_j} \tag{5}$$

and

$$\frac{1}{q_{j+1}+1} < \frac{m_{j+1}}{n_{j+1}} = \frac{m_j-r_j}{n_j(q_j+1)} \leq \frac{1}{q_j(q_j+1)}, \tag{7}$$

where the last inequality is strict unless $j = k$. Thus we have

$$q_{j+1} \geq q_j(q_j+1), \tag{7}$$

as shown by Sylvester [12]. Also, the repeated application of (3) and its successors gives

$$n_k = n_1(q_1+1)(q_2+1)\cdots(q_{k-1}+1) \tag{8}$$

and, in the case where $m_k = 1$, the product on the right of (8) also gives q_k, the denominator of the final term in (4). J. J. Sylvester used the word *sorites* for his representation of a proper fraction as we describe above. It is interesting to read the entry for *sorites* in the Oxford English Dictionary [7] where, following the main part of the entry, there is actually a reference to [12] citing Sylvester's usage of this recondite word.

For the remainder of his paper [12], Sylvester concentrates on what he calls limiting sorites, where the q_j in (4) all satisfy the strict equality $q_{j+1} = q_j(q_j+1)$ in (7). It is easily verified that in this case

$$\frac{1}{q_1} = \frac{1}{q_1+1} + \frac{1}{q_2+1} + \cdots + \frac{1}{q_{k-1}+1} + \frac{1}{q_k} \tag{9}$$

and we can let k tend to infinity in (9) if we please. Sylvester concludes his paper [12] with some number-theoretic properties of the q_j in the case of a limiting sorites, which we will not repeat here. Rather, we discuss in the next section some results concerning Sylvester series which are not mentioned in [12].

2. FURTHER RESULTS ON SYLVESTER SERIES

In the Rhind (or Ahmes) papyrus from Egypt *circa* 1650 B.C. there are representations as a sum of reciprocals of all fractions of the form $2/(2n+1)$, for $2 \leq n \leq 49$. (See Scott [9].) The first three are

$$\frac{2}{5} = \frac{1}{3} + \frac{1}{15}, \ \frac{2}{7} = \frac{1}{4} + \frac{1}{28}, \ \frac{2}{9} = \frac{1}{6} + \frac{1}{18}. \tag{10}$$

Note that the simple identity (see [1, page 40])

$$\frac{2}{2n+1} = \frac{1}{n+1} + \frac{1}{(n+1)(2n+1)} \tag{11}$$

gives the Sylvester series for $2/(2n+1)$. We observe that in (10) the first two fractions are in their Sylvester series form, but the third fraction is not. With 3 as the numerator, there are three outcomes of Sylvester's algorithm (neglecting the trivial case when the denominator is divisible by 3):

$$\frac{3}{6n+1} = \frac{1}{2n+1} + \frac{1}{6n^2+4n+1} + \frac{1}{(2n+1)(6n+1)(6n^2+4n+1)}, \tag{12}$$

$$\frac{3}{3n+2} = \frac{1}{n+1} + \frac{1}{(n+1)(3n+2)}, \tag{13}$$

$$\frac{3}{6n+4} = \frac{1}{2n+2} + \frac{1}{(2n+2)(3n+2)}. \tag{14}$$

Observe that (14) follows readily from (13).

We now present two theorems. The first shows that, for any numerator m, there exist denominators n such that the Sylvester series for m/n has the maximum possible m terms. The second theorem gives a necessary condition on n for the series for m/n to have m terms.

Theorem 1: *the Sylvester series for the function m/n has exactly m terms if $n = km! + 1$, where k is any positive integer.*

Proof: The proof is by induction on m. The result is true for $m = 1$. Assume it holds for some $m = 1 \geq 1$. Now consider a fraction of the form $m/(km!+1)$, where k is any positive integer.

We see that the first reciprocal in its Sylvester series is $1/(k(m-1)!+1)$ and we have

$$\frac{m}{km!+1} = \frac{1}{k(m-1)!+1} + \frac{m-1}{k'(m-1)!+1}, \tag{15}$$

where k' is the positive integer $k(km!+m+1)$. This completes the proof.

Theorem 2: *A necessary condition for m/n to have a Sylvester series with the full m terms is that n is of the form $km+1$, where k is a positive integer.*

Proof: Using the division algorithm, let $n = km+r$, where $0 \leq r < m$. Then, applying the Sylvester algorithm, the leading term in the series for $m/(km+r)$ is $1/(k+1)$ and

$$\frac{m}{km+r} = \frac{1}{k+1} + \frac{m-r}{(k+1)(km+r)}. \tag{16}$$

We require $r=1$ so that the numerator on the right of (16) is $m=1$, and this completes the proof.

We now present a theorem which classifies all proper fractions m/n whose Sylvester series have m terms. (An alternative classification is given by Mays [6].) For this purpose, we now define certain sets of non-negative integers S_m, for $m \geq 3$. We define $S_3 = \{0\}$ and, for each $m \geq 4$, we compute the members of the set S_m, given the set S_{m-1}, as follows. We have $s \in S_m$ if and only if $0 \leq s < (m-1)!$ and

$$ms^2 + (m+1)s \equiv t(m-1) \ (\mathrm{mod}(m-1)!) \tag{17}$$

for some $t \in S_{m-1}$. We note that, for all $m \geq 3$, $0 \in S_m$ and all elements of each S_m are even.

Theorem 3: *For any $m \geq 3$ the fraction m/n has m terms in its Sylvester series if and only if*

$$n = km! + sm + 1,$$

where $s \in S_m$, $k \geq 0$ and k and s are not both zero.

Proof: If the fraction m/n has m terms in its Sylvester series, it follows from Theorem 2 that we may write n in the form $km!+sm+1$, where $0 \leq s < (m-1)!$ The leading term in the series is then $1/(k(m-1)!+s+1)$ and

$$\frac{m}{km!+sm+1} = \frac{1}{k(m-1)!+s+1} + \frac{m-1}{(k(m-1)!+s+1)(km!+sm+1)}. \tag{18}$$

We note that

$$(k(m-1)!+s+1)(km!+sm+1) \equiv (s+1)(sm+1) \ (\mathrm{mod}(m-1)!). \tag{19}$$

We now use an induction argument. In view of (12), (13) and (14), the result is valid for $m=3$. Now we assume that the result holds for some $m-1 \geq 3$. Then from (18) and (19) the result holds also for m provided that s satisfies

$$ms^2(m+1)s \equiv t(m-1) \ (\mathrm{mod}(m-1)!)$$

for some $t \in S_{m-1}$. Since this is how we defined the set S_m, the result holds for m and thus for all $m \geq 3$.

Beginning with the set $S_3 = \{0\}$ and applying the recursive definition, embodying (17), where S_m is constructed from S_{m-1}, we obtain $S_4 = \{0,4\}$, $S_5 = \{0,6,12,18\}$ and $S_6 = \{0,18,30,48,60,78,90,108\}$. Knowing S_m we may, for example, determine the smallest value of n for which m/n has m terms in its Sylvester series. For $3 \leq m \leq 6$ these fractions are 3/7, 4/17, 5/31 and 6/109. Specifically, we have

$$\frac{3}{7} = \frac{1}{3} + \frac{1}{11} + \frac{1}{231}, \quad \frac{4}{17} = \frac{1}{5} + \frac{1}{29} + \frac{1}{1233} + \frac{1}{3039345},$$

$$\frac{5}{31} = \frac{1}{7} + \frac{1}{55} + \frac{1}{3979} + \frac{1}{23744683} + \frac{1}{1127619917796295}$$

and

$$\frac{6}{109} = \frac{1}{19} + \frac{1}{415} + \frac{1}{214867} + \frac{1}{61556888719} + \frac{1}{5683875823083467302723}$$
$$+ \frac{1}{6461288874446552579384137676954062237912 6735}.$$

We remark that the near doubling of the number of digits in successive denominators in the above Sylvester series is consistent with the inequality (7).

3. EXTENSION TO IRRATIONAL NUMBERS

The Sylvester process has been extended to all real numbers between 0 and 1. (See, for example, Sierpinski [10].) Let α denote an irrational number between 0 and 1. There is clearly a smallest positive integer n_1 such that $\alpha - 1/n_1 > 0$. We can replace α by $\alpha - 1/n_1$ and repeat this process. Since the process cannot terminate, we have after k stages

$$s_k = \alpha - \frac{1}{n_1} - \frac{1}{n_2} - \cdots - \frac{1}{n_k} > 0. \tag{20}$$

By definition of n_k, we also have that $s_k + \frac{1}{n_k} - \frac{1}{n_k - 1} < 0$, giving

$$s_k < \frac{1}{n_k(n_k - 1)}. \tag{21}$$

Thus

$$\alpha < \frac{1}{n_1} + \frac{1}{n_2} + \cdots + \frac{1}{n_k} + \frac{1}{n_k(n_k - 1)}$$

and so $n_{k+1} > n_k(n_k - 1)$, which is equivalent to (7). Since $n_1 \geq 2$ we see that the sequence (n_k) is strictly increasing and tends to infinity. It then follows from (20) and (21) that (s_k) is a null sequence and we may write

$$\alpha = \sum_{k=1}^{\infty} \frac{1}{n_k}. \tag{22}$$

We will call the sum on the right of (22) the Sylvester series for α, as we did above for the corresponding finite series when α is rational.

We note that if α has a Sylvester series of the form $\alpha = \sum_{k=1}^{\infty} 1/(rn_k)$ or a finite form of this, where $\alpha < 1/r$, for some positive integer r, then $r\alpha$ has the Sylvester series $r\alpha = \sum_{k=1}^{\infty} 1/n_k$. We will apply this result, which is easily verified, in the next section.

4. FIBONACCI NUMBERS

In the last section, the denominators n_1, n_2, $\cdots$ were chosen appropriately from the set $\{2,3,4,\cdots\}$. We could replace this set by some other suitable set of increasing positive integers. Here we replace it by $S = \{F_3, F_4, F_5, \cdots\}$. For any real α between 0 and 1 we let F_{n_1} denote the smallest number from the set S such that $s_1 = \alpha - 1/F_{n_1} \geq 0$. If $s_1 > 0$ we choose F_{n_2} as the smallest number from the set S such that $s_2 = \alpha - 1/F_{n_1} - 1/F_{n_2} \geq 0$, and so on. If $s_k > 0$, the same arguments deployed in obtained (20) and (21) show that

$$0 < s_k < \frac{1}{F_{n_k-1}} - \frac{1}{F_{n_k}} = \frac{F_{n_k-2}}{F_{n_k-1}F_{n_k}} < \frac{1}{F_{n_k}} \tag{23}$$

and we also obtain

$$\frac{1}{F_{n_k+1}} < \frac{F_{n_k-2}}{F_{n_k-1}F_{n_k}} < \frac{1}{F_{n_k}}.$$

We deduce that the sequence (n_k) is strictly increasing. Thus we can either express α as a finite sum of reciprocals of Fibonacci numbers or we can deduce from (23) that $(s_k)_{k=1}^{\infty}$ is a null sequence and so write $\alpha = \sum_{k=1}^{\infty}(1/F_{n_k})$. In either case we will write this more simply as

$$\alpha = \sum_{n=3}^{\infty} \frac{e_n}{F_n}, \tag{24}$$

where each $e_n \in \{0,1\}$, and the series may be finite or infinite. We will call the sum on the right of (24) the Sylvester-Fibonacci series for α. We now show that at most two consecutive members of the sequence $(e_n)_{n=3}^{\infty}$ can be non-zero. Our proof makes use of the well known identity

$$F_{r+s}F_{r+t} - F_r F_{r+s+t} = (-1)^r F_s F_t, \tag{25}$$

(see Vajda [13]), which is valid for all integers r, s and t.

Theorem 4: *For any real α, $0 < \alpha < 1$, let us construct*

$$\alpha = \sum_{n=3}^{\infty} \frac{e_n}{F_n},$$

as described above, where each $e_n \in \{0,1\}$. Then for any k, $e_{2k} = 1 \Rightarrow e_{2k+1} = 0$.

Proof: We note that

$$\frac{1}{F_{n_k}}+\frac{1}{F_{n_k+1}}-\frac{1}{F_{n_k-1}}=\frac{F_{n_k-1}(F_{n_k+1}+F_{n_k})-F_{n_k+1}}{F_{n_k-1}F_{n_k}F_{n_k+1}}$$

$$=\frac{F_{n_k-1}F_{n_k+2}-F_{n_k}F_{n_k+1}}{F_{n_k-1}F_{n_k}F_{n_k+1}}=\frac{(-1)^{n_k}}{F_{n_k-1}F_{n_k}F_{n_k+1}},$$

the last step following from (25). If n_k is even and $n_{k+1}=n_k+1$, we deduce that

$$\frac{1}{F_{n_k-1}}<\frac{1}{F_{n_k}}+\frac{1}{F_{n_k+1}}=\frac{1}{F_{n_k}}+\frac{1}{F_{n_{k+1}}}$$

and thus, from the definition of n_{k+1},

$$\frac{1}{F_{n_k-1}}<\frac{1}{F_{n_k}}+\alpha-\sum_{j=1}^{k}\frac{1}{F_{n_j}}=\alpha-\sum_{j=1}^{k-1}\frac{1}{F_{n_j}}$$

which contradicts the definition of n_k. It follows that, if n_k is even, $n_{k+1}>n_k+1$ and this completes the proof.

Using (25) again, we obtain

$$\frac{F_{n-k}}{F_nF_k}-\frac{1}{F_{2k}}=(-1)^k\frac{F_{n-2k}}{F_nF_{2k}},$$

for $n\geq 2k>0$. It is also readily verified that

$$\frac{F_{n-k}}{F_nF_k}<\frac{1}{F_{2k-1}} \tag{26}$$

for $n\geq k>0$. Thus, for $n\geq 2k>0$ and k even, $1/F_{2k}$ is the first term in the Sylvester-Fibonacci series for $F_{n-k}/(F_nF_k)$. We may use an induction argument to show that, if $2^sk\leq n<2^{s+1}k$ and k is even, then

$$\frac{F_{n-k}}{F_nF_k}=\frac{1}{F_{2k}}+\frac{1}{F_{4k}}+\cdots+\frac{1}{F_{2^sk}}+\frac{F_{n-2^sk}}{F_nF_{2^sk}}, \tag{27}$$

and the first s terms on the right of (27) are the first s terms of the Sylvester-Fibonacci series for $F_{n-k}/(F_nF_k)$. In fact, we can say more than this, since we can replace (26) by the stronger inequality

$$\frac{F_{n-k}}{F_nF_k}<\frac{1}{F_{2k}-1}, \tag{28}$$

valid for $n\geq k\geq 2$. Thus, repeating the above induction argument, we see that the first s terms on the right of (27) are also those of the Sylvester series, and not merely of the Sylvester-Fibonacci series. To clarify this distinction, note that

$$\frac{13}{40}=\frac{1}{4}+\frac{1}{14}+\frac{1}{280} \quad \text{(Sylvester series)}$$

$$\frac{13}{40}=\frac{1}{5}+\frac{1}{8} \quad \text{(Sylvester-Fibonacci series)}.$$

In particular if we replace k by $2k$ in (27) and put $n = 2^s k$, we obtain the finite Sylvester series

$$\frac{F_{(2^s - 2)k}}{F_{2^s k} F_{2k}} = \frac{1}{F_{4k}} + \frac{1}{F_{8k}} + \cdots + \frac{1}{F_{2^s k}} \qquad (29)$$

valid for $s \geq 2$, $k \geq 1$. It is interesting to compare (29) with the expression given for $\sum_{i=0}^s 1/F_{2^i k}$ by Hoggatt and Bicknell [4]. The identity (29) is particularly simple for the case $k = 1$ and it appears in Good [2] in the modified form

$$\frac{1}{F_1} + \frac{1}{F_2} + \frac{1}{F_4} + \cdots + \frac{1}{F_{2^s}} = 3 - \frac{F_{2^s - 1}}{F_{2^s}}. \qquad (30)$$

Pursuing (29), we can apply the relation $F_{2t} = F_t L_t$ repeatedly to give

$$\frac{F_{2^j k}}{F_{2k}} = \prod_{i=1}^{j-1} L_{2^i k},$$

Thus, using the argument given in the last paragraph of Section 3, we obtain the Sylvester series

$$\frac{F_{(2^s - 2)k}}{F_{2^s k}} = \sum_{j=1}^{s-1} \left(1 / \prod_{i=1}^{j} L_{2^i k} \right), \qquad (31)$$

which holds for all $s \geq 2$ and $k \geq 2$. The latter identity is a special case of an identity of Knopfmacher and Mays [5] which involves Fibonacci and Lucas *polynomials*, rather than *numbers*. However, we emphasize that our interest in (31) is not chiefly with the identity *per se* but rather with our proof that it is a Sylvester series.

The Sylvester series (31) is obviously very special and it is natural to consider the general proper fraction of the form F_m/F_n. We conclude with some remarks about this. It is easily verified that the first term in the Sylvester-Fibonacci series for F_m/F_n, for $3 \leq m < n$, is $1/F_{n-m+2}$. The second term in the series varies with the choice of m and n as follows.

For $m = 3$ and n even and also for the single case where $m = 5$, and $n = 6$, the first two terms of the Sylvester-Fibonacci series are

$$\frac{F_m}{F_n} = \frac{1}{F_{n-m+2}} + \frac{1}{F_{n-m+5}} + \cdots, \qquad (32)$$

where the series (32) terminates after these two terms only in the case of $(m, n) = (5, 6)$.

For $m = 4$ and n even, we have

$$\frac{F_m}{F_n} = \frac{1}{F_{n-m+2}} + \frac{1}{F_{n-m+7}} + \cdots. \qquad (33)$$

For all other cases, that is, for $m = 3$ or 4 and $n \geq 5$ odd, and for all $4 < m < n$ except for $(m, n) = (5, 6)$, we have

$$\frac{F_m}{F_n} = \frac{1}{F_{n-m+2}} + \frac{1}{F_{n-m+6}} + \cdots, \qquad (34)$$

where the series (34) terminates after these two terms only in the case of $(m, n) = (6, 8)$. Note that (34) covers most cases and that it is consistent with (29) when $k = 1$ and $s \geq 3$. Since the proofs of the statements (32)-(34) are elementary and of no intrinsic interest, they are omitted.

REFERENCES

[1]   Eves, H. An Introduction to the History of Mathematics. 5th Edition. Philadelphia: Saunders, 1983.

[2]   Good, I.J. "A reciprocal series of Fibonacci numbers." *The Fibonacci Quarterly*, Vol. *12.4* (1974): p. 346.

[3]   Grimaldi, R.P. Discrete and Combinatorial Mathematics. 3rd Edition. Reading, Massachusetts: Addison-Wesley, 1994.

[4]   Hoggatt, V.E., Jr. and Bicknell, Marjorie. "A reciprocal series of Fibonacci numbers with subscripts $2^n k$." *The Fibonacci Quarterly*, Vol. *14.5* (1976): pp. 453-455.

[5]   Knopfmacher, Arnold and Mays, M.E. "Pierce expansions of ratios of Fibonacci and Lucas numbers and polynomials." *The Fibonacci Quarterly*, Vol. *33.2* (1995): pp. 153-163.

[6]   Mays, Michael E. "A worst case of the Fibonacci-Sylvester expansion." *J. Combin. Math. Combin. Comput.*, Vol. *1* (1987): pp. 141-148.

[7]   Oxford English Dictionary. Oxford, 1971.

[8]   Salzer, H.E. "Further remarks on the approximation of numbers as sums of reciprocals." *Amer. Math. Monthly*, Vol. *55* (1948): pp. 350-356.

[9]   Scott, J.F. A History of Mathematics. London: Taylor & Francis, 1975.

[10]   Sierpinski, W. Elementary Theory of Numbers. Second Edition. Amsterdam: North-Holland, 1988.

[11]   Shallit, J.O. "Metric theory of Pierce expansion." *The Fibonacci Quarterly*, Vol. *24.1* (1986): pp. 122-140.

[12]   Sylvester, J.J. "On a Point in the Theory of Vulgar Fractions." The Collected Mathematical Papers of James Joseph Sylvester 3. Edited by H.F. Baker. Cambridge (1909): pp. 440-445.

[13]   Vajda, S. Fibonacci and Lucas Numbers, and the Golden Section. New York: Wiley, 1989.

AMS Classification Numbers: 11A67, 11B39

ON THE CHARACTERISTIC POLYNOMIAL OF THE j-TH ORDER FIBONACCI SEQUENCE

George W. Grossman and Sivaram K. Narayan[1]

1. INTRODUCTION

The characteristic polynomial of the jth-order Fibonacci sequence or the Fibonacci j-polynomial, is defined to be

$$F_j = x^j - x^{j-1} - x^{j-2} - \cdots - x - 1, \quad j = 2, 3, \cdots. \tag{1.1}$$

F_j is well-known to have two (one positive and one negative) real roots for j even and one positive real root for j odd, (Miller, 1971) [5]. For convenience, in the present paper we denote the positive real root for F_j as ϕ_j and the negative real root as θ_j. Similarly, the real roots (one positive and one negative) of F_j' are denoted γ_j and σ_j, respectively, for j odd. For j even the one positive real root is denoted γ_j. For F_j'', the real roots (one positive and one negative) are denoted ξ_j and η_j, respectively, for j even, and simply ξ_j for j odd. It is also well-known, (Flores, 1967) [3], (Dubeau, 1989)[1] that

$$\lim_{j \to \infty} \phi_j = 2. \tag{1.2}$$

Another elementary proof of this is given in (Grossman, 1997) [4] which also shows that convergence is monotone, see also (Dubeau, 1993) [2]. An approximation to ϕ_j is given by $2 - 1/2^j$, which follows a result of (Dubeau, 1989) [1].

---

[1]The first author acknowledges the support given by Central Michigan University during his sabbatical leave. The second author acknowledges an FRCE Grant from Central Michigan University.

165

The properties of the sequence $\{\phi_j\}_{j=2}^{\infty}$ are known to be related ot the jth-order Fibonacci sequence by

$$\lim_{k \to \infty} \frac{G(j, k+1)}{G(j, k)} = \phi_j, \quad j = 2, 3, \cdots \tag{1.3}$$

where for $k > j$

$$G(j, k) = \sum_{i=1}^{j} G(j, k-i), \quad j = 2, 3, \cdots \tag{1.4}$$

with $G(j, 1) = 1$, $G(j, k) = 2^{k-2}$, $k = 2, 3, \cdots, j$. When j is even the sequence $\{\theta_j\}_{j=2}^{\infty}$ has less obvious relations to the Fibonacci j-sequence, but in the case $j = 2$, $|\phi_2| = |1/\theta_2|$, the Golden Section. It can be shown that

$$\lim_{\substack{j \to \infty \\ j \text{ even}}} \theta_j = -1. \tag{1.5}$$

The paper is divided into four sections including the introduction. Section 2 contains another expression for F_j, evaluations of the functions F_j, F'_j and F''_j at $x = -1, 0, 1, 2$, intervals containing roots, and inequalities of the functions F_j over the intervals $(-\infty, -1)$, $(-1, 0)$, $(0, 2)$, $(2, +\infty)$.

Section 3 establishes the monotone convergence characteristics of the roots. Asymptotic formulae or asymptotic bounds for the roots are also given. Many of the graphical characteristics of the polynomials, F_j, can be shown by use of symbolic calculators such as Maple, Mathematica or Derive, or hand-held calculators, such as the TI-92. Thus, the present analysis has relevance with respect to the use of technology in research. In addition, Maple was used to calculate asymptotic expansions of expressions involving roots.

In section 4 possibilities for further work in the form of generalizations of the methods established in the present paper are discussed.

2. GRAPHICAL PROPERTIES OF F_j, F'_j AND F''_j

Close in graphical form to F_j, is the Golden Polynomial G_j such that $G_{j+2} = xG_{j+1} + G_j$, $G_0 = -1$, $G_1 = x - 1$, (Moore, 1994) [6]. However, considered in the present paper is the simpler recursion $F_{j+1} = xF_j - 1$, $F_0 = 1$, $F_1 = x - 1$. It is known that (Prodinger, 1996) [7]

$$g_n \approx \frac{3}{2} + (-1)^n + \frac{25}{12} 4^{-n}.$$

where g_n, $n = 2, 3, \cdots$ is the greatest real root of G_n.

In the following, the properties are assumed to hold for all values of $j \geq 2$ where F_j is defined, unless otherwise specified. In proofs of lemmas, for ease and clarity of exposition, j is sometimes used for j even, (or odd), and then j is replaced by $2j$ (or $2j+1$) at the conclusion of the proof.

The basic evaluations of F_j are stated:

$$F_j(0) = -1, \; F_j(2) = 1, \; F_{2j}(-1) = 1, \; F_{2j+1}(-1) = -2. \tag{2.1}$$

By employing geometric series, the polynomials F_j, F'_j and F''_j are written as

$$F_j(x) = \begin{cases} \dfrac{-x^{j+1} + 2x^j - 1}{1 - x}, & x \neq 1, \\[2mm] -(j-1) & x = 1. \end{cases} \tag{2.2}$$

$$F'_j(x) = \begin{cases} \dfrac{x^j(1-3j) + jx^{j+1} + 2jx^{j-1} - 1}{(1-x)^2}, & x \neq 1, \\[2mm] \dfrac{j(3-j)}{2}, & x = 1. \end{cases} \tag{2.3}$$

$$F''_j(x) = \begin{cases} \dfrac{x^{j+1}j(1-j) + 2(j-1)(2j-1)x^j + j(7-5j)x^{j-1} + 2j(j-1)x^{j-2} - 2}{(1-x)^3}, & x \neq 1, \\[2mm] \dfrac{j(j-1)(5-j)}{3}, & x = 1. \end{cases} \tag{2.4}$$

L'Hôpital's rule can be used to evaluate F_j, F'_j and F''_j at $x = 1$. From (2.2 2.4) it can be inferred

$$F'_j(0) = -1, \; F''_j(0) = -2, \; (j > 2), \; F'_{2j}(-1) = -3j, \; F'_{2j+1}(-1) = 3j+1, \tag{2.5}$$

$$F''_{2j}(-1) = 2j(3j-2), \; F''_{2j+1}(-1) = -2j(3j+1), \tag{2.6}$$

$$F'_j(2) = 2^j - 1, \; F''_j(2) = (j-2)2^j + 2. \tag{2.7}$$

Employing (2.2-2.4) and Descarte's rule of signs, the interval location of real solutions to $F_j = 0$, $F'_j = 0$ and $F''_j = 0$ can be determined, see table 1. For particular cases $F'_3(1) = 0$,

int/fn	F_{2j}	F_{2j+1}	F'_{2j}	F'_{2j+1}	F''_{2j}	F''_{2j+1}
$(-1,0)$	yes	no	no	yes	yes $(j>1)$	no
$(0,1]$	no	no	yes $(j=1)$	yes $(j=1)$	yes $(j=2)$	yes $(j=1,2)$
$(1,2)$	yes	yes	yes $(j>1)$	yes $(j>1)$	yes $(j>2)$	yes $(j>2)$

Table 1. Does Interval contain a Root, yes or no?

$F'_2(1/2) = 0$, $F''_2 = 2$, $F''_3(1/3) = 0$, $F''_5(1) = 0$, $F''_4((1 + \sqrt{11/3})/4) = 0$. In table 1, the number of negative real roots alternates between 0,1 for j odd and even. It is an open question as to whether this happens for higher derivatives.

The inequalities between F_j, F'_j and F''_j are determined. It can be shown that the intersection points are: $F_j = F_{j+1}$ if $x = 0, 2$ and $F_j = F_{j+2}$ if $x = 0, -1, 2$. Moreover,

$$F_j = xF_{j-1} - 1, \text{ and } F_{j+1} - F_j = x^j(x-2), \tag{2.8}$$

$$F_{j+1} - F_{j-1} = x^{j-1}(x-2)(x+1), \text{ and } F_{j+1} = x^2 F_{j-1} - (x+1) \tag{2.9}$$

which implies

$$sgn(F_{2j+3} - F_{2j+1}) = \begin{cases} -1, & x < -1,\ 0 < x < 2, \\ +1, & -1 < x < 0,\ x > 2. \end{cases} \tag{2.10}$$

$$sgn(F_{2j+2} - F_{2j}) = \begin{cases} -1, & -1 < x < 0,\ 0 < x < 2, \\ +1, & x < -1,\ x > 2. \end{cases} \tag{2.11}$$

In particular, from (2.8)

$$F_{2j+2} > F_{2j+1}, \quad x > 2,\ x < 0,$$

$$F_{2j+1} > F_{2j+2}, \quad 0 < x < 2,$$

$$F_{2j+1} > F_{2j}, \quad x > 2,$$

$$F_{2j} > F_{2j+1}, \quad x < 2.$$

From above, we obtain

Lemma 2.1:

$$F_{2j} > F_{2j+1} > F_{2j+2} > F_{2j+3}, \quad 0 < x < 2,$$

$$F_{2j+3} > F_{2j+2} > F_{2j+1} > F_{2j}, \quad x > 2,$$

$$F_{2j} > F_{2j+2} > F_{2j+3} > F_{2j+1}, \quad -1 < x < 0,$$

$$F_{2j+2} > F_{2j} > F_{2j+1} > F_{2j+3}, \quad x < -1.$$

Figures (1a-c) show the graphs of F_j, F'_j and F''_j $j = 2, 3, \cdots, 6$.

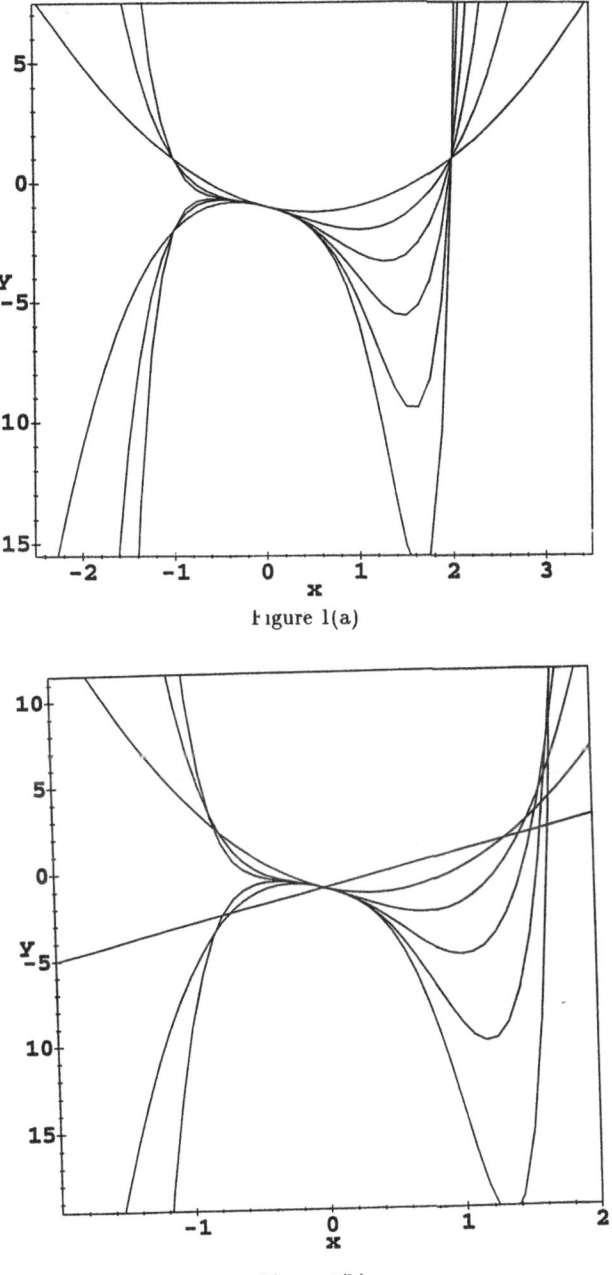

Figure 1(a)

Figure 1(b)

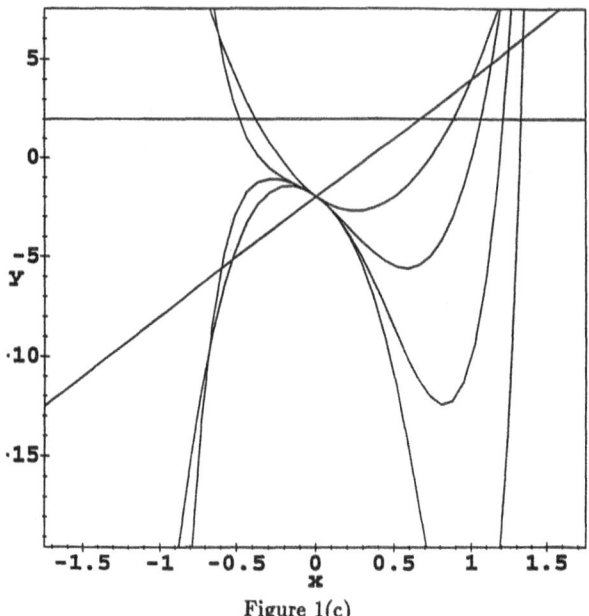

Figure 1(c)

In the next section the problem of asymptotic convergence of the roots of F_j, F'_j and F''_j is solved and extrema for F_j and F'_j are determined.

3. ON THE CONVERGENCE OF ROOTS

From Lemma 2.1 and Table 1 it readily follows that $\phi_j < \phi_{j+1}$ and $\theta_{2j} > \theta_{2j+2}$.

Theorem 3.1: *The negative real roots of F_{2j}, F'_{2j+1} and F''_{2j} given by θ_{2j}, σ_{2j+1} and η_{2j} satisfy*

$$\lim_{j\to\infty} \eta_{2j} = \lim_{j\to\infty} \sigma_{2j+1} = \lim_{j\to\infty} \theta_{2j} = -1. \tag{3.1}$$

Proof: By (2.2) we have for arbitrary j

$$F_j = x^j\left[1 + \frac{1}{1-x}\right] - \frac{1}{1-x}, \quad -1 < x < 0. \tag{3.2}$$

F_j converges uniformly to $-1/(1-x)$ on any closed interval $[x',0]$ where $x' > -1$, which implies the latter result of (3.1) since given $\epsilon > 0$, F_j as given in (3.2) satisfies

$$\left|F_j - \frac{-1}{1-x}\right| \le |2x^j| < \epsilon \tag{3.3}$$

for large enough j and any $x \in [x',0]$. Thus, F_j, as (even) j increases, can only have a limiting root of -1 since $-1/(1-x) \le -1/2$ on $[x',0]$, and by choosing a sequence $\{x'_i\}$ converging to

-1 and letting j increase so that (3.3) is satisfied. By inspection of (2.3-2.4), this type of argument can be extended to the first and second derivatives also. $\square$

Consider the positive roots of F'_j for j odd (the same results as in Lemma 3.4 below hold for j even since by Lemma 2.1 $F_{2j+1} > F_{2j} > F_{2j-1}$).

Lemma 3.2:

$$F'_{2j+1} = 2xF_{2j-1} + x^2 F'_{2j-1} - 1. \tag{3.4}$$

Proof: This result follows from applying the basic recursive relation $F_{j+1} = xF_j - 1$ twice, obtaining the second result of (2.9), taking the derivative and replacing $j \pm 1$ by $2j \pm 1$. $\square$

Corollary 3.3:

$$F'_{2j-1}(\gamma_{2j-1}) = 0 \Rightarrow F'_{2j+1}(\gamma_{2j-1}) < 0. \tag{3.5}$$

The sequence $\{\gamma_{2j+1}\}_{j=1}^{\infty}$ is strictly increasing.

Proof: $F_{2j-1}(0) = -1$, $F_{2j-1}(1/2) = -2 + 3/2^{2j-1}$, $F_{2j-1}(\phi_{2j-1}) = 0$ so that $0 < \gamma_{2j-1} < \phi_{2j-1}$ and $F_{2j-1}(\gamma_{2j-1}) < 0$ and the first result follows by Lemma 3.2. To prove the latter statement: $F'_{2j+1}(\gamma_{2j-1}) < 0 \Rightarrow \gamma_{2j-1} < \gamma_{2j+1} < 2$ since $F_{2j+1}(2) = 1$. $\square$

Lemma 3.4: The sequence of positive real roots, $\{\gamma_j\}_{j=2}^{\infty}$, of $\{F'_j\}_{j=2}^{\infty}$ satisfies

$$\lim_{j \to \infty} \gamma_j = 2, \tag{3.6}$$

$$\lim_{j \to \infty} F_j(\gamma_j) = -\infty. \tag{3.7}$$

Proof: From (2.2-2.3) $F_{2j+1}(1) = -2j$, $F'_{2j+1}(1) = (2j+1)(1-j)$ which implies (3.7). To prove (3.6), from (2.9), for any j

$$F'_{j+1} = F'_{j-1} + x^{j-2}((j+1)x^2 - jx - 2(j-1)), \tag{3.8}$$

then necessarily $(j+1)\gamma_{j-1}^2 - j\gamma_{j-1} - 2(j-1) < 0$ by Corollary 3.3. This implies, since $x^{j-2} > 0$, by considering the real solutions to this quadratic

$$\frac{j - \sqrt{9j^2 - 8}}{2(j+1)} < \gamma_{j-1} < \frac{j + \sqrt{9j^2 - 8}}{2(j+1)}.$$

Employing the *asympt* command in Maple yields (alternately one can expand in powers of j, algebraically, and use Taylor series for $\sqrt{\ }$)

$$-1 + \frac{1}{j} + O\left(\frac{1}{j^2}\right) < \gamma_{j-1} < 2 - \frac{2}{j} + \frac{4}{3j^2} + O\left(\frac{1}{j^3}\right). \tag{3.9}$$

By inspection of (2.3), $F'_j(\gamma_j) = 0$ implies $\gamma_j(1 - 3j) + j\gamma_j^2 + 2j > 0$. Factoring and taking the plus sign give

$$\gamma_j > \frac{3j - 1 + \sqrt{j^2 - 6j + 1}}{2j}$$

or again, employing the *asympt* command gives

$$\gamma_j > 2 - \frac{2}{j} - \frac{2}{j^2} + O\left(\frac{1}{j^3}\right). \tag{3.10}$$

By changing the subscript from j, $j - 1$ to $2j + 1$, obtains from (3.9-3.10)

$$2 - \frac{2}{2j + 1} - \frac{2}{(2j + 1)^2} + O\left(\frac{1}{(2j + 1)^3}\right) < \gamma_{2j + 1} < 2 - \frac{1}{j + 1} + \frac{1}{3}\frac{1}{(j + 1)^2} + O\left(\frac{1}{(j + 1)^3}\right). \quad \Box$$

Similarly, inequalities for γ_{2j} can be obtained.

Example: Using the *solver* on the TI-85 to solve $F'_j = 0$ in (2.3) with $j = 453$, gives $\gamma_{453} = 1.9955751777$, whereas $2 - 2/453 - 2/453^2 = 1.99557524$ and $2 - 1/227 + 1/(3(227^2))$ $= 1.9956011$. Both approximations are accurate to several significant digits.

Lemma 3.5:   *The sequence of positive roots of* $\{F''_{2j}\}$, $\{\xi_{2j}\}_{j=1}^\infty$, *satisfies*

$$\lim_{j \to \infty} \xi_{2j} = 2, \tag{3.11}$$

$$\xi_{2j} < \xi_{2(j + 1)}. \tag{3.12}$$

Proof: The proof is similar to that of Lemma 3.4. $F'_{2j}(x) > 0$ if $x > \gamma_{2j}$ and $F'_{2j}(1) < F'_{2j}(0) < 0$ hence the minimum slope occurs at $\xi_{2j} \in [1, \gamma_{2j})$ since $F''_{2j}(1)F''_{2j}(2) < 0$. Since $F''_{2j + 2} = 2F_{2j} + 4xF'_{2j} + x^2 F''_{2j}$ it follow that $F''_{2(j + 1)}(\xi_{2j}) < 0$ and (3.12) follows, noting $F''_{2j}(2) > 0$. To show (3.11), write for any j, from (3.8)

$$F''_{j + 1} = F''_{j - 1} + x^{j - 3}(j(j + 1)x^2 - j(j - 1)x - 2(j - 1)(j - 2)),$$

so that since $x^{j - 1} > 0$

$$j(j + 1)\xi_{j - 1}^2 - j(j - 1)\xi_{j - 1} - 2(j - 1)(j - 2) < 0.$$

The roots of this quadratic are ($+$ sign)

$$\xi_{j - 1} = \frac{j(j - 1) + \sqrt{j(j - 1)}\sqrt{9j^2 - 9j - 16}}{2j(j + 1)}. \tag{3.13}$$

To second order, the use of Maple yields

$$2 - \frac{4}{j} + \frac{8}{3j^2} + O\left(\frac{1}{j^3}\right) > \xi_{j - 1}. \tag{3.14}$$

Now employing (2.4) implies

$$j(1-j)\xi_j^3 + 2(j-1)(2j-1)\xi_j^2 + j(7-5j)\xi_j + 2j(j-1) < 0,$$

and by Maple, to third order

$$\xi_j > 2 - \frac{4}{j} - \frac{4}{j^2} - \frac{12}{j^3} + O\left(\frac{1}{j^4}\right). \tag{3.15}$$

This calculation was somewhat more complicated, involving the real solution of a cubic and the asymptotic expansion of this solution, which was solved by Maple after some numerical experimentation. Combining (3.14-3.15) and replacing j, $j-1$ by $2j$ gives the approximate formula

$$2 - \frac{4}{2j+1} + \frac{8}{3}\frac{1}{(2j+1)^2} + O\left(\frac{1}{(2j+1)^3}\right) > \xi_{2j} > 2 - \frac{4}{2j} - \frac{4}{(2j)^2} - \frac{12}{(2j)^3} + O\left(\frac{1}{(2j)^4}\right).$$

Similarly, inequalities for ξ_{2j+1} can be obtained.

Example: By using the TI-85 and the *solver* command to solve (2.4) equated to 0 with $j = 454$ gives $\xi_{454} = 1.99116989167$, whereas $2 - 4/455 + 8/(3(455^2)) = 1.991221$ and $2 - 4/454 - 4/454^2 - 12/454^3 = 1.991169893$. Both approximations are good to several significant digits, the latter appears to be more accurate.

Turning to the negative roots of F_{2j}, F'_{2j+1}, the limit portion of Theorem 3.1 is proved by a different method.

Lemma 3.6: *The negative root of F_{2j}, θ_{2j} satisfies*

$$\theta_{2j} \approx -\frac{1}{3^{1/2j}} = -1 + \frac{\ln 3}{2j} + O\left(\frac{1}{j^2}\right), \quad j = 1, 2, \cdots. \tag{3.16}$$

Proof: Substitution θ_{j-1} for x and equating two expressions for F_{j+1} in (2.9) yields

$$\theta_{j-1}^{j-1}(\theta_{j-1} - 2)(\theta_{j-1} + 1) = -(\theta_{j-1} + 1).$$

Dividing by $-(\theta_{j-1} + 1)$ and employing Theorem 3.1 gives $\theta_{j-1} \approx -(1/3)^{1/(j-1)}$. Replacing $j-1$ by $2j$, the expression above can be expanded for large j.   □

Lemma 3.7: *The sequence of negative roots of $\{F'_{2j+1}\}_{j=1}^{\infty}$ given by $\{\sigma_{2j+1}\}_{j=1}^{\infty}$ satisfies*

$$\lim_{j \to \infty} \sigma_{2j+1} = -1, \tag{3.17}$$

$$\sigma_{2j+1} < \sigma_{2j-1}. \tag{3.18}$$

Proof: $F_{2j+1}(\theta_{2j}) = -1$ and $F_{2j+1}(0) = -1$ imply $\theta_{2j} < \sigma_{2j+1}$. (For $j+1$ odd)

$$F'_{j+1}(\sigma_{j-1}) = \sigma_{j-1}^{j-2}((j+1)\sigma_{j-1}^2 - j\sigma_{j-1} - 2(j-1))$$

so that considering the quadratic term in above expression we get

$$-1+\tfrac{1}{j}+O\!\left(\tfrac{1}{j^2}\right)< -1+\tfrac{\ln 3}{j-2}\approx\theta_{j-2}<\sigma_{j-1} \tag{3.19}$$

by Lemma 3.6 and (3.9). Thus, $F'_{j+1}(\sigma_{j-1})<0\Rightarrow F'_{2j+1}(\sigma_{2j-1})<0$. Thus, since $F'_{2j+1}(-1)>0$ (3.18) follows. Also, $F_{2j+1}>F_{2j-1}$ on $(-1,0)$ which is used below. To show (3.17) suppose $\lim_{j\to\infty}\sigma_{2j+1}=\sigma$ where $\sigma\in[-1,0)$. (A limit exists since $F'_{2j+1}(-1)>0$ so that $\{\sigma_{2j+1}\}_{j=1}^{\infty}$ is a bounded decreasing sequence.)

$$\lim_{j\to\infty}F_j(\sigma_{j-2})=\lim_{j\to\infty}F_{j-2}(\sigma_{j-2})$$

for j odd since

$$F_j(\sigma_j)>F_j(\sigma_{j-2})>F_{j-2}(\sigma_{j-2})$$

and $\sup_j F_j(\sigma_j)$ exists. Hence since $F_j=x^2F_{j-2}-x-1$

$$\lim_{j\to\infty}F_j(\sigma_{j-2})=\lim_{j\to\infty}(\sigma_{j-1}^2 F_{j-2}(\sigma_{j-2})-(\sigma_{j-2}+1))$$

which yields after simplification

$$\lim_{j\to\infty}F_j(\sigma_{j-2})=\tfrac{1}{\sigma-1}\le -1.$$

From Lemma 3.2,

$$\lim_{j\to\infty}F'_j(\sigma_{j-2})=\tfrac{2\sigma}{\sigma-1}-1=\tfrac{1+\sigma}{\sigma-1}.$$

From above and (3.8), obtains

$$\lim_{j\to\infty}F'_j(\sigma_{j-2})=\lim_{j\to\infty}\left(\sigma_j^{j-3}\tfrac{j}{2}\!\left(\sigma_{j-2}-\left(-1+\tfrac{1}{j-1}\right)\right)\!\left(\sigma_{j-2}-\left(2-\tfrac{2}{j-1}\right)\right)\right) \tag{3.20}$$

$$=\lim_{j\to\infty}\sigma_j^{j-3}\tfrac{j}{2}(\sigma+1)(\sigma-2)=\tfrac{\sigma+1}{\sigma-1}.$$

From (3.20) yields

$$\lim_{j\to\infty}\sigma_j^{j-3}\tfrac{j}{2}=\tfrac{1}{(\sigma-1)(\sigma-2)}. \tag{3.21}$$

Re-taking the limit in (3.21) after dividing by j and replacing $j-2$ by $2j+1$ gives (3.17). $\qquad\square$

Corollary 3.8:

$$\sigma_{j-2}\approx -\left(\tfrac{1}{6j}\right)^{\frac{1}{j-3}}=-1+\tfrac{\ln 6+\ln j}{j}+O\!\left(\tfrac{(\ln j)^2}{j^2}\right). \tag{3.22}$$

Example: The TI-85 *solver* command was used to solve (2.3) equated to 0 with $j=453$ and gave $\sigma_{453}=-.9826896$ whereas $-(1/6(455))^{1/452}=-.9826477$, accurate to 4 significant digits.

Corollary 3.9:

$$\lim_{j\to\infty} F_{2j+1}(\sigma_{2j+1}) = -\tfrac{1}{2}. \tag{3.23}$$

Corollary 3.10: *If $\{x_{2j+1}\}_{j=1}^{\infty}$ is any sequence converging to -1 such that $\sigma_{2j+1} > x_{2j+1}$ then*

$$\lim_{j\to\infty} F_{2j+1}(x_{2j+1}) = -\tfrac{1}{2}.$$

Proof: For sufficiently large odd j, there is k_j which depends on x_j so that

$$F_j(\sigma_j) \geq F_j(x_j) \geq F_{j-k_j}(\sigma_{j-k_j}).$$

Now let $j\to\infty$ in above. $\Box$

Next we consider η_{2j}, the negative real root of F_{2j}'', which corresponds to maximum of the slope of F_{2j} in the interval $(-1,0)$, since it can be shown that F_{2j}''' does not have a root in this interval.

Lemma 3.11: *The sequence of roots of $\{F_{2j}''\}_{j=1}^{\infty}$ given by $\{\eta_{2j}\}_{j=1}^{\infty}$ satisfies*

$$\lim_{j\to\infty} \eta_{2j} = -1, \tag{3.24}$$

$$\eta_{2j} < \eta_{2j-2}. \tag{3.25}$$

Proof: $\sigma_{2j-1} \leq \eta_{2j}$ since $F_{2j}'(\sigma_{2j-1}) = F_{2j-1}(\sigma_{2j-1}) < -1/2$ and $-1/2 < F_{2j}'(-1/2) \approx -4/9$ by (2.3) for j sufficiently large. Taking the derivative of (3.8) obtains with (3.14) and Maple

$$F_{j+1}''(\eta_{j-1}) \approx \eta_j^{j-3} j(j+1)\left(\eta_{j-1} - 2 + \tfrac{4}{j}\right)\left(\eta_{j-1} + 1 - \tfrac{2}{j}\right) \tag{3.26}$$

which implies with (3.22) that

$$-1 + \tfrac{2}{j} + O\!\left(\tfrac{1}{j^2}\right) < \eta_{j-1}.$$

Hence,

$$F_{j+1}''(\eta_{j-1}) < 0 \Rightarrow F_{2j+2}''(\eta_{2j}) < 0. \tag{3.27}$$

From (3.8-3.10)

$$F_{j+1}'(\eta_{j-1}) \approx F_{j-1}'(\eta_{j-1}) + \eta_j^{j-2}(j+1)\left(\eta_{j-1} - 2 + \tfrac{2}{j}\right)\left(\eta_{j-1} + 1 - \tfrac{1}{j}\right)$$

which is the sum of a negative and positive quantity so that $F_{j+1}'(\eta_{j-1}) > F_{j-1}'(\eta_{j-1})$ and with (3.27) implies (3.25). By this analysis it follows that

$$\lim_{j\to\infty} F_{2j}(\eta_{2j-2}) = \lim_{j\to\infty} F_{2j-2}(\eta_{2j-2}).$$

Assume $\lim_{j\to\infty}\eta_{2j}=\eta$.   Using this with Lemma 3.2 for j even and the inequality
$\theta_{2j-2}<\sigma_{2j-1}<\eta_{2j}$ (from proof of Lemma 3.7) yields

$$\lim_{j\to\infty}F'_{2j}(\eta_{2j})=\lim_{j\to\infty}F'_{2j}(\eta_{2j-2})=\left(\frac{2\eta}{\eta-1}-1\right)\frac{1}{1-\eta^2}=-\frac{1}{(1-\eta)^2}.$$

From (3.26) and replacing $j+1$ by $2j$

$$\lim_{j\to\infty}F''_{2j}(\eta_{2j-2})=\lim_{j\to\infty}\eta_{2j}^{2j-\frac{4}{2}}2j(2j-1)(\eta-2)(\eta+1).\tag{3.28}$$

$\square$

Next we use $F''_{2j}=2F_{2j-2}+4xF'_{2j-2}+x^2F''_{2j-2}$ to obtain

$$\lim_{j\to\infty}F''_{2j}(\eta_{2j-2})=\frac{2}{\eta-1}-\frac{4\eta}{(1-\eta)^2}=-2\frac{1+\eta}{(\eta-1)^2}.\tag{3.29}$$

Equating (3.28) and (3.29) implies (3.24) after dividing by $\eta+1$ and observing $\eta<0$ and taking
the negative root.

Corollary 3.12:

$$\eta_{2j}\approx-\left(\frac{1}{12(j+1)(2j+1)}\right)^{\frac{1}{2(j-1)}}\approx-1+\frac{\ln24/2+\ln j}{j}+O\left(\frac{(\ln j)^2}{j^2}\right).\tag{3.30}$$

Proof: By equating (3.29) and (3.28), without the limit, dividing by $\eta+1$, replacing j by $j+1$
and η by -1. $\square$

Example: The *solver* command was used to solve (2.4) equated to 0 with $j=454$ and gave
$\eta_{454}=-.96953775$ whereas with $j=227$ (3.30) gives $\eta_{454}\approx-.9694272$,   accurate   to   3
significant digits.

Corollary 3.13: *If $\{x_{2j}\}_{j=1}^{\infty}$ is any sequence converging to -1 such that $\eta_{2j}\geq x_{2j}$ then*

$$\lim_{j\to\infty}F'_{2j}(x_{2j})=-\tfrac{1}{4}.$$

Proof: This is similar to Corollary 3.10. $\square$

Lastly, we derive the result for ϕ_j, see also (Flores, 1967) [3], (Dubeau, 1989) [1].

Lemma 3.14: *The positive roots ϕ_j of F_j satisfy*

$$\phi_j\approx2-\frac{1}{2^j}.$$

Proof: This is similar to Lemma 3.6. Letting $\phi_{j-1}=2-\delta_{j-1}$ in (2.9), with $\delta_{j-1}>0$, yields
after simplification $\delta_{j-1}=1/(2-\delta_{j-1})^{j-1}$ and this gives the result by taking a first order
approximation and replacing $j-1$ by j. More precisely, we find by Taylor series that

$$\delta_j=\frac{1}{2^j}+\frac{1}{2}\frac{j\delta_j}{2^j}+O\left(\frac{j^2\delta_j^2}{2^j}\right)$$

and the expression $\delta_j = 1/(2 - \delta_j)^j$ converges to 0 by the Fixed Point Theorem for $0 \leq \delta_j < 1$ and sufficiently large j. $\qquad\qquad\qquad\qquad\qquad\qquad\qquad\qquad\qquad\qquad\qquad\qquad\qquad$ □

4. SUMMARY AND FURTHER IDEAS

In the present paper we have considered the basic properties of characteristic polynomial of the jth-order Fibonacci sequence. This paper also demonstrates the use of modern symbolic calculators, without which the work would have been more time-consuming and complicated. It shows that research possibilities may be facilitated with the use of symbolic calculators. The present analysis also presents possibilities for further fundamental work on the polynomials F_j in terms of finding approximate formulae for the roots of higher order derivatives (> 2) of F_j. This would involve use of (2.8-2.9) and higher order versions of (2.2-2.4). Additionally a powerful symbolic software such as used in the present paper would facilitate the possible solutions.

REFERENCES

[1]   Dubeau, Francçis. On r-Generalized Fibonacci Numbers." *The Fibonacci Quarterly*, Vol. *27.3* (1989): pp. 221-229.

[2]   Dubeau, Francçis. "The Rabbit Problem Revisited." *The Fibonacci Quarterly*, Vol. *31.3* (1993): pp. 268-273.

[3]   Flores, Ivan. "Direct Calculation of k-Generalized Fibonacci Numbers." *The Fibonacci Quarterly*, Vol. *5.3* (1967): pp. 259-266.

[4]   Grossman, George W. "Fractal construction by orthogonal projection using the Fibonacci sequence." *The Fibonacci Quarterly*, Vol. *35.3* (1997): pp. 206-224.

[5]   Miller, M.D. "On Generalized Fibonacci Numbers." *Amer. Math. Monthly*, Vol. *78.10* (1971): pp. 1108-1109.

[6]   Moore, Gregory A. "The Limit of the Golden Numbers is 3/2." *The Fibonacci Quarterly*, Vol. *32.3* (1994): pp. 211-217.

[7]   Prodinger, Helmut. "The asymptotic behaviour of the Golden numbers." *The Fibonacci Quarterly*, Vol. *34.3* (1996): pp. 224-225.

AMS Classification Numbers: 65D99, 11C08, 11B39

QUASI MORGAN-VOYCE POLYNOMIALS AND PELL CONVOLUTIONS

A.F. Horadam

1. PRELUDE

A mathematically fertile class of polynomials $\{R_n^{(r,u)}(x)\}$, introduced recently in [1], is defined recursively by

$$R_n^{(r,u)}(x) = (x+2)R_{n-1}^{(r,u)}(x) - R_{n-2}^{(r,u)}(x) \quad (n \geq 2) \tag{1.1}$$

with

$$R_0^{(r,u)}(x) = u, \quad R_1^{(r,u)}(x) = x + r + u \tag{1.2}$$

in which r, u are thought of as integers.

Aspects of these polynomials are found to involve many kinds of polynomials and their numerical specializations: Fibonacci, Lucas, Chebyshev, and Morgan-Voyce.

Curiosity prompts us to ask: what are the consequences of replacing the $-$ sign in (1.1) by $+$? To the polynomials thus generated we will assign the adjective *quasi*.

Though the occurrence of Pell numbers $P_n(P_0 = 0, 1, 2, 5, 12, 29, \cdots)$ and Pell-Lucas numbers $Q_n(Q_0 = 2, 2, 6, 14, 34, 82, \cdots)$ should not surprise us since $x = 0$ in (3.1) with $r = 1, u = 0$ and $r = 0, u = 2$ give rise to them, it is pleasant (*mirabile dictu!*) to encounter Pell convolutions.

Relevant information on the polynomials $P_n(x)$ and $Q_n(x)$, from which P_n and Q_n are derived when $x = 1$, may be found in [2], while data on the convolutions $P_n^{(m)}(x)$ appear in [3].

This paper is in final form and no version of it will be submitted for publication elsewhere.

2. PELL CONVOLUTIONS

The *mth Pell convolution sequence* $\{P_n^{(m)}(x)\}$ where $m = 0, 1, 2, \cdots; n = 1, 2, 3, \cdots$, is defined in [3, (7.1)]. Ordinary Pell polynomials $P_n(x)$ are specified when $m = 0$. When, further, $x = 1$, the Pell numbers $P_n(1) = P_n$ emerge. Take $P_0^{(m)}(x) = 0$.

For ease of subsequent reference, we record in the restricted Table 1 the simplest *Pell convolution numbers* $P_n^{(m)}(1) = P_n^{(m)}$, i.e., $x = 1$ in [3, pp. 70-71]:

n \ m	0	1	2	3	4
1	1	1	1	1	1
2	2	4	6	8	10
3	5	14	27	44	65
4	12	44	104	200	340
5	29	131	366	810	1555

Table 1. Pell Convolution Numbers $P_n^{(m)}$.

Certain fundamental properties of $\{P_n^{(m)}\}$ are required in the development of our theme. Among these are [3, (7.3) - (7.5) p. 71]

$$P_n^{(m)} = 2P_{n-1}^{(m)} + P_{n-2}^{(m)} + P_n^{(m-1)} \quad (\textit{recurrence relation}) \tag{2.1}$$

$$(n-1)P_n^{(m)} = (m+1)\sum_{i=1}^{n-1} Q_i P_{n-i}^{(m)}; \tag{2.2}$$

$$= 2(m+1)\left\{ P_{n-1}^{(m+1)} + P_{n-2}^{(m+1)} \right\} \tag{2.3}$$

which occur in [3] as equations (7.3), (7.5), and (7.4), respectively ($x = 1$).

Combining (2.1) and (2.3) leads readily to the valuable result, not deduced in [3],

$$P_{n+1-m}^{(m)} + P_{n-1-m}^{(m)} = \frac{n}{m}P_{n+1-m}^{(m-1)}. \tag{2.4}$$

Moreover, a further new result, by (2.1), (2.3), is

$$P_n^{(m)} - P_{n-1}^{(m)} = \frac{n+2m-1}{2m}P_n^{(m-1)}. \tag{2.5}$$

3. THE SEQUENCES $\left\{ S_n^{(r,u)}(x) \right\}$

Definition:

Define the polynomials

$$S_n^{(r,u)}(x) = (x+2)S_{n-1}^{(r,u)}(x) + S_{n-2}^{(r,u)}(x) \quad (n \geq 2) \tag{3.1}$$

with

$$S_0^{(r,u)}(x) = u, \ S_1^{(r,u)}(x) = x + r + u \tag{3.2}$$

where r, u are integers.

Then there exist integers $d_{n,k}^{(r,u)}$:

$$S_n^{(r,u)}(x) = \sum_{k \geq 0} d_{n,k}^{(r,u)} x^k \tag{3.3}$$

with

$$\begin{cases} d_{n,k}^{(r,u)} = 0 \text{ if } k > n \\[2ex] d_{n,n}^{(r,u)} = 1 \text{ if } n \geq 1. \end{cases} \tag{3.4}$$

Recurrences:

Obviously, by (3.1), $d_{n,0}^{(r,u)} = S_n^{(r,u)}(0)$ satisfies the Pellian recurrence

$$d_{n,0}^{(r,u)} = 2d_{n-1,0}^{(r,u)} + d_{n-2,0}^{(r,u)} \quad (n \geq 2) \tag{3.5}$$

with

$$\begin{cases} d_{0,0}^{(r,u)} = u \quad\ = P_0 r + \tfrac{1}{2}Q_0 u \\[2ex] d_{1,0}^{(r,u)} = r + u = P_1 r + \tfrac{1}{2}Q_1 u. \end{cases} \tag{3.6}$$

Immediately, we surmise, and prove (for $k = 0$)

Theorem 1: $d_{n,0}^{(r,u)} = P_n r + \tfrac{1}{2}Q_n u \ \left(\tfrac{1}{2}Q_n = P_n + P_{n-1}\right).$

Proof: Induction on n, with (3.5), produces the result.

Comparison of coefficients of x^k in (3.1) reveals the recurrence ($n \geq 2$, $k \geq 1$)

$$d_{n,k}^{(r,u)} = 2d_{n-1,k}^{(r,u)} + d_{n-2,k}^{(r,u)} + d_{n-1,k-1}^{(r,u)}. \tag{3.7}$$

Pell Convolutions and the Coefficients $d_{n,k}^{(r,u)}$:

Three stages are to be considered, namely, when $k = 1$, $k = 2$, and k is unspecified. Repeated recourses to the law of formation (3.7) of $d_{n,k}^{(r,u)}$ will be essential.

Firstly, therefore, take $k = 1$.

Theorem 2: $d_{n,1}^{(r,u)} = P_n + P_{n-1}^{(1)} r + \dfrac{n-1}{2}P_n u.$

Proof: The theorem is readily verified for $n = 1, 2, 3$.

Assume, for some M, that it is also true for all positive integers $n = N \leq M$, i.e., assume

$$d_{N,1}^{(r,u)} = P_N + P_{N-1}^{(1)}r + \frac{N-1}{2}P_N u. \qquad (I)$$

Now

$$
\begin{aligned}
d_{N+1,1}^{(r,u)} &= 2d_{N,1}^{(r,u)} + d_{N-1,1}^{(r,u)} + d_{N,0}^{(r,u)} \qquad \text{by (3.7)} \\
&= (2P_N + P_{N-1}) + (2P_{N-1}^{(1)} + P_{N-2}^{(1)} + P_N)r \\
&\quad + \{(N-1)P_N + \frac{(N-2)}{2}P_{N-1} + P_N + P_{N-1}\}u \qquad \text{by (I), Theorem 1} \\
&= P_{N+1} + P_N^{(1)}r + \frac{N}{2}P_{N+1}u.
\end{aligned}
$$

Thus, the theorem is valid for $n = N + 1$, and so for all n.

Theorem 3: $d_{n,2}^{(r,u)} = P_{n-1}^{(1)} + P_{n-2}^{(2)}r + \frac{n-2}{4}P_{n-1}^{(1)}u.$

Proof: Proceed by induction as in Theorem 2.

Eventually, we may establish the general situation by induction, namely,

Theorem 4: $d_{n,k}^{(r,u)} = P_{n-k+1}^{(k-1)} + P_{n-k}^{(k)}r + \frac{n-2}{2^k}P_{n-k+1}^{(k-1)}u.$

Example: $d_{5,2}^{(r,u)} = P_4^{(1)} + P_3^{(2)}r + \frac{3}{4}P_4^{(1)}u$

$$= 44 + 27r + 33u \qquad \text{by Table 1,}$$

in agreement with Table 2.

n \ k	0	1	2	3	4	5	6
0	u						
1	$u+r$	1					
2	$3u+2r$	$2+u+r$	1				
3	$7u+5r$	$5+5u+4r$	$4+u+r$	1			
4	$17u+12r$	$12+18u+14r$	$14+7u+6r$	$6+u+r$	1		
5	$41u+29r$	$29+58u+44r$	$44+33u+27r$	$27+9u+8r$	$8+u+r$	1	
6	$99u+70r$	$70+175u+131r$	$131+131u+104r$	$104+52u+44r$	$44+11u+10r$	$10+u+r$	1

Table 2. The coefficients $d_{n,k}^{(r,u)}$.

Combining (3.3), with (3.4), and Theorem 4, we can express the polynomials $S_n^{(r,u)}(x)$ in terms of Pell convolutions. Surprising as the appearance of these convolutions may be, their existence here is a quite natural phenomenon. A detailed analysis of their genesis in simple numerical cases is a worthwhile experience.

Properties of $S_n^{(r,u)}(x)$:

Generating Function:

Standard techniques enable us to derive the generating function

$$\sum_{i=0}^{\infty} S_i^{(r,u)}(x)y^i = \frac{u + \{r - u + (1-u)x\}y}{1 - \{(x+2)y + y^2\}} \tag{3.8}$$

Binet Form:

Associated with (3.1) is the characteristic equation

$$\lambda^2 - (x+2)\lambda - 1 = 0 \tag{3.9}$$

with roots $\alpha(x) \equiv \alpha$, $\beta(x) \equiv \beta$ given by

$$\begin{cases} \alpha = \dfrac{x + 2 + \sqrt{(x+2)^2 + 4}}{2} \\[3mm] \beta = \dfrac{x + 2 - \sqrt{(x+2)^2 + 4}}{2} \end{cases} \tag{3.10}$$

so that

$$\begin{cases} \alpha\beta = -1 \\[2mm] \alpha + \beta = x + 2 \\[2mm] \alpha - \beta = \sqrt{(x+2)^2 + 4} = \Delta. \end{cases} \tag{3.11}$$

Proceeding in the usual way, we deduce the *Binet form*

$$S_n^{(r,u)}(x) = \frac{\{x + r + (1-\beta)u\}\alpha^n - \{x + r + (1-\alpha)u\}\beta^n}{\Delta}. \tag{3.12}$$

Applying (3.12), we establish the *Simson formula*

$$\begin{cases} S_{n+1}^{(r,u)}(x)S_{n-1}^{(r,u)}(x) - \left(S_n^{(r,u)}(x)\right)^2 = (-1)^n\{(x+r+u)[x(1-u)+r-u] - u^2\} \\[4mm] \qquad\qquad = (-1)^n\{S_1^{(r,u)}(x)S_{-1}^{(r,u)}(x) - \left(S_0^{(r,u)}(x)\right)^2 \end{cases} \tag{3.13}$$

since

$$
\left\{
\begin{aligned}
S_{-1}^{(r,u)}(x) &= r - u + (1-u)x \quad \text{by (3.1), (3,2)} \\[2ex]
&= -R_{-1}^{(r,u)}(x) \quad \text{by [1, (5.9)]}
\end{aligned}
\right.
\tag{3.14}
$$

on extending n to include negative numbers. Or, by (1.1), (1.2), (3.1), and (3.2), with superscripts temporarily omitted,

$$
S_{-1}(x) = S_1(x) - (x+2)S_0(x) = R_1(x) - (x+2)R_0(x) = -R_{-1}(x).
$$

4. THE QUASI MORGAN-VOYCE POLYNOMIALS

Definitions:

The *quasi Morgan-Voyce polynomials* $B_n(x)$, $b_n(x)$ and their associated polynomials $C_n(x)$, $c_n(x)$ are specified thus:

$$
B_{n+1}(x) = S_n^{(1,1)}(x)
\tag{4.1}
$$

$$
b_{n+1}(x) = S_n^{(2,1)}(x)
\tag{4.2}
$$

$$
C_n(x) = S_n^{(0,2)}(x)
\tag{4.3}
$$

$$
c_{n+1}(x) = S_n^{(0,1)}(x)
\tag{4.4}
$$

with

$$
S_n^{(0,0)}(x) = xB_n(x).
\tag{4.5}
$$

The first few terms of (4.1)-(4.4) are:

$$
B_0(x) = 0, \; B_1(x) = 1, \; B_2(x) = 2+x, \; B_3(x) = 5 + 4x + x^2, \cdots;
\tag{4.6}
$$

$$
b_0(x) = 1, \; b_1(x) = 1, \; b_2(x) = 3+x, \; b_3(x) = 7 + 5x + x^2, \cdots;
\tag{4.7}
$$

$$
C_0(x) = 2, \; C_1(x) = 2+x, \; C_2(x) = 6 + 4x + x^2, \; C_3(x) = 14 + 15x + 6x^2 + x^3, \cdots;
\tag{4.8}
$$

$$
c_0(x) = -1, \; c_1(x) = 1, \; c_2(x) = 1+x, \; c_3(x) = 3 + 3x + x^2, \cdots.
\tag{4.9}
$$

Readers are encouraged to generate further expressions when $n = 4, 5, 6, \cdots$. Numerical values originate readily for $x = 1$. Then we write $B_n(1) \equiv B_n, \cdots, c_n(1) \equiv c_n$. Recurrence relations involving $B_n, \cdots, c_n$ emerge from (3.1), (3.2) in conjunction with (4.1)-(4.4).

Binet Forms:

From (3.12) and (4.1)-(4.4), the particular *Binet forms* are:

$$B_n(x) = \frac{\alpha^n - \beta^n}{\alpha - \beta} \tag{4.10}$$

$$b_n(x) = \frac{(1-\beta)\alpha^n - (1-\alpha)\beta^n}{\alpha - \beta} = B_n(x) + B_{n-1}(x) \text{ by (4.10)} \tag{4.11}$$

$$C_n(x) = \alpha^n + \beta^n = B_{n+1}(x) + B_{n-1}(x) \text{ by (4.10)} \tag{4.12}$$

$$c_n(x) = \frac{(1+\beta)\alpha^n - (1+\alpha)\beta^n}{\alpha - \beta} = B_n(x) - B_{n-1}(x) \text{ by (4.10).} \tag{4.13}$$

Moreover, from (3.12) and (4.10), with (3.2), we derive

$$S_n^{(r,u)}(x) = S_1^{(r,u)}(x)B_n(x) + S_0^{(r,u)}(x)B_{n-1}(x). \tag{4.14}$$

Properties of $B_n(x), \cdots$:

From amongst many relationships, we select the following few, which ought to be compared with $B_n(x), \cdots$ [4] for possible analogues:

$$B_n(x)C_n(x) = B_{2n}(x) \tag{4.15}$$

$$\frac{dC_n(x)}{dx} = nB_n(x) \tag{4.16}$$

$$B_n(x) - B_{n-2}(x) = b_n(x) - b_{n-1}(x) \tag{4.17}$$

$$C_{n+1}(x) + C_{n-1}(x) = \Delta^2 B_n(x) \tag{4.18}$$

$$\sum_{i=1}^{n} (-1)^{i+1} b_i(x) = (-1)^{n+1} B_n(x) \tag{4.19}$$

$$\sum_{i=1}^{n} c_i(x) = B_n(x) \tag{4.20}$$

and the neat quartet

$$(x+1)B_n(x) + B_{n-1}(x) = c_{n+1}(x) \tag{4.21}$$

$$(x+2)B_n(x) + B_{n-1}(x) = B_{n+1}(x) \text{ (definition)} \tag{4.22}$$

$$(x+3)B_n(x) + B_{n-1}(x) = b_{n+1}(x) \tag{4.23}$$

$$(x+2)B_n(x) + 2B_{n-1}(x) = C_n(x). \tag{4.24}$$

Integration Properties:

$$\int_{-4}^{0} B_{2n}(x)dx = \int_{-4}^{0} C_{2n+1}(x)dx = 0 \tag{4.25}$$

$$\int_{-4}^{0} B_{2n+1}(x)dx = 2\frac{Q_{2n+1}(1)}{2n+1} \tag{4.26}$$

$$\int_{-4}^{0} C_{2n}(x)dx = 2\left(\frac{Q_{2n-1}(1)}{2n-1} + \frac{Q_{2n+1}(1)}{2n+1}\right) \tag{4.27}$$

$$\int_{-4}^{0} b_{2n}(x)dx = \int_{-4}^{0} B_{2n-1}(x)dx \tag{4.28}$$

$$\int_{-4}^{0} b_{2n+1}(x)dx = \int_{-4}^{0} B_{2n+1}(x)dx \tag{4.29}$$

$$\int_{-4}^{0} c_{2n}(x)dx = -\int_{-4}^{0} b_{2n-1}(x)dx \tag{4.30}$$

$$\int_{-4}^{0} c_{2n+1}(x)dx = \int_{-4}^{0} b_{2n+1}(x)dx. \tag{4.31}$$

Generating Functions. Simson Formulas:

We set this information out tabularly for (3.8) and (3.13), thus:

r	u	$S_n^{(r,u)}(x)$	Numerator in (3.8)	R.H.S. of (3.13)
0	1	$c_{n+1}(x)$	$(1-y)$	$(-1)^{n+1}(\alpha+\beta)$
1	1	$B_{n+1}(x)$	1	$(-1)^{n+1}$
2	1	$b_{n+1}(x)$	$(1+y)$	$(-1)^n(\alpha+\beta)$
0	2	$C_n(x)$	$[2-(2+x)y]$	$(-1)^{n+1}\Delta^2$

(4.32)

r, u values for $R_n^{(r,u)}(x), S_n^{(r,u)}(x)$:

Observe the following correspondences:

$$\begin{cases} \begin{array}{cc|cc} r & u & R_n^{(r,u)}(x) & S_n^{(r,u)}(x) \\ \hline 1 & 1 & B_{n+1}(x) \longleftrightarrow B_{n+1}(x) \\ 0 & 2 & C_n(x) \longleftrightarrow C_n(x) \\ 0 & 1 & b_{n+1}(x) \;\; c_{n+1}(x) \\ 2 & 1 & c_{n+1}(x) \;\; b_{n+1}(x) \end{array} \end{cases} \tag{4.33}$$

i.e.

(i) $r+u = 2$ (even) produces "direct" correspondences,

(ii) $r+u = 1,3$ (odd) produce "cross" correspondences.

Connection with P_n nd Q_n:

Putting $x = 0$ in (4.6)-(4.9), or in (4.10)-(4.13), we get

$$B_n(0) = P_n \tag{4.34}$$

$$C_n(0) = Q_n = P_{n+1} + P_{n-1} \tag{4.35}$$

$$b_n(0) = P_n + P_{n-1} = \tfrac{1}{2}Q_n \quad \text{(cf. Theorem 1)} \tag{4.36}$$

$$c_n(0) = P_n - P_{n-1} = \tfrac{1}{2}Q_{n-1} = b_{n-1}(0) \quad \text{by (4.36).} \tag{4.37}$$

Associated Sequences:

Define:

$$B_n^{(l)}(x) = B_{n+1}^{(l-1)}(x) + B_{n-1}^{(l-1)}(x) \tag{4.38}$$

and

$$C_n^{(l)}(x) = C_{n+1}^{(l-1)}(x) + C_{n-1}^{(l-1)}(x) \tag{4.39}$$

with

$$B_n^{(0)}(x) = B_n(x), \quad C_n^{(0)}(x) = C_n(x). \tag{4.40}$$

Repeated application of (4.12) and (4.18) leads to the pleasing outcomes:

$$B_n^{(2l)}(x) = \Delta^{2l}B_n(x) = C_n^{(2l-1)}(x) \tag{4.41}$$

$$B_n^{(2l+1)}(x) = \Delta^{2l}C_n(x) = C_n^{(2l)}(x). \tag{4.42}$$

Polynomial pairs such as $B_n(x)$ and $C_n(x)$ which are linked by properties (4.41) and (4.42) may be called *associated*. Clearly, $b_n(x)$ and $c_n(x)$ are not associated. ($B_n(x)$ and $C_n(x)$ are associated [4], but $b_n(x)$ and $c_n(x)$ are not.)

5. RELATION BETWEEN $S_n^{(r,u)}(x)$ AND $R_n^{(r,u)}(x)$

Temporarily, write $S_n^{(r,u)}(x) \equiv S_n$ and $R_n^{(r,u)}(x) \equiv R_n$ for visual ease. Repeated use of (1.1) with (1.2) and (3.1) with (3.2) will be necessary. Extension of (1.1) and (3.1) to negative subscripts will also be essential.

S_n in terms of R_n:

Clearly, $S_0 = R_0$, $S_1 = R_1$, and we readily deduce that $S_2 = R_2 + 2R_0$, $S_3 = R_3 + 4R_1 + 2R_{-1}$, $S_4 = R_4 + 6R_2 + 8R_0 + 2R_{-2}, \cdots$.

Let $s_n = \sum$ (coefficients of the R's in S_n). E.g., $s_4 = 1 + 6 + 8 + 2 = 17 = \tfrac{1}{2}Q_4$.

Theorem 5: $s_n = \tfrac{1}{2}Q_n$ (always odd).

Proof: Obviously the theorem is true for small values 1,2,3,4 of n. Suppose it is true for $n = k$, i.e., suppose $s_k = \frac{1}{2}Q_k$.

Now

$$S_{k+1} = (x+2)S_k + S_{k-1} \qquad \text{by (3.1)}$$

$$\text{Therefore} \quad s_{k+1} = 2 \cdot \frac{1}{2}Q_k + \frac{1}{2}Q_{k-1} \qquad \text{by hypothesis}$$

$$= \frac{1}{2}Q_{k+1} \qquad \text{by the definition of } Q_n.$$

Therefore the theorem is true for all n.

The explanation for the term $2 \cdot \frac{1}{2}Q_k$ is that each $(x+2)R_i$ in the expansion of S_k is replaceable by 2 R's, R_{i+1} and R_{i-1}, by (1.1).

There does not appear to be an explicit summation expression for S_n as $\sum_{i=0}^{n-1} a_{n-2i}R_{n-2i}$, though a little effort reveals that $a_n = 1$, $a_{n-2} = 2(n-1)$, $a_{n-4} = 2(n-2)^2$, $\cdots$, $a_{-n+2} = 2$. A recurring pattern throughout a tabulation of S_n in terms of R's is that, if $s_{n,t} = $ coefficient of R_t in S_n, then

$$s_{n,t} = s_{n-2,t} + s_{n-1,t-1} + s_{n-1,t+1} \qquad (5.1)$$

e.g. $s_{5,1} = s_{3,1} + s_{4,0} + s_{4,2} = 4 + 8 + 6 = 18$.

R_n in terms of S_n:

It is simple to deduce that $R_2 = S_2 - 2S_0$, $R_3 = S_3 - 4S_1 + 2S_{-1}$, $R_4 = S_4 - 6S_2 + 8S_0 - 2S_{-2}, \cdots$.

Let $r_n = \sum$ (coefficients of the S's in R_n). Then

Theorem 6:

$$r_n = \begin{cases} 1 & \text{for } n = 4m,\ 4m+1, \\ -1 & \text{for } n = 4m+2,\ 4m+3. \end{cases}$$

Proof: Induction may be used.

Similar remarks apply for R_n in terms of S's as for S_n in terms of R's. The presence of the negative signs does complicate things marginally.

Examples:

(i)

$$B_4(x) = B_4(x) + 6B_2(x) + 8B_0(x) + 2B_{-2}(x)$$

$$= 12 + 14x + 6x^2 + x^3 \quad \text{on calculation.}$$

(ii) $$C_4(x) = C_4(x) - 6C_2(x) + 8C_0(x) - 2C_{-2}(x)$$

$$= 2 + 16x + 20x^2 + 8x^3 + x^4 \quad \text{on simplification.}$$

Negative Subscripts:

Start from $S_{-1} = -R_{-1}$ (3.14). One easily has $S_{-2} = R_{-2} + 2R_0$, $S_{-3} = -R_{-3} - 4R_{-1} - 2R_1$, $S_{-4} = R_{-4} + 6R_{-2} + 8R_0 + 2R_2, \cdots$.

Likewise, $\qquad R_{-2} = S_{-2} - 2S_0$, $\qquad\qquad R_{-3} = -S_{-3} + 4S_{-1} - 2S_1$, $R_{-4} = S_{-4} - 6S_{-2} + 8S_0 - 2S_4, \cdots$.

For an instructive comparison and contrast, we conclude with the pattern $(n = 5)$

$$S_5 = R_5 + 8R_3 + 18R_1 + 12R_{-1} + 2R_{-3},$$

$$S_{-5} = -R_{-5} - 8R_{-3} - 18R_{-1} - 12R_1 - 2R_3,$$

$$R_5 = S_5 - 8S_3 + 18S_1 - 12S_{-1} + 2S_{-3},$$

$$R_{-5} = -S_{-5} + 8S_{-3} - 18S_{-1} + 12S_1 - 2S_3.$$

One does not have to be clairvoyant to perceive that the numerical illustrations in this Section imply a quartet of harmoniously interlinked results (Theorems 7(a) - (d)). Induction will be used with the crucial aid of the empirically-derived formula (5.1).

Theorem 7:

(a): If $S_{2n+1} = \displaystyle\sum_{t=-2n+1}^{2n+1} s_{2n+1,t} R_t$, then $S_{-(2n+1)} = -\displaystyle\sum_{t=-2n+1}^{2n+1} s_{2n+1,t} R_{-t}$.

(b): If $S_{2n} = \displaystyle\sum_{t=-2n+2}^{2n} s_{2n,t} R_t$, then $S_{-2n} = \displaystyle\sum_{t=-2n+2}^{2n} s_{2n,t} R_{-t}$.

Proof: Earlier, we showed that (a) and (b) are true for $n = 0, 1, 2$. Assume that for some M, both (a) and (b) are also true for all positive integers $n = N \leq M$, i.e., assume (A) and (B) are true where (A) and (B) mean that in (a) and (b) respectively we have replaced n by N.

Our first objective is then to show that (b) is true for $n = N + 1$. In the summations below we will work from the upper bound to the lower bound for t. Now, by (1.1) and (3.1) extended to negative subscripts,

$$S_{-2(N+2)} = -(x+2)S_{-(2N+1)} + S_{-2N}$$

$$= \sum_{t=-2N+1}^{2N+1} (x+2)s_{2N+1,t}R_{-t} + S_{-2N} \quad \text{by (A)}$$

$$= s_{2N+1,2N+1}(R_{-(2N+2)} + R_{-2N}) + s_{2N+1,2N-1}(R_{-2N} + R_{-(2N-2)})$$

$$\qquad + s_{2N+1,2N-3}(R_{-(2N-2)} + R_{-(2N-4)}) + \cdots$$

$$\qquad + s_{2N+1,-2N+1}(R_{2N-2} + R_{2N}) + (s_{2N,2N}R_{-2N} + s_{2N,2N-2}R_{-(2N-2)}$$

$$\qquad + \cdots + s_{2N,-2N+2}R_{2N-2}) \qquad \text{by (B)}$$

$$= \sum_{t=-2N}^{2N+2} s_{2N+2,t}R_{-t}.$$

Therefore, (b) is true for $n = N+1$ and hence for all n. Similarly, it can now be shown that (a) is true for all n. Thus, the validity of the theorem is established.

In the proof, we have specifically applied the following ideas.

(i) $s_{2N+1,2N+1} = s_{2N+2,2N+2} = 1$ (i.e., the first coefficient, having subscripts equal, is always unity, as was observed earlier).

(ii) $s_{2N+1,2N+1} = s_{2N+2,-2N} = 2$ (i.e., irrespective of the subscript symbolism, the last coefficients is always 2 - noted earlier - due to the presence of the number 2 in the factor $(x+2)$ in definitions (1.1) and (3.1).

(iii) Except for (i) and (ii), all other coefficients in the final summation are the sum of three constituent coefficients to which (5.1) is applicable. No generality is therefore destroyed if, for convenience, we choose as representative the third member R_{-2N+2} the sum of whose coefficients is

$$s_{2N,2N-2} + s_{2N+1,2N-3} + s_{2N+1,2N-1} = s_{2N+2,2N-2}$$

by (5.1), in which $n = 2N+2, t = 2N-2$.

Using inductive processes, we may likewise assert the truth of the following results (complementing those in Theorems 7(a), (b)), where $r_{n,t} = $ coefficient of S_t in R_n.

Theorem 7:

(c): If $R_{2n+1} = \sum_{t=-2n+1}^{2n+1} (-1)^{\frac{t-1}{2}} r_{2n+1,t} S_t$, then

$$R_{-(2n+1)} = -\sum_{t=-2n+1}^{2n+1} (-1)^{\frac{t-1}{2}} r_{2n+1,t} S_{-t}.$$

(d): If $R_{2n} = \sum\limits_{t=-2n+2}^{2n} (-1)^{\frac{t}{2}} r_{2n,t} S_t$, then

$$R_{-2n} = \sum_{t=-2n+2}^{2n} (-1)^{\frac{t}{2}} r_{2n,t} S_{-t}.$$

Accordingly, Theorems 7(a) - (d) have exhibited the existence for R_n and S_n of a pleasing mutual mathematical symbiosis.

6. RISING DIAGONAL FUNCTIONS $S_n^{(r,u)}(x)$

Results below will only be stated as the proofs parallel those in [1].

Upward-slanting diagonals for $S_n^{(r,u)}(x)$ in Table 2, imagined in the mind's eye, produce the *rising diagonal functions* $S_n^{(r,u)}(x) \equiv S_n$ for brevity. Initially,

$$S_0 = r + u, \; S_1 = x + 2r + 3u. \tag{6.1}$$

Generally,

$$S_n = \sum_{k=0}^{[\frac{n+1}{2}]} d_{n+1-k,k}^{(r,u)} x^k \tag{6.2}$$

i.e.,

$$S_4 = d_{5,0}^{(r,u)} + d_{4,1}^{(r,u)} x + d_{3,2}^{(r,u)} x^2$$

$$= 29r + 41u + (12 + 14r + 18u)x + (4 + r + u)x^2.$$

For n even,

$$d_{\frac{n}{2}+1,\frac{n}{2}}^{(r,u)} = n + r + u. \tag{6.3}$$

Theorem 8: $\qquad S_n = 2S_{n-1} + (x+1)S_{n-2} \quad$ (recurrence). $\tag{6.4}$

Corollary 1: $\qquad S_n(-1) = 2^{n-1}(-1 + 2r + 3u). \tag{6.5}$

Corollary 2: $\qquad S_{n+1}(-1)S_{n-1}(-1) = S_n^2(-1). \tag{6.6}$

Determination of the generating function, the Binet form, and the Simson formula relevant to S_n are left to the initiative of the reader (see [1]). From all the above there may be deduced the particular results for the special cases $S_n^{(r,u)}(x) = B_{n+1}(x), \; b_{n+1}(x), \; C_n(x), \; c_{n+1}(x).$

Next, let us re-orient the triangular arrangement of the non-zero $B_n(x)$ by a counter-clockwise rotation into a rectangular array $B_{k,m}$ where k refers to rows and m (coefficients of x^m, $m = 0,1,2,\cdots$) to the columns. Then the first row $(k=1)$ consists entirely of units $1,1,1,1,\cdots$ while the second row $(k=2)$ consists of even numbers $2,4,6,8,\cdots$. A neat nexus now exists between the arrays for $B_{k,m}$ and $P_n^{(m)}$; in fact, they are identical $(k=n)$:

$$\begin{cases} B_{n,m} = P_n^{(m)}, \\ B_{n,0} = P_n \quad (m=0). \end{cases} \tag{6.7}$$

No such very simple attractive connection as (6.7) exists between the Pell convolutions and $b_{n,m}$, $C_{n,m}$, and $c_{n,m}$ when a similar re-orientation is effected.

With some effort, careful inspection reveals that

$$\begin{cases} b_{n,m} = \frac{n}{2m} P_{n+1}^{(m-1)}, \\ b_{n,0} = \frac{1}{2} Q_n, \end{cases} \tag{6.8}$$

$$\begin{cases} C_{n,m} = \frac{n+m-1}{m} P_n^{(m-1)} \\ C_{n,0} = Q_{n-1}, \end{cases} \tag{6.9}$$

$$\begin{cases} c_{n,m} = \frac{n+2m-1}{2m} P_n^{(m-1)} \\ c_{n,0} = \frac{1}{2} Q_{n-1}. \end{cases} \tag{6.10}$$

Examples:

$$B_{5,2} = 366 \quad = P_5^{(2)},$$
$$b_{5,2} = 470 \quad = \frac{5}{4} P_6^{(1)},$$
$$C_{5,2} = 393 \quad = 3 P_5^{(1)},$$
$$c_{5,2} = 262 \quad = 2 P_5^{(1)}.$$

Use Table 1 in conjunction with (2.1) for $P_6^{(1)}$.

Generalization:

Designate by $S_{n,m}$ the re-orientation of the triangular array in Table 2 into a rectangular array where n,m refer to rows and columns, respectively. Observation of the resulting transformed pattern leads to the conclusion that

$$\begin{cases} S_{n,m} = P_n^{(m-1)} + \frac{n-1}{2m} P_n^{(m-1)} u + P_{n-1}^{(m)} r \\ S_{n,0} = \frac{1}{2} Q_{n-1} u + P_{n-1} r \quad (m=0). \end{cases} \tag{6.11}$$

Then the special cases (6.7)-(6.10) may all be verified on applying the relevant values of r and u displayed in (4.32). Algebraic manipulations involved in these verifications require reference to the convolution properties (2.1) and (2.3).

Illustration of (6.11):

$$C_{5,3} = P_5^{(2)} + \frac{4}{6}P_5^{(2)} \cdot 2 + P_4^{(3)} \cdot 0 \quad (u = 2, \ r = 0)$$

$$= 854 \quad \text{by Table 1.}$$

Expressions for the columns in the array in Table 2, relating to (3.3), will bear a formal resemblance to those in (6.11).

It is quite enchanting to behold the emerging structural unity of our theme as it unfolds like a beautiful flower.

7. CONCLUSION

Situations regarding orthogonality and zeros of the polynomials $B_n(x)$, $b_n(x)$, $C_n(x)$, and $c_n(x)$ do not arise in this paper (cf. [4]) as there is no direct correlation between these polynomials and the Chebyshev polynomials.

Further developments of the theory come readily to mind, e.g., extension to negative subscripts of $B_n(x), \cdots$, and consideration of the consequences of adding a constant to the right-hand sides of (1.1) and (3.1). Hidden among the rising diagonal functions, there are possibly differential equations of interest, as is often the case in such situations. All in all, we feel there are many Miltonic "fresh woods and pastures new" to be explored.

REFERENCES

[1]   Horadam, A.F. "A Composite of Morgan-Voyce Generalizations." *The Fibonacci Quarterly*, Vol. *35.3* (1997): pp. 233-239.

[2]   Horadam, A.F. and Mahon, Bro J.M. "Pell and Pell-Lucas Polynomials." *The Fibonacci Quarterly*, Vol. *23.1* (1985): pp. 7-20.

[3]   Horadam, A.F. and Mahon, Br. J.M. "Convolutions for Pell Polynomials." Applications of Fibonacci Numbers, Volume 1. Edited by G.E. Bergum, A.F. Horadam and A.N. Philippou. Kluwer Academic Publishers, Dordrecht, The Netherlands, 1986: pp. 55-80.

[4]   Horadam, A.F. "New Aspects of Morgan-Voyce Polynomials." Applications of Fibonacci Numbers, Volume 7. Edited by G.E. Bergum, A.F. Horadam and A.N. Philippou. Kluwer Academic Publishers, Dordrecht, The Netherlands, 1996: pp. 161-176.

AMS Classification Numbers: 11B39

ON AN ASYMPTOTIC MAXIMALITY OF THE FIBONACCI TREE

Yasuichi Horibe

1. INTRODUCTION

Continuing a previous paper [2], a certain asymptotically maximal property of the Fibonacci tree in the class of the self-similar binary trees will be shown, beginning with a review of the relevant parts of [2] in the next section.

2. SELF-SIMILAR TREES AND THEIR CHARACTERISTIC FUNCTIONS

For given relatively prime integers a, b such that $1 \leq a \leq b$, consider the sequence of binary trees

$$S_k = S_k(a, b), \quad k = 1, 2, \cdots,$$

defined recursively as follows: $S_1, \cdots, S_b$ are just one-leaf trees, and the left subtree of S_k, $k \geq b + 1$, is given by S_{k-a} and the right by S_{k-b}. This recursion defines a tree-growth rule (let us call (a, b)-growth) by viewing $k = 1, 2, \cdots$ as discrete times: the initial production of two children occurs at time $b + 1$ and, if a node is produced at time k, then its left child will be born at time $k + a$ and its right child at time $k + b$. The (a, b)-growing tree looks *self-similar* in an obvious sense, because every subtree is also (a, b)-growing by definition.

The $S_k(1, 2)$ has been called *Fibonacci tree* (of order k). The purpose of this note is to show a certain asymptotic maximality (when $k \to \infty$) of the Fibonacci tree in the class of various (i.e., by varying a, b) self-similar trees thus defined. In order to introduce a new measure in whose sense we will be considering "maximality", we need several characteristic notions associated with S_k.

This paper is in final form and no version of it will be submitted for publication elsewhere.

The *complexity* $K(T)$ of a general binary tree T is defined recursively as follows [2].

$$\begin{cases} K(T) = 0 \ \ \text{for } T \text{ with } n(T) = 1 \\ \\ K(T) = 1 + \min_{U} \ \max\{K(U), K(U')\} \ \text{for } T \text{ with } n(T) \geq 2, \end{cases} \tag{1}$$

where $n(T)$ is the number of leaves in T and the minimization is done by varying proper subtree U of T (with its complementary tree U', the tree obtained by deleting U from T and by contracting the edge incident with the parent of the root of U and with the sibling of that root). (See [2] for a "leaf-searching" interpretation of the complexity.)

Let $h(T)$ be the height of T. We have

$$\log n(T) \leq K(T) \leq h(T).$$

(log is for $\log_2$ and ln for $\log_e$.) The second inequality is obvious and the first is easily shown by induction on the number of leaves as follows: Suppose $n(T) \geq 2$, and let U be an arbitrary proper subtree of T. We have $n(U) + n(U') = n(T)$, and $K(U) \geq \log n(U)$ and $K(U') \geq \log n(U')$ by induction hypothesis. Hence $\max\{K(U), K(U')\} \geq \log\left(\frac{1}{2}n(T)\right)$, so that $K(T) \geq 1 + \log\left(\frac{1}{2}n(T)\right) = \log n(T)$.

We have, therefore,

$$\log n_k \leq K(S_k) \leq h_k. \tag{2}$$

$$(n_k = n(S_k), \ \ h_k = h(S_k))$$

The limit of the *normalized height*, $\lim\limits_{k\to\infty} h_k/\log n_k$, denoted by $g(c)$, was called *growth coefficient* ([2]) and shown to be

$$g(c) = (-\log \lambda(c))^{-1}, \ \ c = \frac{b}{a}, \tag{3}$$

where $\lambda(c)$ is the (unique) positive root of the equation $x^c = 1 - x$, that is,

$$(\lambda(c))^c = 1 - \lambda(c) \ \text{or} \ c = \frac{\log(1 - \lambda(c))}{\log \lambda(c)}. \tag{4}$$

Note that h_k is given simply by

$$h_k = \begin{cases} 0 & (1 \leq k \leq b) \\ \\ \left\lceil \dfrac{k-b}{a} \right\rceil & (b \leq k), \end{cases} \tag{5}$$

since $h_k = i$ when $(i-1)a + b < k \leq ia + b \ (i \geq 1)$. We also note

$$\begin{cases} \frac{1}{2} \le \lambda(c) < 1, \quad \lambda(1) = \frac{1}{2}, \quad \lambda(2) = \frac{\sqrt{5}-1}{2} \\ \\ 1 \le g(c) < \infty, \quad g(1) = 1, \quad g(2) = \dfrac{1}{\log\left(\dfrac{\sqrt{5}+1}{2}\right)} = 1.44. \end{cases} \tag{6}$$

3. NORMALIZED COMPLEXITY

Differentiating (4) with respect to c, regarding c as a continuous variable, we get

$$\lambda(c)(1-\lambda(c))(-\ln\lambda(c)) = \{\lambda(c)+c(1-\lambda(c))\}\lambda'(c). \tag{7}$$

Straightforward but rather tedious calculations, using (4) and the inequality $-\ln x \ge 1-x$, give us

$$\lambda'(c) > 0, \quad \lambda''(c) < 0, \tag{8}$$

$$g'(c) > 0, \quad g''(c) < 0. \tag{9}$$

It is interesting to observe here that the inverse of the growth coefficient can be written (from (4)) as

$$(g(c))^{-1} = \frac{-\lambda(c)\log\lambda(c) - (1-\lambda(c))\log(1-\lambda(c))}{\lambda(c)+c(1-\lambda(c))}.$$

Considering 1 and $c(a:b = 1 \cdot c)$ as "costs" (time costs in our tree-growth model) assigned to left and right branches, respectively, and noting ([2]) that

$$n_{k-a}/n_k \to \lambda(c), \quad n_{k-b}/n_k \to 1 - \lambda(c) \quad (k \to \infty),$$

the denominator above is the average cost (per branching) and the numerator is the entropy (uniformity) of the distribution $(\lambda(c), 1-\lambda(c))$, so that the ratio indicates the "uniformity of branching per cost." The same form

$$\frac{-x\log x - (1-x)\log(1-x)}{x+c(1-x)}$$

appeared in [1], but, here in the case of the self-similar trees, x depends on c.

In this note, however, we are much more interested in the following quantity rather than the related $g(c)$.

$$\kappa(c) = \varlimsup_{k \to \infty} \frac{K(S_k)}{\log n_k}, \text{ the limit superior of the } \textit{normalized complexity}.$$

Remark: We know that $\log n_k \approx (-\log\lambda(c))k/a$ ([2]) and $h_k \approx k/a$ for large k. Hence, from (2), $K(S_k)/\log n_k$ is approximately between $(-\log\lambda(c))k/a$ and k/a for large k.

From $1 \le K(S_k)/\log n_k \le h_k/\log n_k$ we have $1 \le \kappa(c) \le g(c)$. We will show in the next section that $\kappa(c)$ is maximal at $c = 2$ and $\kappa(2) = g(2)$, although $g(c) \uparrow \infty (c \to \infty)$ by (4) and (9).

4. $\kappa(c)$ IS MAXIMAL AT $c = 2$.

Lemma 1: (Theorem 2(A) in [2]): $K(S_k) = h_k$ if $c \le 2$.

From this lemma we see

$$\kappa(c) = g(c) \text{ if } c \le 2. \tag{10}$$

So $\kappa(c)$ is strictly monotone increasing up to $\kappa(2) = g(2)$ according to (9). How does $\kappa(c)$ behave when c exceeds 2?

Lemma 2: $K(S_k) \le 2k/b$ if $2 \le c \le 4$.

Proof: We prove the lemma by induction on k. When $1 \le k \le b$, we see $0 = K(S_k) \le 2k/b$. When $b + 1 \le k \le a + b$, S_k is the two-leaf tree, hence $1 = K(S_k) \le 2k/b$ holds. Now let $k \ge a + b + 1$. The root of S_k has the left-most grandchild u:

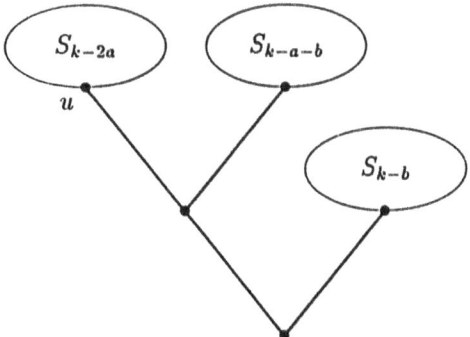

Take $U = S_{k-2a}$ in (1), then the tree U' complementary to U is given by the tree with S_{k-a-b} and S_{k-b} as its left and right subtrees. And by (1) we have

$$K(S_k) \le 1 + \max\{K(U), K(U')\}.$$

By the induction hypothesis

$$K(U) = K(S_{k-2a}) \le \frac{2(k - 2a)}{b}.$$

We also see from (1), using the induction hypothesis, that

$$K(U') \leq 1 + \max\{K(S_{k-a-b}), K(S_{k-b})\}$$

$$\leq 1 + \max\left\{\frac{2(k-a-b)}{b}, \frac{2(k-b)}{b}\right\}$$

$$= 1 + \frac{2(k-b)}{b}.$$

Hence we deduce

$$K(S_k) \leq 1 + \max\left\{\frac{2(k-2a)}{b}, 1 + \frac{2(k-b)}{b}\right\}$$

$$= \max\left\{\frac{2k-(4a-b)}{b}, \frac{2k}{b}\right\}$$

$$= \frac{2k}{b}$$

from the assumption $b \leq 4a$. $\square$

Lemma 2 shows that

$$\frac{K(S_k)}{\log n_k} \leq \frac{2k}{b \log n_k} = \frac{2}{b} \cdot \frac{k}{h_k} \cdot \frac{h_k}{\log n_k}, \quad 2 \leq c \leq 4.$$

Hence, from (5), we have

$$\kappa(c) \leq \frac{2}{b} \cdot a \cdot g(c)$$

$$= \frac{2g(c)}{c}, \quad 2 \leq c \leq 4.$$

Now consider the upper-bounding function $2g(c)/c$.

Lemma 3: $g(c)/c$, $1 \leq c < \infty$, is strictly monotone decreasing.

Proof: Differentiate

$$\frac{g(c)}{c} = \frac{\ln 2}{-c \ln \lambda(c)}.$$

Then

$$\left(\frac{g(c)}{c}\right)' = (\ln 2)\frac{\ln \lambda(c) + c\frac{\lambda'(c)}{\lambda(c)}}{(c \ln \lambda(c))^2}$$

$$= -(\ln 2)\frac{\lambda'(c)}{(1-\lambda(c))(c \ln \lambda(c))^2} < 0,$$

where (7) was used. $\square$

By checking the second derivative we can also show that $g(c)/c$ is a convex-downward function.

From Lemma 1, Lemma 3, and the fact that $2g(c)/c$ is equal to $g(2)$ when $c = 2$, we have proved that $\kappa(c)$ is maximal at $c = 2$, that is, the Fibonacci tree has the maximal normalized complexity asymptotically in the class of the self-similar trees.

A conjecture that naturally arises from the above observations is that $\kappa(c)$ itself will be monotone decreasing from $g(2)$ to 1 for $2 \leq c < \infty$, because, if c becomes large, S_k will be almost "linear" and hence $K(S_k)/\log n_k \approx 1$. Note that $2g(c)/c$ is already 1.07 (near 1) when $c = 4$.

REFERENCES

[1]   Horibe, Y. "An Entropy View of Fibonacci Trees." *The Fibonacci Quarterly*, Vol. *20* (1982): pp. 168-178.

[2]   Horibe, Y. "Growing a Self-Similar Tree." Applications of Fibonacci Numbers, Volume 7. Edited by G.E. Bergum, A.F. Horadam and A.N. Philippou. Kluwer Academic Publishers, Dordrecht, The Netherlands, 1998: pp. 177-184.

AMS Classification Numbers: 05C05

GENERALIZATIONS OF A FIBONACCI IDENTITY

F. T. Howard

1. INTRODUCTION

A well-known identity is

$$F_{m+n} = L_m F_n + (-1)^{m-1} F_{n-m} \tag{1.1}$$

where F_k and L_k are the Fibonacci and Lucas numbers, respectively. With the definitions $F_{-k} = (-1)^{k+1} F_k$ and $L_{-k} = (-1)^k L_k$, formula (1.1) is true for all integer m and n. The identity is easy to prove, and it is evidently useful; Rokach [11], for example, proved that (1.1) is about 2.88 times more efficient than the usual Fibonacci recurrence for computing the Fibonacci numbers.

In this paper we prove some generalizations of (1.1). In particular, we prove the following theorem:

Theorem 1.1: Let e_1, e_2, $\cdots$, e_k be arbitrary complex numbers, with $e_k \neq 0$, and define the sequence $\{a_n\}$ as follows: a_0, a_1, $\cdots$, a_{k-1} are arbitrary complex numbers, and for $n \geq k$,

$$a_n = e_1 a_{n-1} + e_2 a_{n-2} + \cdots + e_k a_{n-k}.$$

Then for $n \geq m \geq 0$,

$$a_{(k-1)m+n} = \sum_{j=1}^{k} (-1)^{j-1} c_{m,jm} a_{(k-j-1)m+n}. \tag{1.2}$$

The numbers $c_{m,jm}$ are defined by

$$\prod_{r=0}^{m-1} (1 - e_1(\theta^r x) - e_2(\theta^r x)^2 - \cdots - e_k(\theta^r x)^k) = 1 + \sum_{j=1}^{k} (-1)^j c_{m,jm} x^{jm},$$

This paper is in final form and no version of it will be submitted for publication elsewhere.

where θ is a primitive m'th root of unity.

Note that if $n = 0$ in Theorem 1.1, then (1.2) is the recurrence for multiples.

One application of Theorem 1.1 in this paper is a generalization of equation (1.1) (Theorem 1.2 below): Let a, b, p and q be arbitrary complex numbers, with $q \neq 0$, and define the sequence $\{w_n\} = \{w_n(a, b; p, q)\}$ by:

$$w_0 = a, \ w_1 = b, \text{ and for } n \geq 2, \ w_n = pw_{n-1} - qw_{n-2}. \tag{1.3}$$

Let
$$v_n = w_n(2, p; p, q) \tag{1.4}$$

and for $n \geq 0$ define

$$w_{-n} = -q^{-n}(w_n - av_n). \tag{1.5}$$

With (1.5), we can routinely verify that the recurrence of (1.3) holds for all integers n. In section 4 we prove the following generalization of (1.1).

Theorem 1.2: Let w_n, v_n, and w_{-n} be defined by (1.3), (1.4) and (1.5), respectively. Then for all integers m and n,

$$w_{m+n} = v_m w_n - q^m w_{n-m}.$$

We should mention here that Horadam [5], [7] has systematically investigated the sequence $\{w_n\}$, and he has proved Theorem 1.2. To the writer's knowledge, however, Theorem 1.1 is a new result, and he hopes it will provide new insights and new techniques for $\{w_n\}$ and more general sequences.

A summary by sections follows. In section 2 we state and prove a theorem which forms much of the theoretical basis for this paper, and in section 3 we use that theorem to give a proof of Theorem 1.1. In section 4 we prove Theorem 1.2. In section 5 we list some of the generalized Fibonacci numbers for which theorem 1.2 is applicable, and we state some formulas that follow from Theorem 1.2. In section 6 we apply Theorem 1.1 to the squares of the generalized Fibonacci numbers and to the Tribonacci sequence.

2. A USEFUL THEOREM

Most of our results are based on the following theorem.

Theorem 2.1: Let $f(x) = \sum_{n=0}^{\infty} a_n x^n$ be a formal power series, and let θ be a primitive m'th root of unity (for example, θ could be $e^{2\pi i/m}$). Let

$$f(x)f(\theta x) \cdots f(\theta^{m-1} x) = \sum_{j=0}^{\infty} b_{m,j} x^j.$$

Then $b_{m,j} = 0$ unless $j = mk$ for some k. That is,

$$f(x)f(\theta x)\cdots f(\theta^{m-1}x) = \sum_{j=0}^{\infty} b_{m,mj} x^{mj}. \tag{2.1}$$

Proof: Let $H(x) = f(x)f(\theta x)\cdots f(\theta^{m-1}x)$.   Clearly $H(x) = H(\theta x)$, so $b_{m,j} = \theta^j b_{m,j}$ for $j = 0,1,2,\cdots$. Since $\theta^j = 1$ only when m divides j, we see that formula (2.1) holds. This completes the proof.   □

Theorem 2.1 is evidently not well-known; at least the writer has not been able to find references for it. In [9] the writer stated and proved a special case of Theorem 2.1, and the method of that paper can easily be generalized. The above proof, however, was suggested by Arnold Adelberg in a private communication.

3. PROOF OF THEOREM 1.1

Using standard techniques [14, p. 8], we can prove that a generating function for $\{a_n\}$ is

$$\frac{P(x)}{1 - e_1 x - e_2 x^2 - \cdots - e_k x^k} = \sum_{j=0}^{\infty} a_j x^j, \tag{3.1}$$

where $P(x)$ is a polynomial of degree less than k. Let θ be a primitive m'th root of unity. Then by Theorem 2.1,

$$\prod_{r=0}^{m-1}(1 - e_1(\theta^r x) - e_2(\theta^r x)^2 - \cdots - e_k(\theta^r x)^k) = 1 + \sum_{j=1}^{k}(-1)^j c_{m,jm} x^{jm}, \tag{3.2}$$

for some numbers $c_{m,jm}$. By (3.1) and (3.2) we have

$$P(x)\prod_{r=1}^{m-1}(1 - e_1(\theta^r x) - e_2(\theta^r x)^2 - \cdots - e_k(\theta^r x)^k) =$$

$$\left(1 + \sum_{j=1}^{k}(-1)^j c_{m,jm} x^{jm}\right)\sum_{j=0}^{\infty} a_j x^j. \tag{3.3}$$

The left side of (3.3) is a polynomial of degree less than km. Thus we compare coefficients of $x^{(k-1)m+n}$ in (3.3) to obtain, for $n \geq m$,

$$0 = a_{(k-1)m+n} - \sum_{j=1}^{k}(-1)^{j-1} c_{m,jm} a_{(k-j-1)m+n}. \tag{3.4}$$

Obviously (3.4) is equivalent to (1.2), so the proof is complete.   □

It is easy to see that $c_{m,km} = (-1)^{(k+1)m} e_k^m$; the main difficulty is computing the numbers $c_{m,jm}$ in general. If we let

$$G(x) = \prod_{r=0}^{m-1}(1 - e_1(\theta^r x) - e_2(\theta^r x)^2 - \cdots - e_k(\theta^r x)^k) \tag{3.5}$$

then, with $e_0 = -1$ (for convenience of notation in the following formulas),

$$x^{km}G\left(\tfrac{1}{x}\right) = \prod_{r=0}^{m-1}\theta^{rk}\left(-e_k - e_{k-1}\theta^{m-r}x - e_{k-2}(\theta^{m-r}x)^2 - \cdots - e_0(\theta^{m-r}x)^k\right)$$

$$= (-1)^{(m-1)k}\prod_{r=0}^{m-1}\left(-e_k - e_{k-1}\theta^{m-r}x - e_{k-2}(\theta^{m-r}x)^2 - \cdots - e_0(\theta^{m-r}x)^k\right).$$

$$= (-1)^{(m-1)k}\prod_{r=1}^{m}\left(-e_k - e_{k-1}\theta^{r}x - e_{k-2}(\theta^{r}x)^2 - \cdots - e_0(\theta^{r}x)^k\right).$$

Thus

$$x^{km}G\left(\tfrac{1}{x}\right) = \begin{cases} (-1)^{(m-1)k}G(x) & \text{if } e_j = e_{k-j} \text{ for } j = 0,\, 1,\, \cdots,\, k \\[2mm] (-1)^{m+(m-1)k}G(x) & \text{if } e_j = -e_{k-j} \text{ for } j = 0,\, 1,\, \cdots,\, k \end{cases} \tag{3.6}$$

Since

$$x^{km}G\left(\tfrac{1}{x}\right) = x^{km} + \sum_{j=1}^{k}(-1)^j c_{m,jm}x^{km-jm},$$

we have

$$c_{m,jm} = \begin{cases} (-1)^{km}c_{m,(k-j)m} & \text{if } e_j = e_{k-j} \text{ for } j = 0,\, 1,\, \cdots,\, k \\[2mm] (-1)^{(k+1)m}c_{m,(k-j)m} & \text{if } e_j = -e_{k-j} \text{ for } j = 0,\, 1,\, \cdots,\, k \end{cases} \tag{3.7}$$

We will use (3.7) in section 6 when we consider the squares of the Fibonacci numbers.

4. PROOF OF THEOREM 1.2

Using the notation of Horadam [5], we define the sequence $\{w_n\} = \{w_n(a,\, b;\, p,\, q)\}$ by means of (1.3). The number $a,\, b,\, p,\, q$ are not restricted to the integers, however; they can be arbitrary complex numbers. We also use the notation

$$u_n = u_n(p,\, q) = w_n(0,\, 1;\, p,\, q) \tag{4.1}$$

$$v_n = v_n(p,\, q) = w_n(2,\, p;\, p,\, q).$$

It is easy to see [5] that we have the generating function

$$\frac{a + (b - ap)x}{1 - px + qx^2} = \sum_{n=0}^{\infty} w_n x^n, \tag{4.2}$$

so the following relationships are clear. For $n \geq 1$,

$$w_n = au_{n+1} + (b - ap)u_n = bu_n - aqu_{n-1}, \tag{4.3}$$

$$v_n = 2u_{n+1} - pu_n = u_{n+1} - qu_{n-1}. \tag{4.4}$$

In fact, using (1.5) we can easily verify that (4.3) and (4.4) hold for all integers n.

We need the following lemma, which might be of interest in its own right.

Lemma 4.1: Let a, b, p, q be arbitrary complex numbers, let m be a positive integer, let θ be a primitive m'th root of unity, and define $A_{m,k}$ by means of

$$\prod_{r=1}^{m-1}(1 - p(\theta^r x) + q(\theta^r x)^2) = \sum_{k=0}^{2m-2} A_{m,k} x^k. \tag{4.5}$$

Then

$$A_{m,k} = \begin{cases} u_{k+1} & \text{for } k = 0, 1, \cdots, m-1 \\ \\ q^{k-m+1}u_{2m-k-1} & \text{for } k = m, m+1, \cdots, 2m-2 \end{cases}$$

where u_k is defined by (4.1).

Proof: For convenience we will use the notation $A_k = A_{m,k}$, keeping in mind that the value of A_k depends on m. From (4.2) and Theorem 2.1, we have, for some number c_m,

$$\sum_{k=0}^{2m-2} A_k x^{k+1} = \prod_{r=0}^{m-1}(1 - p(\theta^r x) + q(\theta^r x)^2)\frac{x}{1 - px + qx^2}$$

$$= (1 - c_m x^m + q^m x^{2m})\frac{x}{1 - px + qx^2}$$

$$= (1 - c_m x^m + q^m x^{2m})\sum_{k=0}^{\infty} u_k x^k. \tag{4.6}$$

Thus

$$A_k = u_{k+1} \quad (k = 0, 1, \cdots, m-1), \tag{4.7}$$

$$A_m = u_{m+1} - c_m. \tag{4.8}$$

Since

$$(1 - px + qx^2)\sum_{k=0}^{2m-2} A_k x^k = (1 - c_m x^m + q^m x^{2m}),$$

we see that

$$A_{2m-2} = q^{m-1} = q^{m-1}u_1,$$

$$A_{2m-3} = \frac{p}{q}A_{2m-2} = pq^{m-2} = q^{m-2}u_2,$$

and, in general, for $k = 2, 3, \cdots, m+1$:

$$A_{2m-k} = \frac{p}{q}A_{2m-k+1} - \frac{1}{q}A_{2m-k+2} = q^{m-k+1}(pu_{k-2} - qu_{k-3}) = q^{m-k+1}u_{k-1}. \tag{4.9}$$

By (4.7) and (4.9), the proof of Lemma 4.1 is complete. $\square$

Proof of Theorem 1.2: Let $A_{m,k}$ be defined by (4.5). We first note that by (4.8) and (4.9),

$$A_{m,m} = u_{m+1} - c_m = qu_{m-1}, \qquad (4.10)$$

so by (4.4) and (4.10),

$$c_m = u_{m+1} - qu_{m-1} = v_m.$$

It follows immediately from Theorem 1.1 that for $n \geq m$,

$$w_{m+n} = v_m w_n - q^m w_{n-m},$$

and Theorem 1.2 is proved for nonnegative subscripts. The fact that Theorem 1.2 is true for all integers m and n can be worked out, with effort, from (1.5), (4.2) and Lemma 4.1. However, the referee has pointed out a remarkable theorem [1] that has the following as a special case: If an identity involving generalized Fibonacci and Lucas numbers (w_n and v_n, respectively) is true for all positive subscripts, then it must also be true for all negative subscripts as well. Thus the proof of Theorem 1.2 is complete.   □

The writer is not claiming this is the easiest proof of Theorem 1.2, but he hopes it is an interesting alternate approach that might have other applications.

5. IDENTITIES DERIVED FROM THEOREM 1.2

We first list some of the special numbers and polynomials for which Theorem 1.2 is applicable. We use the definitions of w_n, $u_n(p, q)$ and $v_n(p, q)$ given by (1.3) and (4.1).

Some special cases of interest are the following:

Generalized Fibonacci [12, p. 10] and ordinary Lucas numbers: $G_n = w_n(a, b; 1, -1)$,

$$L_n = v_n(1, -1);$$

Generalized Fibonacci and ordinary Lucas polynomials [3], [4]: $w_n(a, b; x, -1)$, $v_n(x, -1)$;

Generalized Pell and ordinary Pell-Lucas polynomials [8]: $w_n(a, b; 2x, -1)$, $v_n(2x, -1)$;

Generalized Chebyshev polynomials of the second kind and ordinary Chebyshev polynomials of the first kind [2, 49-50]: $w_n(a, b; 2x, 1)$, $\frac{1}{2} v_n(2x, 1)$;

Generalized Jacobsthal and ordinary Jacobsthal-Lucas polynomials [6]: $w_n(a, b; 1, -2x)$,

$$v_n(1, -2x).$$

Theorem 1.2 is useful in proving many other identities. For example, using the notation of (1.3) and (4.1), define s_n and t_n, for any integer n, by means of

$$s_n = bv_n - aqv_{n-1},$$

$$t_n = bw_n - aqw_{n-1}.$$

Then, with the aid of Theorem 1.2, we can easily prove the following: For all integers m and n,

$$w_{n+1} - qw_{n-1} = s_n, \tag{5.1}$$

$$s_{n+1} - qs_n = (p^2 - 4q)w_n, \tag{5.2}$$

$$s_{n+m} - q^m s_{n-m} = (p^2 - 4q)w_n u_m, \tag{5.3}$$

$$w_{n+1} u_{n-1} = w_n u_n - bq^{n-1}, \tag{5.4}$$

$$w_n^2 = bu_{n-1} t_{n+1} + b^2 q^{n-1}. \tag{5.5}$$

Certainly many more formulas can be proved. Equations (5.1)-(5.5) are just a small sample to illustrate the importance of Theorem 1.2. We omit the proofs, but we note that (5.1)-(5.5) can all be verified by the algorithm "LucasSimplify", which is due to Stan Rabinowitz [10].

Equations (5.1)-(5.5) furnish identities for the special numbers and polynomials defined at the beginning of this section. We observe that if $a = 0$ and $b = 1$, then $s_n = v_n$ and $t_n = w_n = u_n$. If $a = 0$, $b = 1$, $p = 1$, $q = -1$, then (5.1)-(5.5) are familiar identities involving Fibonacci and Lucas numbers [12, pages 176-178].

6. FIBONACCI SQUARES AND THE TRIBONACCI SEQUENCE

In this section we give another example of Theorem 1.1 involving the squares of the generalized Fibonacci numbers. We use the following generating function [5]:

$$H(x) = \frac{h(x)}{1 + (q - p^2)x + q(p^2 - q)x^2 - q^3 x^3} = \sum_{n=0}^{\infty} w_n^2 x^n \tag{6.1}$$

where $h(x) = a^2 + \{b^2 - a^2(p^2 - q)\}x + q(b - pa)^2 x^2$.

Theorem 6.1: If w_n and v_n are defined by (1.3) and (1.4), then for $n \geq m \geq 0$:

$$w_{n+2m}^2 = (v_m^2 - q^m)w_{n+m}^2 - q^m(v_m^2 - q^m)w_n^2 + q^{3m}w_{n-m}^2.$$

Proof: We first note that if $q = 0$, then Theorem 1.2 gives $w_{n+2m}^2 = (v_m w_{n+m})^2$, so the theorem actually holds in this "degenerate" case. If $q \neq 0$, then by (6.1),

$$H\left(\frac{-x}{q}\right) = \frac{h_1(x)}{1 - yx - yx^2 + x^3} = \sum_{n=0}^{\infty} r_n x^n \tag{6.2}$$

where $y = \frac{q - p^2}{q}$, $h_1(x)$ is a polynomial of degree ≤ 2, and $r_n = (-q)^{-n}w_n^2$. Thus by Theorem 1.1 and (3.7),

$$r_{n+2m} = c_{m,m}r_{n+m} + (-1)^{m-1}c_{m,m}r_n + (-1)^m r_{n-m} \tag{6.3}$$

with $c_{m,m}$ defined by

$$\prod_{r=0}^{m-1}(1 - y(\theta^r x) - y(\theta^r x)^2 + (\theta^r x)^3) = 1 - c_{m,m}x^m + (-1)^m c_{m,m}x^{2m} + (-1)^{m-1}x^{3m} \tag{6.4}$$

where θ is a primitive m'th root of unity. Our goal is to determine $c_{m,m}$ from (6.4). We first define the numbers b_n by means of the generating function

$$\frac{1}{1 - yx - yx^2 + x^3} = \sum_{n=0}^{\infty} b_n x^n. \tag{6.5}$$

By (6.1) we know that

$$\frac{-x + x^2}{1 - yx - yx^2 + x^3} = \sum_{n=0}^{\infty} (-q)^{-n+1}u_n^2 x^n \tag{6.6}$$

so clearly for $n = 1, 2, 3, \cdots$

$$b_n - b_{n-1} = (-q)^{-n}u_{n+1}^2 \tag{6.7}$$

where u_n is defined by (4.1). Now for fixed $m > 1$, define $Y_{m,k}$ by means of

$$\prod_{r=1}^{m-1}(1 - y(\theta^r x) - y(\theta^r x)^2 + (\theta^r x)^3) = \sum_{k=0}^{3m-3} Y_{m,k}x^k \tag{6.8}$$

where θ is a primitive m'th root of unity. For convenience we shall use the notation $Y_k = Y_{m,k}$, keeping in mind that Y_k depends on m. Then by (6.4) and (6.8) we have

$$\sum_{k=0}^{3m-3} Y_k x^k = (1 - c_{m,m}x^m + c_{m,m}x^{2m} + (-1)^{m-1}x^{3m})\sum_{j=0}^{\infty} b_j x^j, \tag{6.9}$$

$$(1 - yx - yx^2 + x^3)\sum_{k=0}^{3m-3} Y_k x^k = 1 - c_{m,m}x^m + c_{m,m}x^{2m} + (-1)^{m-1}x^{3m}. \tag{6.10}$$

By (6.9) we have

$$Y_k = b_k \ (k = 0, 1, \cdots, m-1), \tag{6.11}$$

$$Y_m = b_m - c_{m,m},$$

$$Y_{m+1} = b_{m+1} - Y_1 c_{m,m} = b_{m+1} - b_1 c_{m,m},$$

$$Y_{m+2} = b_{m+2} - Y_2 c_{m,m} = b_{m+2} - b_2 c_{m,m},$$

and in general,

$$Y_{m+k} = b_{m+k} - b_k c_{m,m} \ (k = 0, 1, \cdots, m-1). \tag{6.12}$$

By (6.10) we have

$$Y_{3m-3} = (-1)^{m-1} = (-1)^{m-1}b_0,$$

$$Y_{3m-4} = yY_{3m-3} = (-1)^{m-1}b_1,$$

$$Y_{3m-5} = yY_{3m-4} + yY_{3m-5} = (-1)^{m-1}(y^2 + y) = (-1)^{m-1}b_2,$$

and in general, for $k = 0, 1, \cdots, m-1,$

$$Y_{3m-k-3} = (-1)^{m-1}(yb_{k-1} + yb_{k-2} - b_{k-3}) = (-1)^{m-1}b_k. \tag{6.13}$$

Now, by (6.12) and (6.13) we have

$$Y_{2m-1} = b_{2m-1} - b_{m-1}c_{m,m} = (-1)^{m-1}b_{m-2}, \tag{6.14}$$

$$Y_{2m-2} = b_{2m-2} - b_{m-2}c_{m,m} = (-1)^{m-1}b_{m-1}. \tag{6.15}$$

Subtracting (6.14) from (6.15) and solving for $c_{m,m}$, we have

$$c_{m,m} = (-1)^{m-1} + \frac{b_{2m-1} - b_{2m-2}}{b_{m-1} - b_{m-2}},$$

so by (6.7) and Theorem 1.2,

$$c_{m,m} = (-1)^{m-1} + \frac{(-q)^{-2m-1}u_{2m}^2}{(-q)^{-m-1}u_m^2} = (-1)^{m-1} + (-q)^{-m}v_m^2. \tag{6.16}$$

Substituting (6.16) into (6.3) gives

$$w_{n+2m}^2 = (v_m^2 - q^m)w_{n+m}^2 - q^m(v_m^2 - q''')w_n^2 + q^{3m}w_{n-m}^2,$$

and the proof of Theorem 6.1 is complete. ☐

The proof of Theorem 6.1 illustrates the main technique of this paper, and perhaps it indicates how the technique can be useful in proving other formulas. Incidentally, it can be shown that Theorem 6.1 is valid for all integer m and n.

As a final application, consider the generalized Tribonacci sequence $\{P_n\}$ defined by

$$P_n = P_{n-1} + P_{n-2} + P_{n-3}$$

with P_0, P_1, P_2 arbitrary complex numbers. Define $\{K_n\}$ to be the Tribonacci sequence with $P_0 = K_0 = 0$, $P_1 = K_1 = 1$, $P_2 = K_2 = 1$. Here we are using the notation of [13].

Let θ be a primitive m'th root of unity. We know by Theorem 1.1 that

$$\prod_{r=0}^{m-1}(1 - \theta^r x - (\theta^r x)^2 - (\theta^r x)^3) = 1 - c_{m,m}x^m + c_{m,2m}x^{2m} - x^{3m},$$

for some numbers $c_{m,m}$ and $c_{m,2m}$, and using the methods of the first proof of this section, we can easily show that

$$\prod_{r=1}^{m-1}(1 - \theta^r x - (\theta^r x)^2 - (\theta^r x)^3) = \sum_{j=0}^{3m-3} y_j x^j,$$

where

$$y_j = K_{j+1} \quad (j = 0, \cdots, m-1),$$

$$y_{m+j} = K_{m+1+j} - K_{j+1} c_{m,m} \quad (j = 0, \cdots, m-1),$$

$$y_{3m-j-3} = K_{-(j+2)} \quad (j = 0, \cdots, m-1),$$

$$y_{2m-3} = K_{-(m+2)} - c_{m,2m}.$$

This gives

$$c_{m,m} = \frac{K_{2m} - K_{-m}}{K_m}, \tag{6.17}$$

$$c_{m,2m} = K_{-(m+2)} + c_{m,m} K_{m-2} - K_{2m-2}. \tag{6.18}$$

Theorem 6.2: Let $\{P_n\}$ be defined by

$$P_n = P_{n-1} + P_{n-2} + P_{n-3},$$

with P_0, P_1, P_2 arbitrary complex numbers. For $n \geq m \geq 1$,

$$P_{n+2m} = c_{m,m} P_{n+m} - c_{m,2m} P_n + P_{n-m},$$

where $c_{m,m}$ and $c_{m,2m}$ are defined by (6.17) and (6.18), respectively.

For example,

$$P_{n+4} = 3P_{n+2} + P_n + P_{n-2},$$

$$P_{n+6} = 7P_{n+3} - 5P_n + P_{n-3},$$

$$P_{n+8} = 11P_{n+4} + 5P_n + P_{n-4},$$

$$P_{n+10} = 21P_{n+5} - P_n + P_{n-5},$$

and so on. In fact, computational evidence indicates that if $K_{-m} \neq 0$, then

$$c_{m,2m} = \frac{K_{-2m} - K_m}{K_{-m}}.$$

There is also evidence that $c_{m,m} = S_m$ and $c_{m,2m} = S_{-m}$, where $\{S_n\}$ is a sort of generalized Lucas sequence defined by: $S_0 = 3$, $S_1 = 1$, $S_2 = 3$, and $S_n = S_{n-1} + S_{n-2} + S_{n-3}$.

Much work remains to be done here; for example, evidence indicates that Theorem 6.2 actually holds for all integers m and n, and it would certainly be of interest to establish that $c_{m,m} = S_m$ and $c_{m,2m} = S_{-m}$. Results like Theorem 6.2 for fourth order (and higher order) linear recursive sequences would also be interesting. We leave these problems, and many others, for future work.

REFERENCES

[1]   Bruckman, P.S. "Solution to Problem H-487 (Proposed by S. Rabinowitz)." *The Fibonacci Quarterly*, Vol. *33* (1995): p. 382.

[2]   Comtet, L. Advanced Combinatorics. Dordrecht: Reidel, 1974.

[3]   Filipponi, P. and Horadam, A.F. "Derivative Sequences of Fibonacci and Lucas Polynomials." Applications of Fibonacci Numbers, Volume 4. Edited by G.E. Bergum, A.F. Horadam and A.N. Philippou. Kluwer Academic Publishers, Dordrecht, The Netherlands, 1991: pp. 99-108.

[4]   Filipponi, P. and Horadam, A.F. "Integration Sequences of Fibonacci and Lucas Polynomials." Applications of Fibonacci Numbers, Volume 5. Edited by G.E. Bergum, A.F. Horadam and A.N. Philippou. Kluwer Academic Publishers, Dordrecht, The Netherlands, 1993: pp. 317-330.

[5]   Horadam, A.F. "Generating Functions for Powers of a Certain Generalized Sequence of Numbers." *Duke Math. Journal*, Vol. *32* (1965): pp. 437-446.

[6]   Horadam, A.F. "Jacobsthal Representation Polynomials." *The Fibonacci Quarterly*, Vol. *35* (1997): pp. 137-148.

[7]   Horadam, A.F. "Basic Properties of a Certain Generalized Sequence of Numbers." *The Fibonacci Quarterly*, Vol. *3* (1965): pp. 161-176.

[8]   Horadam, A.F. and Mahon, J.M. "Pell and Pell-Lucas Polynomials." *The Fibonacci Quarterly*, Vol. *23* (1985): pp. 7-20.

[9]   Howard, F.T. "Lacunary Recurrences for Sums of Powers of Integers." *The Fibonacci Quarterly*, Vol. *36* (1998): pp. 435-442.

[10]   Rabinowitz, S. "Algorithmic Manipulation of Second Order Linear Recurrences." *The Fibonacci Quarterly* (to appear).

[11]   Rakoch, A. "Optimal Computations, by Computer, of Fibonacci Numbers." *The Fibonacci Quarterly*, Vol. *34* (1996): pp. 436-439.

[12]   Vajda, S. Fibonacci and Lucas Numbers, and the Golden Section. Ellis Horwood Limited, Chichester, 1989.

[13]   Waddill, M.E. and Sacks, L. "Another Generalized Fibonacci Sequence." *The Fibonacci Quarterly*, Vol. *5* (1967): pp. 209-227.

[14]   Wilf, H.S. Generatingfunctionology. Academic Press, Inc., San Diego, 1990.

AMS Classification Numbers: 11B39, 11B37

SOME GENERALIZATIONS OF WOLSTENHOLME'S THEOREM

William A. Kimball and William A. Webb

The theorem known as Wolstenholme's Theorem states that $\sum a_i^{-1}$ is divisible by p^2 for $p > 3$ if the a_i are the numbers $1, 2, \cdots, p-1$. Recently, Alkan noted that similar sums are divisible by p when the a_i run over such sets as $1^2, 2^2, \cdots, (p-1)^2$; the quadratic residues or the quadratic nonresidues modulo p; or a reduced residue system modulo m where $p \mid m$. [1] We will see below how these and many other such results can be combined into some general observations, note the importance of the difference between congruences (mod p) and those which also hold (mod p^2), and offer a new way to generalize Wolstenholme's Theorem with (mod p^2) congruences.

SUMS IN RINGS

Let R be any finite commutative ring with identity 1 and U the set of units in R. We will say that a multiset $S \subseteq U$ is a Wolstenholme set if $\sum_{a \in S} a^{-1} = 0$.

Suppose G is a nontrivial multiplicative subgroup of U. Let $a_1 \neq 1$ be a fixed element of G, and consider the sum $\sum_{a \in G} a^{-1} = a^*$. Then $a^* = \sum_{a \in G} (a_1 a)^{-1} = a_1^{-1} a^*$ so a^* must be a nonunit, although a^* need not equal 0. However, if $R = Z_p$ then every subgroup, such as the quadratic residues, must be a Wolstenholme set since 0 is the only nonunit. Examples such as the squares $1^2, \cdots, (p-1)^2$ are multisets where each element of a subgroup is repeated a fixed number of times.

In a more general ring R with characteristic $\neq 2$, although not every subgroup of U is a Wolstenholme set, the group U itself is. Also, when summing over a subgroup G,
$$\sum a = \sum a^{-1}.$$

This paper is in final form and no version of it will be submitted for publication elsewhere.

If R is a ring with characteristic not equal to 2 then $a \neq -a$ for every $a \in U$. Hence, the elements of $\sum_{a \in U} a$ occur in disjoint pairs a and $-a$ and so $\sum_{a \in U} a = 0$. In particular in Z_m, $\sum_{(a,m)=1} a$ is divisible by m if $m \neq 2$. (Since m can be divisible by p^k for $k > 1$ this is stronger than being divisible by just p.)

Actually, if $(m,6) = 1$ then Leudesdorf has shown that $\sum_{(a,m)=1, 1 \leq a < m} a^{-1}$ is divisible by m^2. The cases when $(m,6) \neq 1$ have also be classified and this sum will be divisible by m^2/d where $d = 2, 3, 4$ or 6. [3, Theorem 128].

Some other easy observations are that if S is a Wolstenholme set so is $U - S$ and aS for any $a \in U$.

The condition that R not have characteristic 2 is necessary in order to conclude $\sum_{a \in U} a = 0$. For example, consider the ring Z_2 or the quotient ring of polynomials over Z_2 modulo x^2. The only units are 1 in the former case and 1 and $x + 1$ which sum to x, in the later case.

It is essentially trivial that $\sum_{1 \leq k < p} k^{-1} = 0 \pmod{p}$. What is really noteworthy about Wolstenholme's Theorem is that the same congruence holds $\pmod{p^2}$ as well. Congruences $\pmod{p}$ such as Fermat's Theorem $a^{p-1} \equiv 1 \pmod{p}$, Wilson's Theorem $(p-1)! \equiv -1 \pmod{p}$ etc. are the meat and potatoes of elementary number theory. Trying to generalize to higher order congruences is quite another matter. When is $2^{p-1} \equiv 1 \pmod{p^2}$? This congruence has many well known consequences, such as to Fermat's Last Theorem. Specifically, Wieferich showed that $x^p + y^p = z^p$ has no solutions where $p \nmid xyz$ provided $2^{p-1} \not\equiv 1 \pmod{p^2}$. A similar statement holds for 3^{p-1} as well. Even though Wiles has finally proved Fermat's Last Theorem, it is not known if $2^{p-1} \equiv 1 \pmod{p^2}$ has infinitely many solutions, although 1093 and 3511 are the only ones less than 3×10^9. [6]

The previously mentioned Wolstenholme sets should really be called mod p sets, since they are in general congruent to 0 $\pmod{p}$ but not $\pmod{p^2}$. As such, they are not really all that remarkable. However, there are a number of related congruences such as those involving certain homogeneous symmetric functions of the numbers $1, 2, \cdots, p - 1$ which are true $\pmod{p^2}$. [3]

SECOND ORDER RECURRENCE SEQUENCES

We will now look at another interesting higher order congruence which has Wolstenholme's Theorem as a special case.

Let $p > 3$ be a prime and consider the second order recurrence sequences satisfying $x_{n+2} = ax_{n+1} + bx_n$, where a and b are integers. We will look at both the degenerate case where $D = a^2 + 4b = 0$ and the nondegenerate case when $D \neq 0$. If $D = 0$ the characteristic equation $x^2 - ax - b$ has a double root of $a/2$. Since a must be even we will write $c = a/2$. The two basic sequences satisfying the recurrence are $u_n = nc^n$ and $v_n = c^n$ which have initial conditions $u_0 = 0$, $u_1 = c$ and $v_0 = 1$, $v_1 = c$, respectively.

If $D \neq 0$, there are distinct roots α and β which are easily seen to satisfy $\alpha + \beta = a$ and $\alpha\beta = -b$. The two basic sequences are now $u_n = \dfrac{\alpha^n - \beta^n}{\alpha - \beta}$ and $v_n = \alpha^n + \beta^n$. (In the case $a = b = 1$ we get the familiar Fibonacci and Lucas sequences where similar results are known. [4]) These formulas define u_n and v_n for all $n \in Z$ if $b \neq 0$.

For any recurrence sequence $\{x_n\}$ and any positive integer m, the rank of apparition (or rank of appearance or entry point) of m is the smallest $n > 0$ such that $m \mid x_n$. We are interested in the rank of apparation r only for the specific sequences u_n defined above and only when $m = p$, a prime. If $D = 0$ and $u_n = nc^n$ then $r = p$ if $p \nmid c$ and $r = 1$ if $p \mid c$. If $D \neq 0$ and $u_n = (\alpha^n - \beta^n)/(\alpha - \beta)$ then it is known that r divides either $p - 1$ or $p + 1$. More specifically, r divides $p - \left(\dfrac{D}{p}\right)$ where $\left(\dfrac{D}{p}\right)$ is the Legendre symbol. [2] [5] If $p \nmid bD$ then r is the order of α/β in the group of units of $Z_p(\sqrt{D})$.

We will consider sums of the form $\sum\limits_{1 \leq n < r} \dfrac{v_n}{u_n}$. For any degenerate sequence $v_n/u_n = 1/n$ and the sum reduces to the usual Wolstenholme sum. The remarkable fact is that even in the nondegenerate case this sum is often divisible by p^2.

Specifically, we have the following theorem.

Theorem: Let $\{u_n\}$ and $\{v_n\}$ denote second order recurrence sequences with discriminant D as defined above. Let p be a prime, $p > 3$, and let r denote the rank of apparition of p in the sequence $\{u_n\}$. If S denotes the sum $\sum\limits_{1 \leq k < r} \dfrac{v_n}{u_n}$ then:

(i) $S \equiv 0 \pmod{p}$.

(ii) If $D = 0$ then $S \equiv 0 \pmod{p^2}$.

(iii) If $D \neq 0$ and $r = p \pm 1$ then $S \equiv 0 \pmod{p^2}$.

Proof: As noted above, if $D = 0$ then S reduces to the sum in Wolstenholme's theorem. The following standard identities for second order recurrences are easy to prove.

$$v_n = u_{n+1} + bu_{n-1} \tag{1}$$

$$v_{n+k} = v_n v_k - (-b)^k v_{n-k} \tag{2}$$

$$2u_{n+k} = u_n v_k + u_k v_n \tag{3}$$

$$2v_{n+k} = v_n v_k + D u_n u_k. \tag{4}$$

In the following $\sum$ will always denote a sum over k where $1 \le k < r$.

If $p \mid D$ then the sequences are degenerate mod p and so $r = p$. Hence, we may assume $p \nmid D$. Similarly, if $p \mid ab$ then r is either 2 or undefined, so we may assume $p \nmid ab$.

To prove (i), note that

$$2S = \sum \left(\frac{v_k}{u_k} + \frac{v_{r-k}}{u_{r-k}} \right) = \sum \frac{2u_r}{u_k u_{r-k}}$$

by (3). Thus $p \mid S$ as claimed. Also, to now prove (iii) it suffices to show that $\sum \frac{2}{u_k u_{r-k}} \equiv 0$ (mod p), or since $p \nmid v_r$ that $\sum \frac{2v_r}{u_k u_{r-k}} \equiv 0$ (mod p).

The following four claims will be used now to prove (iii), and will be established later.

$$\sum \frac{2v_r}{u_r u_{r-k}} \equiv D(r-1) - \sum \left(\frac{v_k}{u_k} \right)^2 \text{ (mod } p\text{)}. \tag{5}$$

$$\left\{ \frac{v_k}{u_k} \right\} \text{ are incongruent modulo } p \text{ for } 1 \le k \le r-1. \tag{6}$$

$$\left(\frac{v_k}{u_k} \right)^2 \not\equiv D \text{ (mod } p\text{) for } 1 \le k \le r-1. \tag{7}$$

$$D \text{ is a quadratic residue modulo } p \text{ if } r \mid p-1. \tag{8}$$

<u>Case 1:</u> $r = p+1$.

Since $p = r-1$ and $\left\{ \frac{v_k}{u_k} \right\}$ for $1 \le k \le r-1$ is a complete residue system modulo p by (6), using (5) we have:

$$\sum \frac{2v_r}{u_r u_{r-k}} \equiv -\sum \left(\frac{v_k}{u_k} \right)^2 \equiv -\sum_{j=1}^{p} j^2 \equiv 0 \text{ (mod } p\text{)}.$$

<u>Case 2:</u> $r = p-1$.

By (5), we have $\sum \frac{2v_r}{u_r u_{r-k}} \equiv -2D - \sum \left(\frac{v_k}{u_k} \right)^2$ (mod p).

But the right side is again $-\sum_{j=1}^{p} j^2$, since D is a quadratic residue modulo p by (8), and numbers $(v_k/u_k)^2$ run over all of the squares modulo p except D by (6) and (7).

To prove the statements in the claim we see that:

$$2 \sum \left(\frac{v_k}{u_k} \right)^2 = \sum \frac{(v_k u_{r-k})^2 + (v_{r-k} u_k)^2}{(u_k u_{r-k})^2} \equiv \sum \frac{-2(v_k u_{r-k} v_{r-k} u_k)}{(u_k u_{r-k})^2}$$

$$\equiv -2 \sum \frac{v_k}{u_k} \frac{v_{r-k}}{u_{r-k}} \equiv 2 \sum \left(D - \frac{2v_r}{u_k u_{r-k}} \right) \text{ (mod } p\text{)}$$

where we have used $(2u_r)^2 = (v_k u_{r-k} + v_{r-k} u_k)^2$ by (3), and $v_k v_{r-k} = 2v_r - D u_k u_{r-k}$ by (4). This proves (5).

To prove (6), assume $\frac{v_k}{u_k} \equiv \frac{v_j}{u_j}$ (mod p) for some j and k where $1 \leq j < k \leq r-1$. This implies

$$0 \equiv u_j v_k - u_k v_j \equiv -(-b)^j u_{-j} v_k - u_k(-b)^j v_{-j} \equiv -2(-b)^j u_{k-j} \pmod{p}$$

by (3) and the fact that clearly $u_n = -(-b)^n u_n$, $v_n = (-b)^n v_n$ for all n. Both this is a contradiction since $1 \leq k - j \leq r - 2$.

Next, if $\left(\frac{v_k}{u_k}\right)^2 \equiv D$ (mod p) then by (4) and (2)

$$2v_k^2 \equiv v_k^2 + Du_k^2 \equiv 2v_{2k} \equiv 2(v_k^2 - (-b)^k v_0) \pmod{p}.$$

Thus, $4(-b)^k \equiv 0$ (mod p) contradicting the fact that $p \nmid b$.

To prove (8), suppose that $r \mid p - 1$. We want to show that $\sqrt{D} \in Z_p$. Since $u_r \equiv 0$ (mod p) we must have $u_{p-1} \equiv 0$ (mod p) so $(\alpha/\beta)^{p-1} = 1$ in Z_p or in a quadratic extension of Z_p. This implies $\alpha/\beta \in Z_p$ and since

$$\frac{\alpha}{\beta} = \frac{a + \sqrt{D}}{a - \sqrt{D}} = \frac{a^2 + D + 2a\sqrt{D}}{a^2 - D}$$

we have $\sqrt{D} \in Z_p$.

The following examples show that various behaviors are possible when $r \neq p \pm 1$.

(I) A typical example for F_n and L_n is $p = 13$. Here $r = 7$ so $r \mid p + 1$.

$$\sum_{1 \leq n \leq 6} \frac{L_n}{F_n} = \frac{767}{60}$$ which is divisible by 13 but not 13^2.

(II) Very simple examples can be constructed for other recurrences.

If $x_n = 2x_{n-1} + 3x_{n-2}$ and $p = 7$ then

u_n	0	1	2	7	$\cdots$
v_n	2	2	10	26	$\cdots$

so $r = 3$ and $r \mid p - 1$.

$$\sum_{1 \leq n \leq 2} \frac{v_n}{u_n} = 7.$$

On the other hand if $x_n = 7x_{n-1} + x_{n-2}$ and $p = 5$ then

u_n	0	1	7	50	$\cdots$
v_n	2	7	51	364	$\cdots$

so $r = 3$.

$$\sum_{1 \leq n \leq 2} \frac{v_n}{u_n} = \frac{100}{7} \text{ which is divisible by } 5^2.$$

REFERENCES

[1] Alkan, Emre. "Variations on Wolstenholme's Theorem." *Amer. Math. Monthly*, Vol. *101* (1994): pp. 1001-1004.

[2] Carmichael, R.D. "On the Numerical Factors of the Arithmetic Forms $\alpha^n \pm \beta^n$." *Annals of Math.*, Vol. *15* (1913-4): pp. 30-70.

[3] Hardy, G.H. and Wright, E.M. An Introduction to the Theory of Numbers, 5th Edition. Oxford University Press, 1979.

[4] Kimball, W.A. and Webb, W.A. "A Congruence for Fibonomial Coefficients Modulo p^3." *The Fibonacci Quarterly*, Vol. *33.4* (1995): pp. 290-297.

[5] Lucas, E. "Theorie der Fonctions Numeriques Simplement Periodiques." *Amer. J. Math.* (1878): pp. 184-240 and 289-321.

[6] Ribenboim, Paulo. 13 Lectures on Fermat's Last Theorem. Springer-Verlag, New York, 1979.

AMS Classification Numbers: 11B39, 11B50, 11A07

CARD SORTING RELATED TO FIBONACCI NUMBERS

Clark Kimberling

1. INTRODUCTION

Imaging holding a deck of n cards in your hand, numbered from 1 to n but not in that order. Let $\alpha = (1 + \sqrt{5})/2$ and $\sigma = (1, 3, 4, 6, 8, 9, 11, 12, 14, 16, 17, 19, \cdots)$, where $\sigma(n) = \lfloor \alpha n \rfloor$ for $n \geq 1$. The sequence σ will be used to select and remove cards from your hand.

To begin, place the 1st card on a table, the 2nd under the others in your hand, the 3rd atop the 1st, the 4th atop the 3rd, the 5th under the 2nd, the 6th atop the 3rd, the 7th under the 5th, and so on. What's happening here is that σ tells which cards to remove, and its complement tells which ones to recycle. *If the procedure continues until all n cards are on the table, and if the card numbers from top to bottom are then $1, 2, \cdots, n$, what was the initial ordering, d_n, of the card numbers in your hand?* This problem stems from a special case proposed by Harris S. Schultz and discussed in [1].

In general, an infinite increasing sequence of positive integers having infinite complement, or any finite increasing sequence of positive integers, is called a *selection sequence*. If σ is an infinite selection sequence, then the determination of the initial order $d_n = (d_n(1), d_n(2), \cdots, d_n(n))$ for *all* deck sizes $n \geq 1$ is the *infinite σ-cardsort problem*. If σ is a finite selection sequence, let n be its length; the *finite σ-cardsort problem* is to determine the initial ordering d_n of the n cards.

Returning now to the special case in which the selection sequence is given by $\sigma(n) = \lfloor (1 + \sqrt{5})n/2 \rfloor$, we introduce by example a *tableau method* for solving the σ-cardsort problem. Take $n = 7$, and write the numbers $1, 2, 3, \cdots, \sigma(n)$ in a row. Then beneath those that are terms of σ, write, in order, $7, 6, 5, 4, 3, 2, 1$, like this:

This paper is in final form and no version of it will be submitted for publication elsewhere.

1	2	3	4	5	6	7	8	9	10	11
7		6	5		4		3	2		1.

Next, append a partial third row of n symbols x_i, as shown:

1	2	3	4	5	6	7	8	9	10	11
7		6	5		4		3	2		1

$$x_1 \quad x_2 \quad x_3 \quad x_4 \quad x_5 \quad x_6 \quad x_7.$$

This tableau indicates that four of the x_i have been assigned values, namely $x_1 = 7$, $x_3 = 6$, $x_4 = 5$, $x_6 = 4$. Next, rewrite those x_i not yet assigned, beginning under position 8:

1	2	3	4	5	6	7	8	9	10	11
7		6	5		4		3	2		1

$$x_1 \quad x_2 \quad x_3 \quad x_4 \quad x_5 \quad x_6 \quad x_7 \quad x_2 \quad x_5 \quad x_7.$$

In this manner, we have picked up $x_2 = 3$ and $x_5 = 2$. Only x_7 remains unassigned. Write it under position 11, and write the now completed tableau as

1	2	3	4	5	6	7	8	9	10	11
7	3	6	5	2	4	1	3	2	1	1

$$x_1 \quad x_2 \quad x_3 \quad x_4 \quad x_5 \quad x_6 \quad x_7 \quad x_2 \quad x_5 \quad x_7 \quad x_7.$$

All seven x_i are now assigned; the left-most seven assignments form the *initial deck*, $d_7 = (7, 3, 6, 5, 2, 4, 1)$, and the remaining four form the *extension*, $e_7 = (3, 2, 1, 1)$. It is easy to check that for the *extended deck*

$$d_7 \oplus e_7 := (7, 3, 6, 5, 2, 4, 1, 3, 2, 1, 1),$$

the positions $1, 3, 4, 6, 8, 9, 11$ are respectively occupied by cards numbered $7, 6, 5, 4, 3, 2, 1$, as required. The method just used for $n = 7$ applies to $n = 1, 2, \cdots, 16$ to give these results:

n	d_n	e_n
1	1	(none)
2	2, 1	1
3	3, 1, 2	1
4	4, 1, 3, 2	1, 1
5	5, 2, 4, 3, 1	2, 1, 1
6	6, 1, 5, 4, 2, 3	1, 2, 1
7	7, 3, 6, 5, 2, 4, 1	3, 2, 1, 1
8	8, 3, 7, 6, 1, 5, 2, 4	3, 1, 2, 1
9	9, 1, 8, 7, 3, 6, 2, 5, 4	1, 3, 2, 1, 1
10	10, 4, 9, 8, 3, 7, 1, 6, 5, 2	4, 3, 1, 2, 1, 1
11	11, 4, 10, 9, 2, 8, 3, 7, 6, 1, 5	4, 2, 3, 1, 2, 1
12	12, 2, 11, 10, 4, 9, 1, 8, 7, 3, 6, 5	2, 4, 1, 3, 2, 1, 1
13	13, 5, 12, 11, 2, 10, 4, 9, 8, 3, 7, 6, 1	5, 2, 4, 3, 1, 2, 1, 1
14	14, 1, 13, 12, 5, 11, 4, 10, 9, 2, 8, 7, 3, 6	1, 5, 4, 2, 3, 1, 2, 1
15	15, 6, 14, 13, 5, 12, 2, 11, 10, 4, 9, 8, 1, 7, 3	6, 5, 2, 4, 1, 3, 2, 1, 1
16	16, 6, 15, 14, 1, 13, 5, 12, 11, 2, 10, 9, 4, 8, 3, 7	6, 1, 5, 2, 4, 3, 1, 2, 1

Table 1. Initial decks and their extensions, using $\sigma(n) = \lfloor n(1 + \sqrt{5})/2 \rfloor$

An inspection of Table 1 suggests that for each n_1 there exists n_2 such that the juxtaposition

$$D_{n_1} := d_{n_1} \oplus e_{n_1}$$

is a tail-segment of $D_{n_2} := d_{n_2} \oplus e_{n_2}$; e.g.,

$$D_4 = (4, 1, 3, 2, 1, 1) \subset (7, 3, 6, 5, 2, 4, 1, 3, 2, 1, 1) = D_7.$$

Indeed,

$$D_1 \subset D_2 \subset D_4 \subset D_7 \subset D_{12} \subset \cdots. \tag{1}$$

Also,

$$D_1 \subset D_3 \subset D_8 \subset D_{21} \subset \cdots \text{ (note the Fibonacci numbers)}, \tag{2}$$

and

$$D_2 \subset D_5 \subset D_{13} \subset D_{34} \subset \cdots \text{ (note the Fibonacci numbers)}. \tag{3}$$

On the other hand, it is easy to devise selection sequences for which the above procedure gives chains that do not extend to the right indefinitely. For example, let σ^* consist of all the positive integers not of the form $(n^2 + 5n - 2)/2$, for $n = 1, 2, 3, \cdots$. The sequence σ^* begins with

$$1, 3, 4, 5, 7, 8, 9, 10, 12, 13, 14, 15, 16, 18,$$

and from this we obtain decks shown in Table 2. From this table, one sees that the chain

$$D_1 \subset D_2 \subset D_3 \subset D_5 \tag{4}$$

has a longest member, D_5, that is not a tail-segment of D_n, for $6 \le n \le 9$. Methods to be developed below will show that D_5 is not contained in *any* longer D_n.

n	d_n	e_n
1	1	(none)
2	1, 2	1
3	3, 1, 2	1
4	4, 1, 3, 2	1
5	5, 2, 4, 3, 1	2, 1
6	6, 2, 5, 4, 3, 1	2, 1
7	7, 2, 6, 5, 4, 1, 3	2, 1
8	8, 2, 7, 6, 5, 1, 4, 3	2, 1
9	9, 2, 8, 7, 6, 1, 5, 4, 3	2, 1, 1

Table 2. Initial decks and extensions, using σ^*

2. AMENABLE SELECTION SEQUENCES — THE PERIODIC CASE

In view of chains (1)-(4), a selection sequence σ is deemed *amenable* if every D_n is a member of an infinite chain

$$D_{n_1} \subset D_{n_2} \subset D_{n_3} \subset \cdots$$

induced by σ and the tableau method. In this paper we are primarily interest in amenable selection sequences.

For the sake of convenience, we define $\sigma(0) = 0$.

Theorem 1: Suppose σ is a strictly increasing sequence of positive integers for which the difference sequence Δ defined by

$$\Delta(n) = \sigma(n) - \sigma(n-1)$$

for $n = 1, 2, \cdots$ is periodic. Then σ is an amenable selection sequence.

Proof: Let p be the period of Δ, and let $g = \Delta(1) + \Delta(2) + \cdots + \Delta(p)$. Suppose that $1 \le r \le p$, that $h \ge 1$, and, as an induction hypothesis, that

$$D_r \subset D_{p+r} \subset \cdots \subset D_{(h-1)p+r}.$$

Suppose you hold an initial deck of $n = hp + r$ cards, where $h \ge 1$ and $1 \le r \le p$. From the g cards in positions $1, 2, \cdots, g$, select and remove those in positions $\sigma(1), \sigma(2), \cdots, \sigma(p)$. The remaining $g - p$ cards are placed, in their original order, behind the other cards in your hand. You now hold $n - p = (h-1)p + r$ cards and will select for removal those whose positions in the original deck were

$$\sigma(p+1), \sigma(p+2), \cdots, \sigma(n);$$

the positions in the presently held cards numbered $1, 2, \cdots, n-p$ are

$$\sigma(p+1) - g = \sigma(1), \quad \sigma(p+2) - g = \sigma(2), \quad \cdots, \quad \sigma(n) - g = \sigma(p).$$

By induction hypothesis, the extended deck for these $n-p$ cards is $D_{(h-1)p+r}$. Therefore,

$$D_{(h-1)p+r} \subset D_{hp+r}.$$

As the argument just given holds for $h=1$ (as a first induction step), a separate argument is not needed. By induction, D_n is a member of the infinite chain

$$D_r \subset D_{p+r} \subset D_{2p+r} \subset \cdots. \qquad\qquad \square$$

An alternative proof of Theorem 1, involving dispersions and fractal sequences, is given in [1]. Note that Theorem 1 does not cover the introductory example, since there, the number α is irrational. In order to cover such cases, it will be helpful to employ some notation customarily associated with continued fractions.

3. AMENDABLE SELECTION SEQUENCES — NON-PERIODIC CASES

Suppose α has continued fraction $[a_0, a_1, a_2, \cdots]$, and let

$$p_{-2} = 0, \quad p_{-1} = 1, \quad p_i = a_i p_{i-1} + p_{i-2}$$

and

$$q_{-2} = 1, \quad q_{-1} = 0, \quad q_i = a_i q_{i-1} + q_{i-2}$$

for $i \geq 0$. The *principal convergents of* α are the rational numbers p_i/q_i for $i \geq 0$. The numbers $q_0, q_2, q_4, \cdots$ are the *even-indexed denominators*, and the numbers $q_1, q_3, q_5, \cdots$ are the *odd-indexed denominators*. We shall use the notation $((\ ))$ for the fractional-part function, defined by

$$((x)) = x - \lfloor x \rfloor. \qquad\qquad (5)$$

Theorem 2: Suppose α is an irrational number greater than 1, and $\sigma(n) = \lfloor n\alpha \rfloor$ for $n = 1, 2, 3, \cdots$. Then for every positive integer n, there exists a positive integer m such that the sum

$$\sigma(m+k) - \sigma(k)$$

is invariant of k for $k = 1, 2, \cdots, n$.

Proof: Let n be a positive integer, let μ be the greatest index i for which the convergent p_i/q_i to α satisfies $q_i \leq n$, and let $m = q_\mu$. It is know (e.g., [3, theorem 1]) that

$$((\,(m+k)\alpha)) = \begin{cases} ((m\alpha)) + ((k\alpha)) - 1 & \text{if } m \text{ is an even-indexed denominator} \\[2ex] ((m\alpha)) + ((k\alpha)) & \text{if } m \text{ is an odd-indexed denominator,} \end{cases}$$

for $k = 1, 2, \cdots, n$. Consequently,

$$((\,(m+k)\alpha)) - ((k\alpha)) = ((m\alpha + \alpha)) - ((\alpha)),$$

so that by (5),

$$\lfloor (m+k)\alpha \rfloor - \lfloor k\alpha \rfloor = \lfloor m\alpha + \alpha \rfloor - \lfloor \alpha \rfloor,$$

and

$$\sigma(m+k) - \sigma(k) = \sigma(m+1) - \sigma(1),$$

as required.   □

Theorem 3: If $\sigma(n) = \lfloor \alpha n \rfloor$ for some irrational number $\alpha > 1$, for $n = 1, 2, 3, \cdots$, then σ is amenable selection sequence.

Proof: The tableau for n has the form

1 2 3 4 5 6 7		$n+n'-1$ $n+n'$
(n $n-1$		3 2 1)

the n-tableau

where n matches (i.e., is directly beneath) $\sigma(1)$, $n-1$ matches $\sigma(2), \cdots$, 1 matches $\sigma(n)$. Let μ be the greatest index i for which the convergent p_i/q_i to α satisfies $q_i \leq n$, and let $m = q_\mu$, as in the proof of Theorem 2. The tableau for $n + m$ has the form

1 2 3 4 5 6 7		$n+n''-1$ $n+n''$
$(n+m$ $n+m-1$	$n+1$ n $n-1$ 3 2 1)	

the $(n+m)$-tableau

where $n + m$ matches $\sigma(1)$, $n + m - 1$ matches $\sigma(2)$, $\cdots$, $m + 1$ matches $\sigma(n)$; also,

$$n \qquad \text{matches } \sigma(m+1)$$

$$n-1 \qquad \text{matches } \sigma(m+2)$$

$$\vdots$$

$$1 \qquad \text{matches } \sigma(m+n)$$

By Theorem 2, this last set of matchings can be written as follows:

$$n \qquad \text{matches } \sigma(1)+K$$

$$n-1 \qquad \text{matches } \sigma(2)+K$$

$$\vdots$$

$$1 \qquad \text{matches } \sigma(n)+K,$$

where $K = \sigma(m+1) - \sigma(1)$. Thus, starting from the *right-hand side* of the $(n+m)$-tableau, the positions of cards numbered $1, 2, \cdots, n$ are respectively identical to the positions of these same cards in the n-tableau, as required. $\qquad\qquad\square$

4. CARD SORTING BY DISPERSION

Suppose σ is a strictly increasing sequence of positive integers. Let σ' be the complement of σ. The dispersion, A, of σ' is the array consisting of all the positive integers, each occurring exactly once, with row 1 consisting of

$$a(1,1) = 1, \quad a(1,2) = \sigma'(1), \quad a(1,3) = \sigma'(a(1,2)), \quad \cdots, \quad a(1,j) = \sigma'(a(1,j-1)), \quad \cdots,$$

row 2 consisting of

$$a(2,1) = \text{least positive integer not in row 1}, \quad a(2,2) = \sigma'(a(2,1)), \quad a(2,3) = \sigma'(a(2,2)), \cdots,$$

row 3 consisting of

$$a(3,1) = \text{least positive integer not in row 1 or row 2}, \; a(3,2) = \sigma'(a(3,1)), \; a(3,3) = \sigma'(a(3,2)), \cdots,$$

and so on. (For example, the Zeckendorf-Wythoff array is a dispersion.)

With the notion of dispersion in mind, we turn now to a second method of obtaining chains of the form (1)-(4), beginning with the example in Section 1. There, recall, the selection sequence is given by

$$\sigma(k) = \lfloor k\alpha \rfloor, \text{ where } \alpha = (1+\sqrt{5})/2,$$

so that

$$\sigma = (1,3,4,6,8,9,11,12,14,16,17,19,\cdots)$$

and

$$\sigma' = (2,5,7,10,13,15,18,\cdots).$$

One can easily obtain a northwest corner of the dispersion, A, of σ':

1	2	5	13	34
3	7	18	47	
4	10	26	68	
6	15	39		
8	20	52		
9	23	60		
11	28	73		
12	31	81		
14	36			
16	41			
17	44			
19	49			

Let s be the sequence defined by

$$s(k) = \text{the number of the row of } A \text{ that contains } k:$$
$$s = (1,1,2,3,1,4,2,5,6,3,7,8,1,9,4,10,11,2,12,\cdots).$$

Let $(n_1,n_2,n_3,\cdots) = (1,2,4,7,\cdots)$ be the sequence of indices in chain (1), and note that the terms of chain (1) are reversed initial segments of s:

$$D_1 = (1)$$
$$D_2 = (2,1,1)$$
$$D_4 = (4,1,3,2,1,1)$$
$$D_7 = (7,3,6,5,2,4,1,3,2,1,1)$$
$$D_{12} = (12,2,11,10,4,9,1,8,7,3,6,5,2,4,1,3,2,1,1).$$

Since s is a fractal sequence (introduced in [4] and defined just below), the above example suggests that fractal sequences are closely related to cardsort problems. A more precise expression of this relationship comprises the next two theorems. We begin with the definition of fractal sequence.

Definitions: (from [2]). A sequence $s = s(n)$ is an *infinitive sequence* if for every i,

(F1) $s(n) = i$ for infinitely many n.

Let $a(i, j)$ be the jth index n for which $s(n) = i$. The array $A = \{a(i, j)\}$ is the *associated array* of x. An infinitive sequence s is a *fractal sequence* if two conditions hold:

(F2) if $k + 1 = s(n)$, then there exists $m < n$ such that $k = s(m)$;

(F3) if $h < i$ then for every j there is exactly one k such that $a(i, j) < a(h, k) < a(i, j + 1)$.

According to (F2), the first occurrence of each $k \geq 2$ in s must be preceded at least once by each of the numbers 1, 2, $\cdots$, $k - 1$, and according to (F3), between consecutive occurrences of i in s, each h less than i occurs exactly once.

The reason for the use of the adjective "fractal" is that any sequence satisfying (F1)-(F3) contains itself as a proper subsequence (infinitely many times); for example, if the first occurrence of each positive integer in a fractal sequence s is removed, then the remaining sequence is s. It is proved in [2] that the array A defined in (F1) is a dispersion, namely that of the complement of the first column of A, and conversely, that any dispersion yields a fractal sequence via (F1).

Theorem 4: Suppose s is a fractal sequence. Define a sequence σ as follows: for $k = 1, 2, 3, \cdots$, let $\sigma(k)$ be the position in σ of the first occurrence of k. Then for $n = 1, 2, 3, \cdots$, there exists a finite selection sequence σ_n such that the σ_n-cardsort problem has solution D_n given by

$$D_n = (s(\sigma(n)), s(\sigma(n) - 1), \cdots, s(2), s(1)). \tag{6}$$

Such a finite selection sequence is given by

$$\sigma_n(k) = 1 + \sigma(n) - \sigma(n - k + 1) \tag{7}$$

for $k = 1, 2, \cdots, n$.

Proof: Let $\sigma(k)$ be the position in s of the first occurrence of k. Suppose $n \geq 1$, and let σ_n satisfy (7) for $1 \leq k \leq n$. For such k, starting at $s(1)$ and counting to the right in s, the number in position $\sigma(k)$ is k. Consequently, starting at $n = s(\sigma(n))$ and counting to the left, k is in position $1 + \sigma(n) - \sigma(n - k + 1)$, which equals $\sigma_n(k)$. Since the positions in s occupied by the numbers $\sigma(n), \sigma(n - 1), \cdots, \sigma(1)$ in s agree with the requirement of the σ_n-cardsort problem, and since these positions determine D_n uniquely, the extended deck is D_n. (The initial order of the deck of n cards in your hand is therefore given by the first n terms of D_n.) $\qquad \square$

Theorem 4 shows how to manufacture a cardsort problem, with solution for each n, from an essentially arbitrary selection sequence σ. (That is, one could start with arbitrary σ, obtain s as the fractal sequence of the dispersion of the complement of σ, and then apply Theorem 4.) However, it is natural to ask if for certain values of n the first n terms of σ already comprise a finite selection sequence for which D_n is a solution. Next, we shall see that there are indeed such special values of n, and that they are given in terms of palindromes.

Corollary 1: Let s and σ be as in Theorem 4. For $n \geq 2$, the first n terms of σ comprise a finite selection sequence having solution D_n given by (6) if and only if

$$\big(\sigma(n) - \sigma(n-1), \sigma(n-1) - \sigma(n-2), \cdots, \sigma(2) - \sigma(1)\big) = \big(\sigma(2) - \sigma(1), \sigma(3) - \sigma(2), \cdots, \sigma(n) - \sigma(n-1)\big);$$

i.e., the difference sequence $\big(\sigma(2) - \sigma(1), \sigma(3) - \sigma(2), \cdots, \sigma(n) - \sigma(n-1)\big)$ is a palindrome.

Proof: Put $\sigma_n(k) = \sigma(k)$ in (7).   □

Corollary 2: The sequence σ in Theorem 4 is amenable if and only if for every $n \geq 1$ there exists $m \geq 1$ such that

$$\sigma(m+k) - \sigma(k) \tag{8}$$

is invariant for $k = 0, 1, \cdots, n$, or equivalently,

$$\sigma(n+m-k-1) - \sigma(n-k-1) = \sigma(n+m) - \sigma(n) \tag{9}$$

for $k = 0, 1, \cdots, n-1$, where $\sigma(0) := 0$.

Proof: The definition of amenability for σ is clearly equivalent to the following: if $n \geq 1$, then for some $m \geq 1$, the "shift equations"

$$\sigma(n+m) - \sigma(n+m-1) = \sigma(n) - \sigma(n-1),$$

$$\sigma(n+m-1) - \sigma(n+m-2) = \sigma(n-1) - \sigma(n-2),$$

$$\vdots$$

$$\sigma(m+1) - \sigma(m) = \sigma(1) - \sigma(0)$$

are satisfied by σ. On adding the first $k+1$ of these equations and canceling like terms, we obtain equation (9) for $k = 0, 1, \cdots, n-1$.

Conversely, starting with equations (9), we easily obtain the system of shift equations. Clearly, the set of equations (8) is equivalent to the set (9).   □

Theorem 5: Suppose σ is an amenable selection sequence. Let

$$D_{n_1} \subset D_{n_2} \subset D_{n_3} \subset \cdots$$

be an infinite chain induced by σ and the tableau method. Let E_{n_i} be the reversal of D_{n_i}. Then

$$\lim_{i \to \infty} E_{n_i}$$

is a fractal sequence.

Proof: The given selection sequence σ uniquely determines the fractal sequence s having first occurrence of k in position $\sigma(k)$, for $k = 1, 2, 3, \cdots$. One way to specify a deck d_{n_i} obtained by the tableau method is by counting back, in s, starting at $\sigma(n_i)$, using the finite selection sequence σ_{n_i} given by (7) for $k = 1, 2, \cdots, n_i$. The sequence σ_{n_i} thus selects $\sigma(n_i), \sigma(n_i - 1), \cdots, \sigma(1)$ as the respective positions in s of the numbers $n_i, n_i - 1, \cdots, 1$, and these uniquely determine the extended deck D_{n_i} of equation (6). The reversal of D_{n_i} is

$$E_{n_i} = (s(1), s(2), \cdots, s(\sigma(n_i) - 1), s(\sigma(n_i))),$$

and clearly, $\lim_{i \to \infty} E_{n_i} = s$.   □

We are now in a position to summarize the relation between cardsort problems and fractal sequences. According to theorem 4, each fractal sequence s generates a sequence of cardsort problems; if s also satisfies the hypothesis of Corollary 2, then s generates an amenable selection sequence σ. Conversely, by Theorem 5, an amenable selection generates a family of fractal sequences, one for each of the infinite chains of extended decks D_n.

Returning now to the chain (1)

$$D_1 \subset D_2 \subset D_4 \subset D_7 \subset D_{12} \subset \cdots$$

associated with $\alpha = (1 + \sqrt{5})/2$, we shall prove that the subscripts are $F_3 - 1, F_4 - 1, F_5 - 1, \cdots$. Write horizontally the fractal sequence s of the dispersion of the complement of σ; beneath s, write the terms of σ so that $\sigma(n)$ is directly below the first occurrence of n. Then write the difference sequence Δ just below σ, with $\Delta(n)$ situated between (and below) $\sigma(n - 1)$ and $\sigma(n)$. Finally, place the number $\nu(i) := j$ beneath the ith number $\Delta(j)$ for which the segment $(\Delta(2), \Delta(3), \cdots, \Delta(j))$ is a palindrome. The work appears as follows:

s:	1	1	2	3	1	4	2	5	6	3	7	8	1	9	4	10
σ:	1		3	4		6		8	9		11	12		14		16
Δ:		2		1	2		2		1	2		1	2		2	
ν:		2			4				7.							

230   C. KIMBERLING

Additional terms of Δ and ν may be helpful:

Δ: 2 1 2 2 1 2 1 2 2 1 2 2 1 2 1 2 2 1 2 1 2 2 1 2 2 1 2 1 2 2 1 2 2 1 2 1 2 2 1 2 1 2 2 1 2 2 1 2 1 2 2 1 2 1 2 2 1 2 1

ν: 2   4   7   12   20   33.

In order to verify that $\nu(i) = F_{i+1} - 1$ for $i \geq 2$, recall that $\Delta(n) = \lfloor n\alpha \rfloor - \lfloor (n-1)\alpha \rfloor$, so it suffices to show that

$$\lfloor (F_i - h)\alpha \rfloor - \lfloor (F_i - h - 1)\alpha \rfloor = \lfloor (h+1)\alpha \rfloor - \lfloor h\alpha \rfloor \tag{10}$$

for $h = 1, 2, \cdots, F_1 - 2$. As (10) is equivalent to

$$\lfloor (F_i - h)\alpha \rfloor + \lfloor h\alpha \rfloor = \lfloor (F_i - h - 1)\alpha \rfloor + \lfloor (h+1)\alpha \rfloor,$$

it suffices to prove that

$$\lfloor (F_i - h)\alpha \rfloor + \lfloor h\alpha \rfloor$$

is invariant for $h = 1, 2, \cdots, F_1 - 2$. To see that this indeed the case, we begin with the fact that for any real numbers x and y,

$$((x - y)) = \begin{cases} ((x)) - ((y)) & \text{if } ((x)) \geq ((y)) \\ \\ ((x)) - ((y)) + 1 & \text{if } ((x)) < ((y)). \end{cases} \tag{11}$$

Taking $x = F_i \alpha$ and $y = h\alpha$ yields

$$((F_i - h)\alpha)) + ((h\alpha)) = \begin{cases} ((F_i \alpha)) & \text{if } i \text{ is even} \\ \\ ((F_i \alpha)) + 1 & \text{if } i \text{ is odd.} \end{cases} \tag{12}$$

Now $\lfloor (F_i - h)\alpha \rfloor + \lfloor h\alpha \rfloor = F_i \alpha - (((F_i - h)\alpha)) - ((h\alpha))$, so that by (12),

$$\lfloor (F_i - h)\alpha \rfloor + \lfloor h\alpha \rfloor = \begin{cases} \lfloor F_i \alpha \rfloor & \text{if } i \text{ is even} \\ \\ \lfloor F_i \alpha \rfloor + 1 & \text{if } i \text{ is odd;} \end{cases}$$

this proves that the sums (11) are invariant for $h = 1, 2, \cdots, F_i - 2$, as desired.

REFERENCES

[1]   Kimberling, Clark and Schultz, Harris S. "Card sorting by dispersions." *Ars Combinatoria*, to appear.

[2]   Kimberling, Clark. "Fractal sequences and interspersions." *Ars Combinatoria*, Vol. *45* (1997): pp. 157-168.

[3]   Kimberling, Clark. "Palindromic sequences from irrational numbers." *The Fibonacci Quarterly*, Vol. *36* (1998): pp. 171-173.

[4]   Kimberling, Clark. "Numeration sequences and fractal sequences." *Acta Arithmetica*, Vol. *73* (1995): pp. 103-117.

AMS Classification Numbers: 11B39

ON THE INHOMOGENEOUS GEOMETRIC LINE-SEQUENCE

Jack Lee

1. INTRODUCTION

There have been some recent investigations in the inhomogeneity of the second order recurrence sequence, see for example [1], [2], [3], [6] and [7]. We extend the investigation to include the inhomogeneity also in the corresponding geometric sequence. Consider the line-sequence generated by the recurrence relation

$$u_n = cu_{n-2} + bu_{n-1} + k, \quad n \in z, \tag{1.1}$$

where c and b are non-zero integers and k the linear inhomogeneous term. We seek the conditions under which the terms of the line-sequence generated by (1.1) also satisfy the inhomogeneous geometric relation

$$u_n = xu_{n-1} + g, \tag{1.2}$$

where x is the geometric ratio and g the geometric inhomogeneous term.

To indicate the geometric inhomogeneity a line-sequence possesses, we need to modify the format employed in [6] to include the geometric inhomogeneous constant g in the expression of the line-sequence as follows:

$$U_{u_0, u_1}(c, b; k, g): \cdots, u_{-3}, u_{-2}, u_{-1}, [u_0, u_1], u_2, u_3, u_4, \cdots \tag{1.3}$$

where the term u_0 is the origin, and the terms u_0 and u_1 form the generating pair, and so forth. For example, the following pair of line-sequences

---

This paper is in final form and no version of it will be submitted for publication elsewhere.

233

$$U_{0,1+\sqrt{2}}(-2,4;1,1+\sqrt{2}):\cdots, -\frac{\sqrt{2}}{2},[0,1+\sqrt{2}],5+4\sqrt{2},19+14\sqrt{2},\cdots \qquad (1.4)$$

$$U_{0,1-\sqrt{2}}(-2,4;1,1-\sqrt{2}):\cdots, \frac{\sqrt{2}}{2},[0,1-\sqrt{2}],5-4\sqrt{2},19-14\sqrt{2},\cdots \qquad (1.5)$$

not only satisfies relation (1.1), but also (1.2) if we take $x = 2+\sqrt{2}$ for line-sequence (1.4) and the corresponding conjugate for line-sequence (1.5). For reasons to be explained later, we shall refer to this pair as an IA-Golden Pair.

It is clear that a general term u_r satisfying (1.2) is given by the formula

$$u_r = x^r u_0 + \frac{r}{|r|}g \sum_{i=0}^{|r|-1} x^j, \; j = \begin{cases} i, & \text{if } r > 0 \\ -i-1, & \text{if } r < 0 \end{cases} ; r \neq 0. \qquad (1.6)$$

A few terms in the neighborhood of the origin is shown below:

$$u_{-3} = x^{-3}u_0 - g(x^{-3} + x^{-2} + x^{-1})$$

$$u_{-2} = x^{-2}u_0 - g(x^{-2} + x^{-1})$$

$$u_{-1} = x^{-1}u_0 - gx^{-1}$$

$$u_0 = u_0$$

$$u_1 = xu_0 + g$$

$$u_2 = x^2 u_0 + g(1+x)$$

$$u_3 = x^3 u_0 + g(1+x+x^2). \qquad (1.7)$$

Substituting (1.2) into (1.1) leads to the equation

$$(x^2 - bx - c)u_n + (x - b + 1)g - k = 0. \qquad (1.8)$$

Since u_n is an arbitrary term in the line-sequence, its coefficient must vanish, so the geometric ratio x must satisfy the condition,

$$x^2 - bx - c = 0. \qquad (1.9)$$

This is none other than equation (4.4) in [4], so the only permissible values that the geometric ratio x may assume are:

$$\alpha = \frac{b + (b^2 + 4c)^{1/2}}{2}, \quad \beta = \frac{b - (b^2 + 4c)^{1/2}}{2}. \qquad (1.10)$$

Here we use α and β to denote the generalized large and the (negative) generalized small golden ratios, respectively, rather than A and B, which will be used in this report to denote the regular large and the (negative) regular small golden ratios.

Furthermore, the constant term in (1.8) must vanish, so we obtain the condition relating the two inhomogeneous constants:

$$g = \frac{k}{x - b + 1},$$ (1.11)

with the condition

$$x - b + 1 \neq 0.$$ (1.12)

Following are some examples of application.

2. BINET'S FORMULA IN IA-SPACE

The pair of basis vectors of the IA-space are given by (4.9) and (4.10) in [6]:

$$J_{1-k,\,-k}(c, b; k): \cdots, \; \frac{c + b^2}{c^2} - k, \; -\frac{b}{c} - k, [1 - k, \, -k], c - k, cb - k, \cdots$$ (2.1)

$$J_{-k,\,1-k}(c, b; k): \cdots, \; \frac{-b}{c^2} - k, \frac{1}{c} - k, [\, -k, 1 - k], b - k, c + b^2 - k, \cdots$$ (2.2)

subjecting to the condition

$$c + b = 2,$$ (2.3)

see (4.8) in [6]. Note that owing to (2.3), condition (1.12) is automatically satisfied in the IA-space.

Consider the geometrical line-sequence corresponding to (2.1):

$$J_{1-k,\,x(1-k)+g}(c, b: k): \cdots, \; x^{-1}(1 - k - g), \; [1 - k, x(1 - k) + g], x(x(1 - k) + g) + g, \cdots.$$ (2.4)

Using (1.9) and (1.11), we obtain

$$J_{1-k,\,x(1-k)+g}(c, b; k, g) = J_{1-k,\,x-k}(c, b; k, g).$$ (2.5)

The two line-sequences generated from the right hand side of (2.5) are

$$J_{1-k,\,\alpha-k}(c, b; k, g): \cdots, \; \frac{\alpha - b}{c} - k, [1 - k, \alpha - k], c + b\alpha - k, \cdots$$ (2.6)

$$J_{1-k,\,\beta-k}(c, b; k, g): \cdots, \; \frac{\beta - b}{c} - k, [1 - k, \beta - k], c + b\beta - k, \cdots.$$ (2.7)

For $c = b = 1$ and $k = 0$, this pair reduces to the Golden Pair, namely, (2.6) and (2.7) in [5]. We therefore refer to this pair as the IA-Golden Pair. The line-sequences (1.4) and (1.5) appeared previously is an example of such a pair.

Applying the inhomogeneous operations (4.12), (4.4) and (4.5) in [6] to the IA-Golden Pair, we obtain the Binet's formula in the IA-space:

$$J_{-k,\,1-k} = (J_{1-k,\,\alpha-k} - J_{1-k,\,\beta-k})/(\alpha - \beta).$$ (2.8)

We therefore refer to the line-sequence $J_{-k,1-k}$, as the IA-Fibonacci line-sequence. In line-sequential form, it looks as follows:

$$J_{-k,1-k}(c,b;k): \cdots, (-b-c^2k)/c^2, (1-ck)/c, [-k,1-k], b-k, c+b^2-k, \cdots \qquad (2.9)$$

For example, the IA-Fibonacci line-sequence corresponding to the IA-Golden Pair (1.4) and (1.5) looks as follows:

$$J_{-1,0}(-2,4;1): \cdots, -2, 3/-2, [-1,0], 3, 13, \cdots. \qquad (2.10)$$

For $k = 0$, line-sequence (2.9) reduces to the homogeneous line-sequence $G_{0,1}$, see (4.3) in [4]. For $k = 1$ and $b = c = 1$, formula (2.8) reduces to the Binet formula in the IH-space, see (2.17) in [6]. If $k = 0$, and $c = b = 1$, the line-sequence $J_{-k,1-k}$ further reduces to the (regular) Fibonacci line-sequence, $F_{0,1}$, see [4], and formula (2.8) further reduces to the (regular) Binet's formula for Fibonacci line-sequence.

The corresponding Lucas line-sequence in the IA-space is then given by the inhomogeneous addition, (4.4) in [6], of the IA-Golden Pair:

$$J_{2-k,b-k} = J_{1-k,\alpha-k} + J_{1-k,\beta-k}. \qquad (2.11)$$

In line-sequential form, it looks as follows:

$$J_{2-k,b-k}(c,b;k): \cdots, (-b+(b-2)k)/c, [2-k,b-k], 2c+b^2-k, b(3c+b^2)-k, \cdots. \qquad (2.12)$$

Thus the IA-Lucas line-sequence corresponding to the pair (1.4) and (1.5) looks as follows:

$$J_{1,3}(-2,4;1): \cdots, 2, 1, [1,3], 11, 39, \cdots. \qquad (2.13)$$

For $k = 0$, the IA-Lucas line-sequence (2.12) reduces to $G_{2,b}$, the anharmonic Lucas line-sequence, see (4.11) in [4]. If $c = b = 1$, it further reduces to the regular Lucas line-sequence, $F_{2,1}$, see (3.1) in [5].

Let u_a and v_a represent the ath element in the IA-Fibonacci and IA-Lucas line-sequences, respectively. Then it can be verified that the following relation holds:

$$(u_a + k)(v_a + k) = u_{2a} + k. \qquad (2.14)$$

This is the IA version of the well known relation, $F_n L_n = F_{2n}$ between the Fibonacci and Lucas terms.

Alternately, parallel to (2.15) and (2.16) of [6], we could apply operations (4.12), (4.4) and (4.5) in [6] onto the basis pair (2.1) and (2.2), and obtain the Golden Pair in the IA-space:

$$J_{1-k,-k} + \alpha J_{-k,1-k} = J_{1-k,\alpha-k} \qquad (2.15)$$

$$J_{1-k,-k} + \beta J_{-k,1-k} = J_{1-k,\beta-k}. \tag{2.16}$$

Onto this pair applying operations (4.12), (4.4) and (4.5) in [6], once again, we obtain the IA-Binet's formula (2.8).

Now consider the geometrical line-sequence corresponding to (2.2):

$$J_{-k,x(-k)+g}(c,b:k,g):\cdots,\ x^{-1}(-k-g),[-k,x(-k)+g],x(x(-k)+g)+g\cdots. \tag{2.17}$$

Using (1.9) and (1.11), we obtain:

$$J_{-k,x(-k)+g}(c,b;k,g) = J_{-k,-k}(c,b;k). \tag{2.18}$$

The right hand side of (2.18) shows a sequence of $-k$'s, independent of the values of c and b under the condition (2.3):

$$J_{-k,-k}(c,b;k,g):\cdots,\ -k,[-k,-k],-k,-k,\cdots. \tag{2.19}$$

This is just (4.7) of [6], the identity element of addition. However, since x on the left hand side of (2.18) has two values, this additive identity is therefore two-fold degenerated, namely a doublet.

3. BINET'S FORMULA IN HA-SPACE

The set of basis vectors of the HA-space is given by (5.4), (5.5) and (5.6) in [6]:

$$H_{1,0}(c,b;0)\cdots,\ \frac{c+b^2}{c^2},\ -\frac{b}{c},[1,0],c,cb,\cdots\cdots \tag{3.1}$$

$$H_{0,1}(c,b;0)\cdots,\ -\frac{b}{c^2},\frac{1}{c},[0,1],b,c+b^2,\cdots\cdots \tag{3.2}$$

$$H_{0,0}(c,b;1)\cdots,\ \frac{b-c}{c^2},\frac{-1}{c},[0,0],1,b+1,\cdots\cdots. \tag{3.3}$$

Consider the geometric line-sequence corresponding to (3.1):

$$H_{1,x+g}(c,b;k,g):\cdots,\ x^{-1}(1-g),[1,x+g],x(x+g)+g,(x(x(x+g)+g)+g,\cdots. \tag{3.4}$$

The two line-sequences generated from (3.4) are

$$H_{1,\alpha+\delta}(c,b;k,\delta):\cdots,\alpha^{-1}(1-\delta),[1,\alpha+\delta],\alpha(\alpha+\delta)+\delta,\alpha(\alpha(\alpha+\delta)+\delta)+\delta,\cdots \tag{3.5}$$

$$H_{1,\beta+\varepsilon}(c,b;k,\varepsilon):\cdots,\beta^{-1}(1-\varepsilon),[1,\beta+\varepsilon],\beta(\beta+\varepsilon)+\varepsilon,\beta(\beta(\beta+\varepsilon)+\varepsilon)+\varepsilon,\cdots \tag{3.6}$$

where $\delta = \dfrac{k}{a-b+1}$ and $\varepsilon = \dfrac{k}{\beta-b+1}$. This is the HA-Golden Pair. Applying operations (5.1) and (5.2) in [6] to this pair, we obtain

$$H_{0,1+\phi} = (H_{1,\alpha+\delta} - H_{1,\beta+\varepsilon})/(\alpha-\beta) \tag{3.7}$$

where $\phi = \dfrac{k}{c+b-1}$. This is the Binet's formula in the HA-space and its left side is the HA-Fibonacci line-sequence. In line-sequential form, the latter looks:

$$H_{0,1+\phi}(c,b:k):\cdots, c^{-1}(1+\phi-k),[0,1+\phi],b(1+\phi)+k,c(1+\phi)+b^2(1+\phi)+(b+1)k,\cdots.$$

$$(3.8)$$

The one corresponding to the pair (1.4) and (1.5) looks as follows:

$$H_{0,1+\phi}(-2,4;1):\cdots, -1/2,[0,2],9,33,\cdots.$$

For $k = 0$, so $\phi = 0$, line-sequence (3.8) reduces to the homogeneous line-sequence $G_{0,1}$, see (4.3) in [4]. The corresponding Lucas line-sequence is then given by the homogeneous addition, (5.2) in [6], of the HA-Golden Pair:

$$H_{2,b+\phi} = H_{1,\alpha+\delta} + H_{1,\beta+\epsilon}.$$

$$(3.9)$$

In line-sequential form, it looks as follows:

$H_{2,b+\phi}(c,b;k):$

$$\cdots,(-b+\phi-k)/c,[2,b+\phi],2c+b(b+\phi)+k,b(3c+b^2)+(c+b^2)\phi+(b+1)k,\cdots.$$

$$(3.10)$$

The one corresponding to the pair (1.4) and (1.5) looks as follows:

$$H_{2,5}(-2,4;1):\cdots,2,[2,5],17,59,\cdots.$$

The question as to how does the basic relation $F_n L_n = F_{2n}$ look like in the HA-space is still open at this point and further investigation is needed.

Consider the geometrical line-sequence corresponding to (3.2):

$$H_{0,x(0)+g}(c,b;0):\cdots,x^{-1}(-g),[0,x(0)+g],x(x(0)+g)+g,x(x(x(0)+g)+g)+g,\cdots.$$

$$(3.11)$$

Since $k = 0$, so $g = 0$, and $H_{0,x(0)+g}(c,b;0) = H_{0,0}(c,b;0)$. Hence this line-sequence degenerates into a sequence of zeros, which serves as the identity element of addition in the HA-space.

We summarize our results together with those obtained previously in the following table.

Space	HH-	IH-	IA-	HA-
Homogeneity	$k = 0$	$k = 1$	$k \neq 0$	$k \neq 0$
Basic pair	$F_{1,0}; F_{0,1}$	$I_{0,-1}; I_{-1,0}$	$J_{1-k,-k}; J_{-k,1-k}$	$H_{1,0}; H_{0,1}; H_{0,0}$
Identity element	$F_{0,0}$	$I_{-1,-1}$	$J_{-k,-k}$	$H_{0,0}(c,b;0)$
Fibonacci	$F_{0,1}$	$I_{-1,0}$	$J_{-k,1-k}$	$H_{0,1+\phi}$
Lucas	$L_{2,1}$	$I_{1,0}$	$J_{2-k,b-k}$	$H_{2,b+\phi}$
Golden Pair	$F_{1,A}; F_{1,B}$	$I_{0,A-1}; I_{0,B-1}$	$J_{1-k,\alpha-k}; J_{1-k,\beta-k}$	$H_{1,\alpha+\delta}; H_{1,\beta+\epsilon}$
Binet's formula	$\dfrac{F_{1,A}-F_{1,B}}{A-B}$	$\dfrac{I_{0,A-1}-I_{0,B-1}}{A-B}$	$\dfrac{J_{1-k,\alpha-k}-J_{1-k,\beta-k}}{\alpha-\beta}$	$\dfrac{H_{1,\alpha+\delta}-H_{1,\beta+\epsilon}}{\alpha-\beta}$

Table 1. Some basic quantities of geometric line-sequences

The author wishes to express his gratitude to the anonymous referee for his valuable suggestions resulting in the improvement of this report.

REFERENCES

[1] Andrade, Ana and Pethe, S.P. "On the rth-Order Nonhomogeneous Recurrence Relation and Some Generalized Fibonacci Sequences." *The Fibonacci Quarterly*, Vol. *30.3* (1992): pp. 256-262.

[2] Bicknell-Johnson, Marjorie and Bergum, Gerald E. "The Generalized Fibonacci Numbers $\{c_n\}, c_n = c_{n-1} + c_{n-2} + k$." Applications of Fibonacci Numbers, Volume 3. Edited by G.E. Bergum, A.F. Horadam and A.N. Philippou. Kluwer Academic Publishing, Dordrecht, The Netherlands, 1988: pp. 193-205.

[3] Bolian, Liu. "A Matrix Method to Solve Linear Recurrences with Constant Coefficients." *The Fibonacci Quarterly*, Vol. *30.1* (1992): pp. 2-8.

[4] Lee, Jack Y. "Some Basic Properties of the Fibonacci Line-Sequence." Applications of Fibonacci Numbers, Volume 4. Edited by G.E. Bergum, A.F. Horadam and A.N. Philippou. Kluwer Academic Publishing, Dordrecht, The Netherlands, 1991: pp. 203-214.

[5]   Lee, Jack Y. "The Golden-Fibonacci Equivalence." *The Fibonacci Quarterly*, Vol. *30.3*
        (1992): pp. 216-220.

[6]   Lee, Jack Y. "On Some Basic Linear Properties of the Second Order Inhomogeneous
        Line-Sequence." *The Fibonacci Quarterly*, Vol. *35.2* (1997): pp. 111-121.

[7]   Zhang, Zhizheng. "Some Properties of the Generalized Fibonacci Sequences
        $C_n = C_{n-1} + C_{n-2} + r$." *The Fibonacci Quarterly*, Vol. *35.2* (1997): pp. 169-171.

AMS Classification Numbers: 11B39

FIBONACCI NUMBERS OF THE FORM $k^2 + k + 2$

Florian Luca

1. INTRODUCTION

Cohn (see [1]) determined all the Fibonacci and the Lucas numbers which are squares and London & Finkelstein (see [3]) found all the Fibonacci and the Lucas numbers which are cubes. Luo Ming (see [4], [5], [6], [7]) has determined all the Fibonacci and the Lucas numbers which are either triangular or pentagonal and Williams (see [9]) found all the Fibonacci numbers which are of the form $k^2 + 1$ for some integer k. Stark (see [2]), or [8]) asks which Fibonacci numbers are half the difference (or sum) of two cubes.

In this paper we present the following result:

Theorem:

The only Fibonacci numbers which are of the form $k^2 + k + 2$ are 2 and 8.

The method of proof is as follows. Since an integer u is of the form $k^2 + k + 2$ for some integer k if and only if $4u - 7$ is a perfect square, it suffices to find all integers n such that $4F_n - 7$ is a perfect square. To do this, we find, for each nonsquare $4F_n - 7$, an integer W_n, such that the Jacobi symbol

$$\left(\frac{4F_n - 7}{W_n}\right) = -1.$$

Using elementary congruences we can show that, if $4F_n - 7$ is a perfect square, then

$$n \equiv \pm 3, 6 \pmod{2^5 \cdot 3 \cdot 5^2 \cdot 7 \cdot 11}. \tag{1}$$

This work was partially supported by the Alexander von Humboldt Foundation.

This paper is in final form and no version of it will be submitted for publication elsewhere.

241

Following Luo Ming (see [4]) we develop a special Jacobi symbol criterion with which we can further show that each congruence class of (1) above contains exactly one value of n such that $4F_n - 7$ is a perfect square, i.e. $n = \pm 3, 6$.

2. PRELIMINARIES

Let $(L_m)_{m \in \mathbb{Z}}$ be the Lucas sequence. The following relations are well known (see [4]):

$$F_{-n} = (-1)^{n+1} F_n, \quad L_{-n} = (-1)^n L_n; \tag{2}$$

$$2F_{m+n} = F_m L_n + L_n F_m, \quad 2L_{m+n} = 5F_m F_n + L_m L_n; \tag{3}$$

$$F_{2n} = F_n L_n, \quad L_{2n} = L_n^2 + 2(-1)^{n+1}; \tag{4}$$

$$L_n^2 - 5F_n^2 = 4(-1)^n; \tag{5}$$

$$F_m \mid F_n \text{ if } m \mid n; \tag{6}$$

$$L_m \mid L_n \text{ if } m \mid n \text{ and } \tfrac{n}{m} \text{ is odd}; \tag{7}$$

$$F_{2kt+n} \equiv (-1)^t F_n \pmod{L_k} \text{ if } k \equiv \pm 2 \pmod 6. \tag{8}$$

Moreover, since $x = \pm F_n$, $y = \pm L_n$ are the complete set of solutions of the Diophantine equations $5x^2 - y^2 = \pm 4$, the condition $F_n = k^2 + k + 2$ is equivalent to finding all integer solutions of the Diophantine equations

$$5(k^2 + k + 2)^2 - y^2 = \pm 4.$$

From our theorem, it follows easily that the only solutions of the Diophantine equations above are $(k, y) = (0, \pm 4), (-1, \pm 4), (2, \pm 18), (-3, \pm 18)$.

3. A JACOBI SYMBOL CRITERION AND ITS CONSEQUENCES

We first establish a Jacobi symbol criterion that plays a key role in the proof of the theorem and then we give some of its consequences.

Criterion:

If n, a and b are positive integers such that $n \equiv 0 \pmod 4$, $3 \nmid n$, $(a, L_n) = 1$, $b \mid F_n$ and $b \equiv 1 \pmod 2$, then

$$\left(\frac{4aF_{2n} + b}{L_{2n}} \right) = -\left(\frac{b}{L_{2n}} \right) \cdot \left(\frac{8aF_n + bL_n}{64a^2 + 5b^2} \right). \tag{9}$$

Proof: Let $F_n = bG$ for some positive integer G. Since $4 \mid n$ but $3 \nmid n$, it follows that $n \equiv \pm 4 \pmod{12}$ and $2n \equiv \pm 4 \pmod{12}$. Hence, $L_n \equiv L_{2n} \equiv 7 \pmod 8$. In particular, $\left(\frac{2}{L_n} \right) = \left(\frac{2}{L_{2n}} \right) = 1$. It follows that

$$\left(\frac{4aF_{2n}+b}{L_{2n}}\right) = \left(\frac{8aF_{2n}+2b}{L_{2n}}\right) = \left(\frac{8abGL_n + bL_n^2}{L_{2n}}\right) = \left(\frac{b}{L_{2n}}\right) \cdot \left(\frac{L_n}{L_{2n}}\right) \cdot \left(\frac{8aG + L_n}{L_{2n}}\right)$$

$$= -\left(\frac{b}{L_{2n}}\right) \cdot \left(\frac{L_n}{L_{2n}}\right) \cdot \left(\frac{L_{2n}}{8aG + L_n}\right) \text{ (because } L_{2n} \equiv 8aG + L_n \equiv 3 (\text{mod } 4))$$

$$= \left(\frac{b}{L_{2n}}\right) \cdot \left(\frac{L_{2n}}{L_n}\right) \cdot \left(\frac{L_{2n}}{8aG + L_n}\right) = \left(\frac{b}{L_{2n}}\right) \cdot \left(\frac{-2}{L_n}\right) \cdot \left(\frac{L_{2n}}{8aG + L_n}\right) = -\left(\frac{b}{L_{2n}}\right) \cdot \left(\frac{\frac{1}{2}(5F_n^2 + L_n^2)}{8aG + L_n}\right)$$

$$= -\left(\frac{b}{L_{2n}}\right) \cdot \left(\frac{40F_n^2 + 8L_n^2}{8aG + L_n}\right) = -\left(\frac{b}{L_{2n}}\right) \cdot \left(\frac{a}{8aG + L_n}\right) \left(\frac{40aF_n^2 + 8aL_n^2}{8aG + L_n}\right).$$

Now

$$8aL_n^2 = 8aL_n(L_n + 8aG - 8aG) \equiv -64a^2 L_n G \ (\text{mod } (8aG + L_n))$$

and

$$40aF_n^2 = 5bF_n(8aG + L_n - L_n) \equiv -5b^2 L_n G \ (\text{mod } (8aG + L_n)).$$

So,

$$\left(\frac{4aF_{2n}+1}{L_{2n}}\right) = -\left(\frac{b}{L_{2n}}\right) \cdot \left(\frac{a}{8aG + L_n}\right) \cdot \left(\frac{-(5b^2 + 64a^2)L_n G}{8aG + L_n}\right)$$

$$= -\left(\frac{b}{L_{2n}}\right) \cdot \left(\frac{a}{8aG + L_n}\right) \cdot \left(\frac{-1}{8aG + L_n}\right) \left(\frac{64a^2 + 5b^2}{8aG + L_n}\right) \cdot \left(\frac{L_n G}{8aG + L_n}\right).$$

Since $\left(\frac{64a^2 + 5b^2}{8aG + L_n}\right) = \left(\frac{8aG + L_n}{64a^2 + 5b^2}\right)$ and $\left(\frac{-1}{8aG + L_n}\right) = -1$, it follows that

$$\left(\frac{4aF_{2n}+1}{L_{2n}}\right) = \left(\frac{b}{L_{2n}}\right) \cdot \left(\frac{a}{8aG + L_n}\right) \cdot \left(\frac{L_n}{8aG + L_n}\right) \cdot \left(\frac{G}{8aG + L_n}\right) \cdot \left(\frac{8aG + L_n}{64a^2 + 5b^2}\right). \quad (10)$$

Notice that if $G \equiv 1 \ (\text{mod } 4)$, then

$$\left(\frac{G}{8aG + L_n}\right) = \left(\frac{8aG + L_n}{G}\right) = \left(\frac{L_n}{G}\right) = \left(\frac{G}{L_n}\right)$$

and if $G \equiv 3 \ (\text{mod } 4)$, then

$$\left(\frac{G}{8aG + L_n}\right) = -\left(\frac{8aG + L_n}{G}\right) = -\left(\frac{L_n}{G}\right) = \left(\frac{G}{L_n}\right).$$

Hence,

$$\left(\frac{G}{8aG + L_n}\right) = \left(\frac{G}{L_n}\right). \quad (11)$$

Moreover,

$$\left(\frac{L_n}{8aG+L_n}\right) = -\left(\frac{8aG+L_n}{L_n}\right) = -\left(\frac{2}{L_n}\right)\cdot\left(\frac{a}{L_n}\right)\cdot\left(\frac{G}{L_n}\right) = -\left(\frac{a}{L_n}\right)\cdot\left(\frac{G}{L_n}\right). \qquad (12)$$

From formulae (10), (11) and (12), it follows that

$$\left(\frac{4aF_{2n}+1}{L_{2n}}\right) = -\left(\frac{b}{L_{2n}}\right)\cdot\left(\frac{a}{8aGL_n+L_n^2}\right)\cdot\left(\frac{8aG+L_n}{64a^2+5b^2}\right). \qquad (13)$$

Notice that

$$\left(\frac{a}{8aGL_n+L_n^2}\right) = \left(\frac{8aGL_n+L_n^2}{a}\right) = \left(\frac{L_n^2}{a}\right) = 1.$$

Hence, formula (13) becomes

$$\left(\frac{4aF_{2n}+1}{L_{2n}}\right) = -\left(\frac{b}{L_{2n}}\right)\cdot\left(\frac{8aG+L_n}{16a^2+5b^2}\right) = -\left(\frac{b}{L_{2n}}\right)\cdot\left(\frac{8abG+bL_n}{16a^2+5b^2}\right)\cdot\left(\frac{b}{64a^2+5b^2}\right). \qquad (14)$$

Formula (9) follows now from formula (14), the fact that $bG = F_n$ and the fact that

$$\left(\frac{b}{64a^2+5b^2}\right) = \left(\frac{64a^2+5b^2}{b}\right) = \left(\frac{64a^2}{b}\right) = 1. \qquad \square$$

Remark: A slightly more general version of the above criterion for $b=1$ appears in [4].

Case $n \equiv \pm 3 \pmod{2^5\cdot 3\cdot 5\cdot 7\cdot 11}$.

Assume that $4F_n - 7$ is a perfect square. Since $|F_n| = F_{|n|}$, and since $k^2+k+2 \geq 2 > 0$ for any integer k, it follows that we may assume that $n \geq 0$. We show that $n = 3$. Indeed, for if not, assume that $n > 3$. Write

$$n = c + 2^\alpha 3^\beta 5^\gamma 7^\delta 11^\epsilon m$$

where $c \in \{\pm 3\}$, $\alpha \geq 5$, $\beta \geq 1$, $\gamma \geq 1$, $\delta \geq 1$, $\epsilon \geq 1$ and $m \geq 1$. We assume that $(m, 2310) = 1$. Let $b = 2^\alpha 3^\beta 5^\gamma 7^\delta 11^\epsilon m$. From formulae (2) and (3), it follows that

$$2F_n = 2F_{b+c} = F_b L_c + F_c L_b = L_{\pm 3}F_b + F_{\pm 3}L_b = \mathrm{sign}(c)4F_b + 2L_b. \qquad (15)$$

Notice that

$$b = \begin{cases} 2^\alpha\cdot 7\cdot\left(1+2\cdot\left(\dfrac{3^\beta 5^\gamma 7^{\delta-1}11^\epsilon-1}{2}\cdot m+\dfrac{m-1}{2}\right)\right) \text{ or} \\[3mm] 2^\alpha\cdot 5\cdot 7\cdot\left(1+2\cdot\left(\dfrac{3^\beta 5^{\gamma-1}7^{\delta-1}11^\epsilon-1}{2}\cdot m+\dfrac{m-1}{2}\right)\right) \text{ or} \\[3mm] 2^\alpha\cdot 7\cdot 11\cdot\left(1+2\cdot\left(\dfrac{3^\beta 5^\gamma 7^{\delta-1}11^{\epsilon-1}-1}{2}\cdot m+\dfrac{m-1}{2}\right)\right) \text{ or} \\[3mm] 2^\alpha\cdot 5\cdot 7\cdot 11\cdot\left(1+2\cdot\left(\dfrac{3^\beta 5^{\gamma-1}7^{\delta-1}11^{\epsilon-1}-1}{2}\cdot m+\dfrac{m-1}{2}\right)\right). \end{cases} \qquad (16)$$

Let

$$
\left\{
\begin{array}{l}
t_1 = 3^\beta 5^\gamma 7^\delta - {}^1 11^\epsilon - 1, \\[2mm]
t_2 = 3^\beta 5^\gamma - 1 7^\delta - {}^1 11^\epsilon - 1, \\[2mm]
t_3 = 3^\beta 5^\gamma 7^\delta - {}^1 11^{\epsilon - 1} - 1, \\[2mm]
t_4 = 3^\beta 5^\gamma - 1 7^\delta - {}^1 11^{\epsilon - 1} - 1.
\end{array}
\right.
$$

Clearly, $t_1 \equiv t_2 \pmod 4$, $t_3 \equiv t_4 \pmod 4$ but $t_1 \not\equiv t_3 \pmod 4$. Moreover, all four numbers t_1, t_2, t_3 and t_4 are even. Hence, using formulae (8), (16) and the fact that $(m, 6) = 1$, it follows that one can always choose b_1 to be one of the four numbers $2^{\alpha - 1} \cdot 7, 2^{\alpha - 1} \cdot 5 \cdot 7, 2^{\alpha - 1} \cdot 7 \cdot 11$ or $2^{\alpha - 1} \cdot 5 \cdot 7 \cdot 11$, in such a way that

$$ F_b \equiv -\,\mathrm{sign}(c) F_{2b_1} \pmod{L_{2b_1}}. \tag{17} $$

Moreover, since $112 = 2^4 \cdot 7 \mid b_1$ and $5 \equiv 2 \pmod 3$, it follows easily that b_1 can be chosen in such a manner that $b_1 \equiv 112 \pmod{336}$. From formulae (15), (7) and (17), it follows that

$$ 4F_n - 7 = \mathrm{sign}(c) 8 F_b + 4 L_b - 7 \equiv -8 F_{2b_1} - 7 \equiv -(8 F_{2b_1} + 7) \pmod{L_{2b_1}}. $$

Notice that $4 \mid b_1$, $3 \nmid b_1$, $2 \nmid L_{b_1}$ and $7 \mid F_8 \mid F_{b_1}$. Hence, by the criterion for $a = 2$ and $b = 7$, it follows that

$$
\left(\frac{4F_n - 7}{L_{2b_1}} \right) = \left(\frac{-(8 F_{2b_1} + 7)}{L_{2b_1}} \right) = -\left(\frac{-1}{L_{2b_1}} \right) \cdot \left(\frac{7}{L_{2b_1}} \right) \cdot \left(\frac{16 F_{b_1} + 7 L_{b_1}}{64 \cdot 4 + 5 \cdot 49} \right) = \left(\frac{7}{L_{2b_1}} \right) \cdot \left(\frac{16 F_{b_1} + 7 L_{b_1}}{501} \right)
$$

$$ = -\left(\frac{L_{2b_1}}{7} \right) \cdot \left(\frac{16 F_{b_1} + 7 L_{b_1}}{3} \right) \cdot \left(\frac{16 F_{b_1} + 7 L_{b_1}}{167} \right). \tag{18} $$

The sequence $(L_m)_{m \in \mathbf{Z}}$ is periodic modulo 7 with period 16. The sequence $(16 F_m + 7 L_m)_{m \in \mathbf{Z}}$ is periodic modulo 3 with period 8 and periodic modulo 167 with period 336. From equation (18), the fact that $16 \mid b_1$ and $b_1 \equiv 112 \pmod{336}$, it follows that

$$ \left(\frac{4F_n - 7}{L_{2b_1}} \right) = -\left(\frac{L_0}{7} \right) \cdot \left(\frac{16 F_0 + 7 L_0}{3} \right) \cdot \left(\frac{16 F_{112} + 7 L_{112}}{167} \right) = -\left(\frac{2}{7} \right) \cdot \left(\frac{14}{3} \right) \cdot \left(\frac{16 F_{112} + 7 L_{112}}{167} \right) = -1 $$

because

$$ \left(\frac{16 F_{112} + 7 L_{112}}{167} \right) = -1. $$

Hence, $4F_n - 7$ cannot be a perfect square in this case.

Case $n \equiv 6 \pmod{2^5 \cdot 3 \cdot 5^2 \cdot 7 \cdot 11}$.

Assume that $4F_n - 7$ is a perfect square. Since $|F_n| = F_{|n|}$ and since $k^2 + k + 2 \geq 2 > 0$ for any integer k, it follows that we may assume that $n > 0$. We show that

$n = 6$. Indeed, for if not, assume that $n > 6$. Write

$$n = 6 + 2^\alpha 3^\beta 5^\gamma 7^\delta 11^\epsilon m$$

where $\alpha \geq 5$, $\beta \geq 1$, $\gamma \geq 2$, $\delta \geq 1$, $\epsilon \geq 1$ and $m \geq 1$. We assume that $(m, 2310) = 1$. Let $b = 2^\alpha 3^\beta 5^\gamma 7^\delta 11^\epsilon m$. From formula (3), it follows that

$$2F_n = 2F_{b+6} = F_b L_6 + F_6 L_b = 18 F_b + 8 L_b. \tag{19}$$

Notice that

$$b = \begin{cases} 2^\alpha \cdot 5 \cdot 11 \cdot \left(1 + 2 \cdot \left(\dfrac{3^\beta 5^{\gamma-1} 7^\delta 11^{\epsilon-1} - 1}{2} \cdot m + \dfrac{m-1}{2}\right)\right) \text{ or} \\[2mm] 2^\alpha \cdot 5^2 \cdot 11 \cdot \left(1 + 2 \cdot \left(\dfrac{3^\beta 5^{\gamma-2} 7^\delta 11^{\epsilon-1} - 1}{2} \cdot m + \dfrac{m-1}{2}\right)\right) \text{ or} \\[2mm] 2^\alpha \cdot 5 \cdot 7 \cdot 11 \cdot \left(1 + 2 \cdot \left(\dfrac{3^\beta 5^{\gamma-1} 7^{\delta-1} 11^{\epsilon-1} - 1}{2} \cdot m + \dfrac{m-1}{2}\right)\right) \text{ or} \\[2mm] 2^\alpha \cdot 5^2 \cdot 7 \cdot 11 \cdot \left(1 + 2 \cdot \left(\dfrac{3^\beta 5^{\gamma-2} 7^{\delta-1} 11^{\epsilon-1} - 1}{2} \cdot m + \dfrac{m-1}{2}\right)\right). \end{cases} \tag{20}$$

Let

$$\begin{cases} t_1 = 3^\beta 5^{\gamma-1} 7^\delta 11^{\epsilon-1} - 1, \\[1mm] t_2 = 3^\beta 5^{\gamma-2} 7^\delta 11^{\epsilon-1} - 1, \\[1mm] t_3 = 3^\beta 5^{\gamma-1} 7^{\delta-1} 11^{\epsilon-1} - 1, \\[1mm] t_4 = 3^\beta 5^{\gamma-2} 7^{\delta-1} 11^{\epsilon-1} - 1. \end{cases}$$

Clearly, $t_1 \equiv t_2 \pmod 4$, $t_3 \equiv t_4 \pmod 4$ but $t_1 \not\equiv t_3 \pmod 4$. Moreover, all four numbers t_1, t_2, t_3 and t_4 are even. Hence, using formulae (8), (20) and the fact that $(6, m) = 1$, it follows that one can always choose b_1 to be one of the four numbers $2^{\alpha-1} \cdot 5 \cdot 11$, $2^{\alpha-1} \cdot 5^2 \cdot 11$, $2^{\alpha-1} \cdot 5 \cdot 7 \cdot 11$ or $2^{\alpha-1} \cdot 5^2 \cdot 7 \cdot 11$, in such a way that

$$F_b \equiv -F_{2b_1} \pmod{L_{2b_1}}. \tag{21}$$

Moreover, since $20 = 2^2 \cdot 5 \mid b_1$ and $5 \equiv 2 \pmod 3$, it follows easily that b_1 can be chosen in such a manner that $b_1 \equiv 40 \pmod{60}$. From formulae (19), (7) and (21), it follows that

$$4F_n - 7 = 36 F_b + 16 L_b - 7 \equiv -36 F_{2b_1} - 7 = -(36 F_{2b_1} + 7) \pmod{L_{2b_1}}.$$

Notice that $4 \mid b_1$, $3 \nmid b_1$, $3 \nmid L_{b_1}$ and $7 \mid F_8 \mid F_{b_1}$. Hence, by the criterion for $a = 9$ and $b = 7$, it follows that

$$\left(\frac{4F_n - 7}{L_{2b_1}}\right) = \left(\frac{-(36F_{2b_1} + 7)}{L_{2b_1}}\right) = -\left(\frac{-1}{L_{2b_1}}\right) \cdot \left(\frac{7}{L_{2b_1}}\right) \cdot \left(\frac{72F_{b_1} + 7L_{b_1}}{64 \cdot 81 + 5 \cdot 49}\right) = \left(\frac{7}{L_{2b_1}}\right) \cdot \left(\frac{72F_{b_1} + 7L_{b_1}}{5429}\right)$$

$$= -\left(\frac{L_{2b_1}}{7}\right) \cdot \left(\frac{72F_{b_1} + 7L_{b_1}}{89}\right) \cdot \left(\frac{72F_{b_1} + 7L_{b_1}}{61}\right). \tag{22}$$

The sequence $(L_m)_{m \in Z}$ is periodic modulo 7 with period 16. The sequence $(72F_m + 7L_m)_{m \in Z}$ is periodic modulo 89 with period 44 and periodic modulo 61 with period 60. From equation (21), the fact that $2^4 \cdot 11 \mid b_1$ and $b_1 \equiv 40 \pmod{60}$, it follows that

$$\left(\frac{4F_n - 7}{L_{2b_1}}\right) = -\left(\frac{L_0}{7}\right) \cdot \left(\frac{72F_0 + 7L_0}{89}\right) \cdot \left(\frac{72F_{40} + 7L_{40}}{61}\right) = -\left(\frac{2}{7}\right) \cdot \left(\frac{14}{89}\right) \cdot \left(\frac{72F_{40} + 7L_{40}}{61}\right) = -1$$

because

$$\left(\frac{72F_{40} + 7L_{40}}{61}\right) = -1.$$

Hence, $4F_n - 7$ cannot be a perfect square in this case.

4. SOME COMPUTATIONS

We now show that if $4F_n - 7$ is a perfect square, then n must satisfy congruence (1).

Modulo $2^5 \cdot 3 \cdot 5 \cdot 7$.

Let $A = 6720 = 2^6 \cdot 3 \cdot 5 \cdot 7$. The following table lists all prime numbers $p < 30,000$ such that the period d_p of $(4F_m - 7)_{m \in Z}$ modulo p is a divisor of A.

d_p:	8	20	16	10	28	48	14	30	40	32	60	70	168	336	42	448	240	56	84	448	192	70
p:	3	5	7	11	13	23	29	31	41	47	61	71	83	167	211	223	241	281	421	449	769	911

d_p:	96	168	160	80	64	120	160	192	3360	112	1344	240	210	6720
p:	1103	1427	1601	2161	2207	2521	3041	3167	3361	14503	18143	20641	21211	26881

If one wants to study what congruence classes n modulo A can lead to solutions of $F_n = k^2 + k + 2$, it suffices to find all $0 \le h \le A - 1$ such that

$$\left(\frac{4F_h - 7}{p}\right) = 1.$$

for each one of the prime numbers p listed above. One can use Mathematica to check that the only such h are 3, 6, 6717. Hence, $n \equiv \pm 3, 6 \pmod{2^6 \cdot 3 \cdot 5 \cdot 7}$.

Modulo $2^5 \cdot 3 \cdot 5 \cdot 11$.

Let $B = 10560 = 2^6 \cdot 3 \cdot 5 \cdot 11$. The following table lists all prime numbers $p < 30,000$ such that the period d_p of $(4F_m - 7)_{m \in Z}$ modulo p is a divisor of B.

d_p:	8	20	16	10	48	30	40	88	32	60	44	22	240	176	88	110	220	192	176	176
p:	3	5	7	11	23	31	41	43	47	61	89	199	241	263	307	331	661	769	881	967

d_p:	96	1320	704	160	80	64	120	160	192	528	66	132	240	1760
p:	1103	1321	1409	1601	2161	2207	2521	3041	3167	5281	9901	19801	20641	21121

If one wants to study what congruence classes n modulo B can lead to solutions of $F_n = k^2 + k + 2$, it suffices to find all $0 \le h \le B - 1$, such that

$$\left(\frac{4F_h - 7}{p} \right) = 1,$$

for each one of the prime numbers p listed above. One can use Mathematica to check that the only such h are 3, 6, 10557. Hence, $n \equiv \pm 3, 6 \pmod{2^6 \cdot 3 \cdot 5 \cdot 11}$.

Modulo $2^5 \cdot 3 \cdot 5^2$.

Let $C = 4800 = 2^6 \cdot 3 \cdot 5^2$. The following table lists all prime numbers $p < 30,000$ such that the period d_p of $(4F_m - 7)_{m \in Z}$ modulo p is a divisor of C.

d_p:	8	20	16	10	48	30	40	32	60	50	50	240	200	600	192	96	1200	160	80	64
p:	3	5	7	11	23	31	41	47	61	101	151	241	401	601	769	1103	1201	1601	2161	2207

d_p:	120	100	160	192	4800	150	150	240
p:	2521	3001	3041	3167	4801	12301	18451	20641

If one wants to study what congruence classes n modulo C can lead to solutions of $F_n = k^2 + k + 2$, it suffices to find all $0 \le h \le C - 1$, such that

$$\left(\frac{4F_h - 7}{p} \right) = 1,$$

for each one of the prime numbers p listed above. One can use Mathematica to check that the only such h are 3, 6, 2397, 2403, 4797. Hence, $n \equiv \pm 3, 6 \pmod{2^5 \cdot 3 \cdot 5^2}$.

From the previous computations it follows easily that n satisfies congruence (1). The assertion of the theorem follows now from the results of Section 3.

5. ACKNOWLEDGEMENTS

I would like to thank Shafiq Ahmed and Bob Bell for their assistance with Mathematica.

6. REFERENCES

[1]   Cohn, J.H.E. "On square Fibonacci numbers." *J. London Math. Soc.*, Vol. *39* (1964): pp. 537-540.

[2]   Guy, R.K. Unsolved Problems in Number Theory. Springer-Verlag, New York, (1981): p. 106.

[3]   London, H. and Finkelstein, R. "On Fibonacci and Lucas numbers which are perfect powers." *The Fibonacci Quarterly*, Vol. *7* (1969): pp. 476-481, 487. (Errata *The Fibonacci Quarterly*, Vol. *8* (1970): p. 248).

[4]   Ming, L. "On triangular Fibonacci numbers." *The Fibonacci Quarterly*, Vol. *27* (1989): pp. 98-108.

[5]   Ming, L. "On triangular Lucas numbers." Applications of Fibonacci Numbers, Volume 4. Edited by G.E. Bergum, A.F. Horadam and A.N. Philippou. Kluwer Academic Publishers, Dordrecht, The Netherlands, 1991: pp. 98-108.

[6]   Ming, L. "Pentagonal numbers in the Fibonacci sequence." Applications of Fibonacci Numbers, Volume 6. Edited by G.E. Bergum, A.F. Horadam and A.N. Philippou. Kluwer Academic Publishers, Dordrecht, The Netherlands, 1994: pp. 349-354.

[7]   Ming, L. "Pentagonal numbers in the Lucas sequence." *Portugaliae Mat.*, Vol. *53* (1996): pp. 325-329.

[8]   Stark, H.M. Problem 23, Summer Institute on Number Theory, Stony Brook, 1969.

[9]   Williams, H.C. "On Fibonacci numbers of the form $k^2 + 1$." *The Fibonacci Quarterly*, Vol. *13* (1975): pp. 213-214.

AMS Classification Numbers: 11A25, 11B39

ON CERTAIN POLYNOMIALS OF EVEN SUBSCRIPTED LUCAS NUMBERS

R.S. Melham

1. INTRODUCTION

A recent paper of Swamy [4] contains eight finite sums. Three of these are sums discovered by Jennings [3] which express $F_{(2q+1)n}$ as a polynomial in F_n, and F_{mn}/F_n as a polynomial in L_n. Three are sums discovered by Filipponi [2] which express L_{mn} as a polynomial in L_n, and L_{2qn} as a polynomial in F_n. The remaining two are sums discovered by Swamy which express F_{2qn}/L_n as a polynomial in F_n, and $L_{(2q+1)n}/L_n$ as a polynomial in F_n.

To derive his sums, Filipponi made use of Waring's formula

$$X^m + Y^m = \sum_{k=0}^{[m/2]} (-1)^k \frac{m}{m-k} \binom{m-k}{k} (XY)^k (X+Y)^{m-2k}, m \geq 1, \qquad (1.1)$$

(here [ ] denotes the greatest integer function) while Swamy made use of the identity

$$\frac{X^m - Y^m}{X - Y} = \sum_{k=0}^{[(m-1)/2]} (-1)^k \binom{m-k-1}{k} (XY)^k (X+Y)^{m-2k-1}, m \geq 1, \qquad (1.2)$$

which appeared in an earlier paper of Carlitz [1].

In this paper we extend the results of the above mentioned articles by expressing $F_{(2q+1)n}/F_n$, $L_{(2q+1)n}/L_n$, F_{2qn}/F_{2n}, and L_{2qn} as polynomials in L_{2n}. These polynomials are given in the next section in the form of double sums.

2. THE MAIN RESULTS

For notational convenience, in the statement of the theorems which follow, we choose to write, say, $(-1)^n$ instead of $(-1)^{-n}$. In what follows we take α and β to be the roots of $x^2 - x - 1 = 0$.

This paper is in final form and no version of it will be submitted for publication elsewhere.

Theorem 1:

$$\frac{F_{(2q+1)n}}{F_n} = \sum_{r=0}^{q} (-1)^{q+n(q+r)} 2^{-r} L_{2n}^r \sum_{k=r}^{q} (-2)^k \binom{k}{r} \binom{q+k}{2k}, \quad n \geq 1 \; q \geq 0.$$

Proof: Let $X = \alpha^n$, $Y = \beta^n$, and $m = 2q+1$ in (1.2). Since $XY = (\alpha\beta)^n = (-1)^n$ we have

$$\frac{F_{(2q+1)n}}{F_n} = \sum_{k=0}^{q} (-1)^k \binom{2q-k}{k} (-1)^{nk} (\alpha^n + \beta^n)^{2(q-k)},$$

and, with k replaced by $q - k$, the right side becomes

$$\sum_{k=0}^{q} (-1)^{q-k} \binom{q+k}{q-k} (-1)^{n(q-k)} (\alpha^n + \beta^n)^{2k}.$$

Since $\binom{q+k}{q-k} = \binom{q+k}{2k}$ and $(\alpha^n + \beta^n)^2 = L_{2n} + 2(-1)^n$ we rewrite the above as

$$\sum_{k=0}^{q} (-1)^{q+k+nq+nk} \binom{q+k}{2k} \left(L_{2n} + 2(-1)^n \right)^k$$

$$= \sum_{k=0}^{q} (-1)^{q+k+nq+nk} \binom{q+k}{2k} \sum_{r=0}^{k} \binom{k}{r} 2^{k-r} (-1)^{n(k-r)} L_{2n}^r$$

$$= \sum_{k=0}^{q} \sum_{r=0}^{k} (-1)^{q+k+n(q+r)} \binom{q+k}{2k} \binom{k}{r} 2^{k-r} L_{2n}^r.$$

Finally, in order to collect like powers of L_{2n}, we reverse the order of summation and this completes the proof. $\square$

Theorem 2:

$$\frac{L_{(2q+1)n}}{L_n} = \sum_{r=0}^{q} (-1)^{r+n(q+r)} 2^{-r} L_{2n}^r \sum_{k=r}^{q} (-2)^k \binom{k}{r} \binom{q+k}{2k}, \quad n, q \geq 0.$$

This theorem may be established by letting $X = \alpha^n$, $Y = -\beta^n$, and $m = 2q+1$ in (1.2), and following the same steps used to prove Theorem 1. $\square$

Theorem 3:

$$\frac{F_{2qn}}{F_{2n}} = \sum_{r=0}^{q} (-1)^{q+1+n(q+r+1)} 2^{-r} L_{2n}^r \sum_{k=r}^{q-1} (-2)^k \binom{k}{r} \binom{q+k}{2k+1}, \quad n, q \geq 1.$$

Proof: Letting $X = \alpha^n$, $Y = \beta^n$, and $m = 2q$ in (1.2) we obtain

$$\frac{F_{2qn}}{F_n} = \sum_{k=0}^{q-1} (-1)^k \binom{2q-k-1}{k} (-1)^{nk} (\alpha^n + \beta^n)^{2q-2k-1}.$$

If we replace k by $q-1-k$ this becomes

$$\frac{F_{2qn}}{F_n} = \sum_{k=0}^{q-1} (-1)^{(n+1)(q+k+1)} \binom{q+k}{q-k-1} (\alpha^n + \beta^n)^{2k+1}.$$

Dividing both sides by $L_n = \alpha^n + \beta^n$, and noting that $F_{2n} = F_n L_n$ and $\binom{q+k}{q-k-1} = \binom{q+k}{2k+1}$, we have

$$\frac{F_{2qn}}{F_{2n}} = \sum_{k=0}^{q-1} (-1)^{(n+1)(q+k+1)} \binom{q+k}{2k+1} \left(L_{2n} + 2(-1)^n \right)^k.$$

We now proceed as in the proof of Theorem 1. □

Theorem 4:

$$L_{2qn} = 2q \sum_{r=0}^{q} (-1)^{q+n(q+r)} 2^{-r} L_{2n}^r \sum_{k=r}^{q} (-2)^k \frac{1}{q+k} \binom{k}{r} \binom{q+k}{2k}, \quad n \geq 0, q \geq 1.$$

To prove Theorem 4 we let $X = \alpha^n$, $Y = \beta^n$, and $m = 2q$ in (1.1), and follow the same steps used to prove Theorem 1. □

In Theorems 1-4 it is desirable to obtain closed expressions for the inner sums, as this would facilitate any use of these theorems. We give these sums in the next section. However, because our closed expressions are given in a piecemeal manner, depending on the parity of $q - r$, they cannot be incorporated into our statements of Theorems 1-4, and must be given separately.

We mention that we have obtained expressions for $\dfrac{F_{(2q+1)n}}{F_n}$ and $\dfrac{L_{(2q+1)n}}{L_n}$ as polynomials in L_{2n} by using (1.1). However, we have chosen to present the results obtained with the use of (1.2), since, as we soon see, it is simpler to obtain closed expressions for the associated inner sums.

3. THE INNER SUMS

In this section we adhere to the usual conventions for binomial coefficients. This means, for example, that $\sum_{k \geq 0} \binom{k}{r} x^r$ means the same as $\sum_{k \geq r} \binom{k}{r} x^r$, since $\binom{k}{r} = 0$ for $0 \leq k < r$. We require the following.

$$\sum_{k \geq r} \binom{k}{r} x^k = \frac{x^r}{(1-x)^{r+1}}, \quad |x| < 1, \ r \geq 0, \tag{3.1}$$

$$\sum_{k \geq r} \frac{1}{k} \binom{k}{r} x^k = \frac{1}{r} \cdot \frac{x^r}{(1-x)^r}, \quad |x| < 1, \ r \geq 1, \tag{3.2}$$

$$\sum_{n \geq 0} \binom{n+r}{n} x^n = \frac{1}{(1-x)^{r+1}}, \quad |x| < 1, \ r \geq 0, \tag{3.3}$$

$$-\ln(1-x) = x + \frac{x^2}{2} + \frac{x^3}{3} + \cdots, \quad -1 \leq x \leq 0. \tag{3.4}$$

The power series in (3.1) and (3.3) occur as (4.3.1) and (2.5.7) respectively in Wilf [5]. To obtain (3.2), we consider (3.1) for $r \geq 1$, divide by x, and integrate, noting that $\int \frac{x^{r-1} dx}{(1-x)^{r+1}} = \frac{1}{r} \cdot \frac{x^r}{(1-x)^r}$ for $r \geq 1$.

Lemma 1: for $q, r \geq 0$,

$$\sum_{k=r}^{q}(-2)^{k}\binom{k}{r}\binom{q+k}{2k}=\begin{cases}(-1)^{(q+r)/2}2^{r}\binom{(q+r)/2}{r}, & q\equiv r(\text{mod }2),\\[3mm](-1)^{(q+r+1)/2}2^{r}\binom{(q+r-1)/2}{r}, & q\equiv r+1(\text{mod }2).\end{cases}$$

Proof: We use the "Snake Oil" method as described in Section 4.3 in Wilf [5]. Considering the required sum as a function of q, we let $F(x)$ be the generating function defined by

$$F(x)=\sum_{q\geq0}x^{q}\sum_{k\geq r}(-2)^{k}\binom{k}{r}\binom{q+k}{2k}.$$

Changing the order of summation, we have

$$F(x)=\sum_{k\geq r}(-2)^{k}\binom{k}{r}\sum_{q\geq0}\binom{q+k}{2k}x^{q}$$

$$=\sum_{k\geq r}(-2)^{k}\binom{k}{r}x^{-k}\sum_{q\geq0}\binom{q+k}{2k}x^{q+k}$$

$$=\sum_{k\geq r}(-2)^{k}\binom{k}{r}x^{-k}\sum_{s\geq k}\binom{s}{2k}x^{s}$$

$$=\sum_{k\geq r}(-2)^{k}\binom{k}{r}x^{-k}\cdot\frac{x^{2k}}{(1-x)^{2k+1}}\qquad\text{(by (3.1))}$$

$$=\frac{1}{(1-x)}\sum_{k\geq r}\binom{k}{r}\left(\frac{-2x}{(1-x)^{2}}\right)^{k}$$

$$=\frac{\left(\dfrac{-2x}{(1-x)^{2}}\right)^{r}}{(1-x)\left(1+\dfrac{2x}{(1-x)^{2}}\right)^{r+1}}\qquad\text{(by (3.1))}$$

which simplifies to $\dfrac{(-2)^{r}(x^{r}-x^{r+1})}{(1+x^{2})^{r+1}}$. But by (3.3) this can be written as

$$(-2)^{r}(x^{r}-x^{r+1})\sum_{n\geq0}\binom{n+r}{n}(-x^{2})^{n}=\sum_{n\geq0}(-1)^{n+r}2^{r}\binom{n+r}{n}(x^{2n+r}-x^{2n+r+1}).$$

Since the expression on the right is the expansion of $F(x)$ in powers of x, the required sum is the coefficient of x^{q}. If $q\equiv r$ (mod 2) then we can write $q=2\left(\frac{q-r}{2}\right)+r$, and so the

coefficient of x^{q} is $(-1)^{(q-r)/2+r}2^{r}\binom{(q-r)/2+r}{(q-r)/2}$, or $(-1)^{(q+r)/2}2^{r}\binom{(q+r)/2}{r}$. On the

other hand, if $q\equiv r+1$ (mod 2) we can write $q=2\left(\frac{q-r-1}{2}\right)+r+1$, and so the coefficient of

x^q is $(-1)^{(q-r-1)/2+r+1}2^r\binom{(q-r-1)/2+r}{(q-r-1)/2}$, that is $(-1)^{(q+r+1)/2}2^r\binom{(q+r-1)/2}{r}$.

This establishes Lemma 1.   ☐

Lemma 2: For $q \geq 1$ and $r \geq 0$,

$$\sum_{k=r}^{q-1}(-2)^k\binom{k}{r}\binom{q+k}{2k+1}=\begin{cases} 0, & q \equiv r(\text{mod } 2), \\ (-1)^{(q+r-1)/2}2^r\binom{(q+r-1)/2}{r}, & q \equiv r+1(\text{mod } 2). \end{cases}$$

Proof: Proceeding as in the proof of Lemma 1 we write

$$F(x) = \sum_{q \geq 1} x^q \sum_{k \geq r}(-2)^k\binom{k}{r}\binom{q+k}{2k+1}$$

$$= \sum_{k \geq r}(-2)^k\binom{k}{r}\sum_{q \geq 1}\binom{q+k}{2k+1}x^q$$

$$= \sum_{k \geq r}(-2)^k\binom{k}{r}x^{-k}\sum_{q \geq 1}\binom{q+k}{2k+1}x^{q+k}$$

$$= \sum_{k \geq r}(-2)^k\binom{k}{r}x^{-k}\sum_{s \geq k+1}\binom{s}{2k+1}x^{s}.$$

Since the remainder of the proof is similar to that of Lemma 1, we omit the details.   ☐

For our final lemma, which gives the inner sum in Theorem 4, we need to consider separately the cases $r = 0$ (part (a)) and $r \geq 1$ (part (b)).

Lemma 3: Let $q \geq 1$ and $r \geq 1$. Then

$$\sum_{k=0}^{q}(-2)^k\frac{1}{q+k}\binom{q+k}{2k}=\begin{cases} 0, & q \text{ odd}, \\ \dfrac{(-1)^{q/2}}{q} & q \text{ even}, \end{cases} \qquad (a)$$

$$\sum_{k=r}^{q}(-2)^k\frac{1}{q+k}\binom{k}{r}\binom{q+k}{2k}=\begin{cases} 0, & q = r+1 \;(\text{mod } 2), \\ \dfrac{(-1)^{(q+r)/2}2^{r-1}}{r}\binom{\frac{q+r-2}{2}}{r-1}, & q \equiv r \;(\text{mod } 2). \end{cases} \qquad (b)$$

Proof: To prove part (a) set

$$F(x) = \sum_{q \geq 1} x^q \sum_{k \geq 0} (-2)^k \frac{1}{q+k} \binom{q+k}{2k}$$

$$= \sum_{k \geq 0} (-2)^k \sum_{q \geq 1} \frac{1}{q+k} \binom{q+k}{2k} x^q$$

$$= \sum_{k \geq 0} (-2)^k x^{-k} \sum_{q \geq 1} \frac{1}{q+k} \binom{q+k}{2k} x^{q+k}$$

$$= \sum_{k \geq 0} (-2)^k x^{-k} \sum_{s \geq k+1} \frac{1}{s} \binom{s}{2k} x^s.$$

Because, for $k = 0$, $s \geq 2k$ does not imply that $s \geq k+1$, we must consider the cases $k = 0$ and $k \geq 1$ separately. The last sum then splits as

$$\sum_{s \geq 1} \frac{1}{s} \binom{s}{0} x^s + \sum_{k \geq 1} (-2)^k x^{-k} \sum_{s \geq k+1} \frac{1}{s} \binom{s}{2k} x^s$$

$$= \sum_{s \geq 1} \frac{1}{s} x^s + \sum_{k \geq 1} (-2)^k x^{-k} \sum_{s \geq 2k} \frac{1}{s} \binom{s}{2k} x^s$$

$$= -\ln(1-x) + \sum_{k \geq 1} (-2)^k x^{-k} \cdot \frac{1}{2k} \cdot \frac{x^{2k}}{(1-x)^{2k}}$$

$$= -\ln(1-x) + \frac{1}{2} \sum_{k \geq 1} \frac{1}{k} \left(\frac{-2x}{(1-x)^2} \right)^k,$$

where we have used (3.2) and (3.4). Again, using (3.4), the last line becomes

$$-\ln(1-x) - \frac{1}{2} \ln\left(1 + \frac{2x}{(1-x)^2}\right) = -\ln(1-x) - \frac{1}{2} \ln\left(\frac{1+x^2}{(1-x)^2}\right)$$

$$= -\frac{1}{2} \ln(1+x^2)$$

$$= \sum_{n \geq 1} (-1)^n \frac{x^{2n}}{2n}.$$

Hence, if q is odd, the coefficient of x^q is zero, while if q is even, the coefficient of x^q is $\dfrac{(-1)^{q/2}}{q}$, and this proves part (a).

To prove part (b) we consider

$$G(x) = \sum_{q \geq 1} x^q \sum_{k \geq r} (-2)^k \frac{1}{q+k} \binom{k}{r} \binom{q+k}{2k}$$

$$= \sum_{k \geq r} (-2)^k \binom{k}{r} x^{-k} \sum_{q \geq 1} \frac{1}{q+k} \binom{q+k}{2k} x^{q+k}$$

$$= \sum_{k \geq r} (-2)^k \binom{k}{r} x^{-k} \sum_{s \geq k+1} \frac{1}{s} \binom{s}{2k} x^s$$

$$= \sum_{k \geq r} (-2)^k \binom{k}{r} x^{-k} \sum_{s \geq 2k} \frac{1}{s} \binom{s}{2k} x^s$$

$$= \sum_{k \geq r} (-2)^k \binom{k}{r} x^{-k} \cdot \frac{1}{2k} \cdot \frac{x^{2k}}{(1-x)^{2k}} \quad \text{(by (3.2))}$$

$$= \frac{1}{2} \sum_{k \geq r} \frac{1}{k} \binom{k}{r} \left(\frac{-2x}{(1-x)^2} \right)^k$$

$$= \frac{(-1)^r 2^{r-1} x^r}{r} \cdot \frac{1}{(1+x^2)^r},$$

which we obtain by using (3.2) and simplifying. But by (3.3) the last expression is

$$\frac{(-1)^r 2^{r-1} x^r}{r} \sum_{n \geq 0} \binom{n+r-1}{n} (-x^2)^n = \frac{2^{r-1}}{r} \sum_{n \geq 0} \binom{n+r-1}{n} (-1)^{n+r} x^{2n+r}.$$

From this we see that if $q \equiv r+1 \pmod 2$, the coefficient of x^q is zero, and if $q \equiv r \pmod 2$

then we can write $q = 2\left(\frac{q-r}{2}\right) + r$. In this case the coefficient of x^q is

$$\frac{(-1)^{(q-r)/2 + r} 2^{r-1}}{r} \binom{(q-r)/2 + r - 1}{(q-r)/2}, \quad \text{or} \quad \frac{(-1)^{(q+r)/2} 2^{r-1}}{r} \binom{\frac{q+r-2}{2}}{r-1}, \quad \text{and this}$$

establishes Lemma 3.   □

To conclude, we give a simple example. In Theorem 1 let $q = 5$. Then for $r = 0$, 1, 2, 3, 4, and 5, the inner sums evaluated with the use of Lemma 1 are -1, -6, 12, 32, -16 and -32 respectively. Considering the two parities of n, we obtain

$$\frac{F_{11n}}{F_n} = \begin{cases} 1 + 3L_{2n} - 3L_{2n}^2 - 4L_{2n}^3 + L_{2n}^4 + L_{2n}^5, & n \text{ even}, \\[2ex] -1 + 3L_{2n} + 3L_{2n}^2 - 4L_{2n}^3 - L_{2n}^4 + L_{2n}^5, & n \text{ odd}. \end{cases}$$

REFERENCES

[1]   Carlitz, L. "A Fibonacci Array." *The Fibonacci Quarterly*, Vol. *1.2* (1963): pp. 17-27.

[2]   Filipponi, P. "Some Binomial Fibonacci Identities." *The Fibonacci Quarterly*, Vol. *33.3*
       (1995): pp. 251-257.

[3]   Jennings, D. "Some Polynomial Identities for the Fibonacci and Lucas Numbers." *The
       Fibonacci Quarterly*, Vol. *31.2* (1993): pp. 134-137.

[4]   Swamy, M.N.S. "On Certain Identities Involving Fibonacci and Lucas Numbers." *The
       Fibonacci Quarterly*, Vol. *35.3* (1997): pp. 230-232.

[5]   Wilf, Herbert S. Generatingfunctionology. Boston-San Diego-New York: Academic
       Press, 1990.

AMS Classification Numbers: 11B39, 05A19

ON THE RANK OF APPEARANCE OF LUCAS SEQUENCES

Siguna Müller

0. INTRODUCTION

Motivated by the study of the order of a group element and its importance for (cryptographic) applications we suggest a generalization of the order-concept in terms of the rank of appearance for Lucas sequences. After summarizing some important properties of the rank of appearance, we give a short description and portray some relationships to similar concepts involving the periodicity of linear recurrences. The goal of this paper is to establish the number of parameters with the same rank of appearance modulo a prime p. To this end, special cases in varying and counting the parameters will be considered.

1. MOTIVATION AND BACKGROUND

Nearly all procedures and mechanisms of modern cryptography require the provision of large random prime numbers. Therefore algorithms and procedures for the identification of large primes are of essential interest. Basically all of the algorithms used originate in some sense from Fermat's Little Theorem

$$a^n \equiv a \pmod{n},$$

which is known to be true for all $a \in \mathbb{Z}$, when n is a prime. Thus, if this congruence is violated for some a, then n is certainly composite. The problem with this test is the existence of

*Research supported by the Österreichischen Fonds zur Förderung der wissenschaftlichen Forschung, FWF-Project no. P 13088-MAT.

This paper is in final form and no version of it will be submitted for publication elsewhere.

pseudoprimes n which, although composite, pretend to be prime by satisfying the above condition. It follows immediately, that any such composite number n fulfills

$$ord_p(a) \mid (n-1) \tag{1.1}$$

for all prime divisors p of n such that $\gcd(a,p) = 1$. Thus, any parameter a with small multiplicative order modulo p will yield a poor, unreliable 'evidence' for the primality of n. This test can considerably be strengthened in the following way.

The Strong Probable Prime Test: Given an odd integer n, let $n-1 = 2^s d$ with $2 \nmid d, s > 0$. Choose a base $a \in \mathbb{Z}$ with $1 < a < n-1$. If

$$a^d \equiv 1 \pmod{n} \text{ or } a^{2^r d} \equiv -1 \pmod{n} \tag{1.2}$$

for some $0 \leq r < s$, then n is declared prime to base a. A prime will pass the test for all a. Pseudoprimes to this test can be characterized in the following way

Lemma 1.1: *Let n be a positive, odd composite integer. Then n is a strong pseudoprime to base a if and only if $a^{n-1} \equiv 1 \pmod{n}$ and there exists an integer c such that, for every prime p dividing n, $\nu_2(ord_p(a)) = c$ where $\nu_2(m)$ for $m \in \mathbb{N}$ denotes the exponent k of the highest power of 2 which divides m, i.e. $2^k \| m$.*

Contrary to the Fermat test which allows the existence of so-called Carmichael numbers which pass the test of all a, similar numbers to the Strong Probable Prime Test do not exist.

Thus, the reliability of the test seems to depend strongly on values of the (powers of 2 of the) orders of elements modulo p. Actually, the non-existence of Strong Carmichael numbers is based on the following theorem. Indeed, it turns out that most of the number theoretic and algebraic applications to cryptography utilize this very fundamental fact.

Theorem 1.2: *Let p be a prime. If $t \nmid (p-1)$, no integer has order $t \pmod{p}$. If $t \mid (p-1)$, the number of $a \in \mathbb{Z}_p^*$ with $ord_p(a) = t$ is $\phi(t)$.*

The importance of this theorem is twofold. Firstly, it allows the existence of primitive roots, i.e. elements with order equal to $p-1$, and secondly it gives a number of parameters with the same order.

In may cryptographic settings, elements with small orders make the breaking of the system more easily tractable. This makes it therefore essential to either eliminate them by certain choices of the primes and/or give a number or an upper bound for elements with small orders.

As can be seen above, the Fermat test is completely infeasible as a primality testing criterion. Thus, generalizations of probabilistic tests have been proposed which replace the power polynomials with other functions. These suggestions include the use of Lucas sequences.

Let P and Q be integers and $D = P^2 - 4Q$. The Fundamental, respectively Primordial Lucas sequence, associated with the parameters P, Q, is defined by means of the second-order recursion relation $u_{m+1} = Pu_m - Qu_{m-1}$, respectively $v_{m+1} = Pv_m - Qv_{m-1}$, with initial terms $u_0(P,Q) = 0$, $u_1(P,Q) = 1$, $v_0(P,Q) = 2$, and $v_1(P,Q) = P$. To simplify notations, we sometimes write $u_m = u_m(P,Q)$, $v_m = v_m(P,Q)$.

The counterpart to the Fermat test involving the Lucas sequences is based on the following properties of prime numbers.

For an odd prime n it can be shown:

$$v_n(P,Q) \equiv P \pmod{n}, \tag{1.3}$$

$$\text{if } \gcd(n, D) = 1 \text{ then } u_{n - \left(\frac{D}{n}\right)}(P,Q) \equiv 0 \pmod{n}, \tag{1.4}$$

where $\left(\frac{D}{n}\right)$ denotes the Jacobi-respectively Lengendre symbol. If an odd composite number n satisfies property (1.3), respectively (1.4), then it is called a Dickson pseudoprime, respectively Lucas pseudoprime, for P and Q (Lpsp(P,Q)).

Conditions are already known (cf. [5]) for pseudoprimes for the v-test with respect to all parameters P and Q and examples of such numbers have been found (cf. [10, 4]).

It turns out that pseudoprimes to the u-test can be characterized in a very similar way to the Fermat test, when the rank of appearance is employed as the counterpart of the order of a group element.

Definition 1.3: Let $u = u(P,Q)$ be the Lucas sequence of the first kind and let m be any integer. The rank of appearance $\rho(m, P, Q)$ modulo m (or simply $\rho(m)$) is the smallest integer l, if it exists, such that $u_l(P,Q) \equiv 0 \pmod{m}$.

It can be shown that $\rho(m, P, Q)$ exists if $p \nmid Q$ for all prime divisors p of m (cf. [19]). If p is any prime with $\gcd(p, QD) = 1$, where $D = P^2 - 4Q$, then it is known that the rank of appearance is a divisor d of $p - \left(\frac{D}{p}\right)$.

In this vein, a Lucas pseudoprime is a composite number for which

$$\rho(p, P, Q) \mid \left(n - \left(\frac{D}{n}\right) \right) \tag{1.5}$$

for all primes p dividing n.

Similarly to the Strong Probable Prime Test, a stronger version of the Lucas test can be obtained by means of the *Strong Lucas Probable Prime Test*

$$u_d(P,Q) \equiv 0 \pmod{n}, \tag{1.6}$$

or

$$v_{2^r d}(P,Q) \equiv 0 \pmod{n} \text{ for some } r,\ 0 \le r < s, \tag{1.7}$$

where $n - \left(\frac{D}{n}\right) = 2^s d, 2 \nmid d$.

It has been shown in [8] that pseudoprimes to the Strong Lucas Probable Prime Test (in short sLpsp) can be characterized in the following way.

Lemma 1.4: *Let n be a positive, odd composite integer and $Q \in \mathbf{Z}_n^*$. Then n is a $sLpsp(P,Q)$ if and only if it is a $Lpsp(P,Q)$ and $\nu_2(\rho(p,P,Q))$ is a constant value c for all prime factors p of n.*

An analysis of the Lucas Test has shown (cf. [9]) that, if n is composite, congruence (1.4) cannot be fulfilled for all P and Q with $\gcd(n, DQ) = 1$. Thus, by replacing the Fermat-by the Lucas u-test we can avoid the existence of Carmichael numbers. This means that there are no Lucas pseudoprimes with respect to all P and Q. Even more, the author has also shown a stronger result, namely that there are no composite integers n which are Lucas pseudoprimes for a fixed value of $Q \in \mathbf{Z}_n^*$ and all varied values of P (or vice versa) with discriminants $D = P^2 - 4Q$ coprime to n.

This seems to be another indicator that probable prime tests based on Lucas sequences are more reliable than the ones based on power polynomials.

In view of equations (1.1) and (1.5) it is thus of interest to contrast the values of the order of a group element with those of the rank of appearance. Specifically, we are going to establish a formula for the number of parameters with the same rank of appearance.

The question about the number of parameters with the same rank of appearance was first investigated in [21]. It is shown that, if the discriminant D is kept fixed and $d > 1$ divides $\left(p - \left(\frac{D}{p}\right)\right)$ for a fixed odd prime p, then there are exactly $\phi(d)$ distinct values of P modulo p such that there exists a parameter $Q \in \mathbf{Z}_p^*$ with $P^2 - 4Q \equiv D \pmod{p}$ and $\rho(p,P,Q) = d$.

By contrasting the rank of appearance with the order of elements of the group $\mathbf{Z}_p^*$, Williams' result can be seen as the counterpart of Theorem 1.2 involving Lucas sequences.

In Williams' case, the discriminant D is always fixed, which implies $\left(\frac{D}{P}\right) = \varepsilon \in \{-1, 1\}$. Consequently, those parameters are considered, for which the roots of the characteristic polynomial $f(x) = x^2 - Px + Q$ are both in F_p, respectively both in F_{p^2}. Obviously there is no particular reason, why the *discriminant* D needs to be fixed. Alternatively, we are going to consider the cases, where, instead of D, either one of the parameters $P = P_0$ or $Q = Q_0$ is kept fixed. It has been shown in [9] that there exist parameters with the maximal rank of appearance modulo p, $d_0 = p - \left(\frac{D}{P}\right)$ respectively $d_0 = \frac{1}{2}\left(p - \left(\frac{D}{P}\right)\right)$. In particular, for variations of parameters with a fixed Legendre symbol $\left(\frac{P^2 - 4Q}{p}\right) \in \{-1, 1\}$ the number of P for a fixed Q (or vice versa) with this maximal rank d_0 has been established. The goal of this paper is to generalize these results for arbitrary divisors d of $p - \varepsilon$ and to establish the number of parameters with the same rank of appearance for a fixed value of P respectively Q. Most interestingly, although in a completely different setting, our results will turn out to be quite similar to both Williams' and the ones stated in Theorem 1.2.

2. FUNDAMENTAL PROPERTIES AND RELATED CONCEPTS

For our purpose it suffices to consider positive integers m, m_i, respectively odd primes p, with $\gcd(mm_i, QD) = 1$, respectively $\gcd(p, QD) = 1$, where $D = P^2 - 4Q$. Then the rank of appearance is known to have the following properties (cf. [2], [11], [19])

$$m \mid u_n(P, Q) \text{ if and only if } \rho(m, P, Q) \mid n, \tag{2.1}$$

$$\rho(p, P, Q) \mid \left(p - \left(\frac{D}{P}\right)\right), \tag{2.2}$$

$$\rho(p, P', Q') = \rho(p, P, Q) \text{ when } P' \equiv P, \ Q' \equiv Q \pmod{p}, \tag{2.3}$$

$$\rho(p, P, Q) \mid \frac{p - (D/p)}{2} \text{ if and only if } \left(\frac{Q}{p}\right) = 1, \tag{2.4}$$

$$\rho(\text{lcm}(m_1, \cdots, m_k)) = \text{lcm}(\rho(m_1), \cdots, \rho(m_k)), \tag{2.5}$$

$$\text{If } p^c \parallel u_{\rho(p, P, Q)}(P, Q) \text{ then } \rho(p^e, P, Q) = p^{\max(e, c) - e} \rho(p, P, Q). \tag{2.6}$$

In particular, the rank of appearance is closely connected to the period of linear recurrence sequences. Relationships are known for the general case where Q is an arbitrary nonzero integer (cf. [13, 22]).

In general, let $(w) = w(a_1, \cdots, a_k)$ be a k-th order linear recurrence over any finite field F_q satisfying the recursion relation $w_{n+k} = a_1 w_{n+k-1} - a_2 w_{n+k-2} + \cdots + (-1)^{k+1} a_k w_n$ with certain initial terms $w_0, \cdots, w_k$. With the calculations carried out in any finite field F_q, the *period* μ of (w) is defined to be the least positive integer t such that $w_{n+t} = w_n$ for all $n \geq 0$.

The *restricted period* α of (w) is the least positive integer t such that $w_{n+t} = sw_n$ for all non-negative integers n and some non-zero element s, called the multiplier of (w).

We will let the *unit sequence* $w(a_1, \cdots, a_k)$ denote the particular recurrence satisfying the above recursion relation with initial terms $w_0 = w_1 = \cdots = w_{k-2} = 0$, $w_{k-1} = 1$. We shall assume throughout this paper that the parameter a_k is $\neq 0$. In particular, it is known then that $w(a_1, \cdots, a_k)$ is purely periodic.

Specifically, when $k = 2$ and $q = p$, a prime then $\alpha = \rho(p)$, as the following well-known lemma shows (cf. Lemma 1 on page 317 of [13]).

Lemma 2.1:   Let $u_{\rho+1}(P, Q) \equiv s \pmod{p}$, where $\rho = \rho(p, P, Q)$ is the rank of appearance and p is a odd prime. Then s is the multiplier of (u) and ρ equals the restricted period α of $u(P, Q)$ modulo p.

If the multiplicative order of the multiplier s is denoted by β, then it is a well-known fact (cf. [16]) that

$$\beta = \frac{\mu}{\alpha}, \tag{2.7}$$

so that there is obviously also a close connection between the period and the rank of appearance of the u-sequence modulo a prime p.

If $f(x) = x^k - a_1 x^{k-1} + a_2 x^{k-2} + \cdots + (-1)^k a_k$ is the characteristic polynomial of $w(a_1, \cdots, a_k)$ with distinct characteristic roots $r_1, \cdots, r_k$, then it is known (cf. [16]) that the restricted period α is the smallest non-negative integer n such that, in F_q,

$$r_1^n = r_2^n = \cdots = r_k^n = s,$$

where $s \in F_q$ is the multiplier of (w).

For $k = 2$ and $q = p$ there is an interesting connection between the rank of appearance and the order of the roots r_i, $i = 1, 2$. Namely, if f is irreducible, then $\mu = \beta\alpha = \beta\rho(p)$ is the multiplicative order of the roots r_1, r_2 in F_{p^2} (cf. Lemma 1 of [16]).

It is a well-known fact (cf. [7]) that if d is a divisor of $q^k - 1$ with $d \nmid (q^n - 1)$, for $1 \leq n \leq k - 1$ then there are exactly $\frac{\phi(d)}{k}$ monic irreducible polynomials of degree k over F_q, such that each root of the polynomial has multiplicative order d. This result was utilized by L. Somer (cf. [16]) in counting the sequences with the same period μ, when the characteristic polynomial f is assumed to be irreducible over F_q. With the same conditions on d, he has shown that there are exactly $\frac{\phi(d)}{k}$ k-th order unit recurrence sequences with $\mu = d$ (this follows from the proof of the Corollary to Theorem 2 on page 201 of [16] and Theorem 4.3.1 of [15]).

In this paper we are going to count the number of second-order recurrence sequences with the same rank of appearance ρ, irrespective of whether the characteristic polynomial is irreducible or not. We restrict ourselves to considering the problem over F_p, where p is an odd prime.

As the rank of appearance is a function in two variables, P and Q, the following cases will be distinguished.

- Keeping $Q = Q_0$ fixed and varying P.

- Keeping $P = P_0$ fixed and varying Q.

Please note that the case where the discriminant $D = P^2 - 4Q$ is assumed to be fixed, the number of parameters P for which there exists a Q with $P^2 - 4Q \equiv D \pmod{p}$ and $\rho(p, P, Q) = d \mid \left(p - \left(\frac{D}{p} \right) \right)$ for some $Q \in Z_p^*$ has already been investigated in [21].

3. KEEPING $Q = Q_0$ FIXED AND VARYING P.

In what follows, let $Q \in \mathbf{Z}^*$ be fixed and let $P \in \mathbf{Z}$ be arbitrary with $D = P^2 - 4Q$ coprime to a fixed prime p. Whenever we want to stress that Q is a special fixed value, we will write Q_0 instead of Q.

3.1 The rank of appearance for a fixed parameter Q.

3.1.1: $\boxed{\text{Let } \left(\frac{D}{p} \right) = \left(\frac{P^2 - 4Q_0}{p} \right) = -1.}$

We begin with stating some results of k-th order linear recurrences. The general results have been established in [16] over any field F_q, $q = p^t$ for p prime. We will restrict ourselves to the case that $q = p$. In the following, let $q' = (q^k - 1)/(q - 1)$ and $e = \gcd\left(\frac{q-1}{\mathrm{ord}_q(a_k)}, q' \right)$.

Proposition 3.1: (*Theorem 3 of* [16])

Let $w(a_1, \cdots, a_k)$ be a unit sequence such that its characteristic polynomial f is irreducible. Then $\alpha \geq k$ and $\alpha \mid \frac{q^k - 1}{q - 1}$, but $\alpha \nmid \frac{q^n - 1}{q - 1}$ for $1 \leq n \leq k-1$. Further, α is of the form q'/h, where $h \mid q', e \mid h$, and $\gcd\left(\frac{h}{e}, \mathrm{ord}_q(a_k) \right) = 1$.

Proposition 3.2: (*Theorem 4 of* [16])

Let a_k be a fixed non-zero element of F_q. Let $d \mid \frac{q^k - 1}{q - 1}$ such that $d \nmid \frac{q^n - 1}{q - 1}$ for $1 \leq n \leq k-1$, $d \geq k$ and d is of the form q'/h, where $h \mid q'$, $e \mid h$, and $\gcd\left(\frac{h}{e}, \mathrm{ord}_q(a_k) \right) = 1$. Then there exists a k-th order unit sequence $w(a_1, \cdots, a_k)$ such that f is irreducible and $\alpha = d$.

Proposition 3.3: (*Theorem 6 of* [16])

Let $w(a_1, \cdots, a_k)$ *be a unit sequence such that its characteristic polynomial f is irreducible. Then $e_1 \cdot ord_q(a_k) \mid \beta$ and $\beta \mid e \cdot ord_q(a_k)$, where $e = e_1 e_2$ and e_2 is the largest factor of e relatively prime to $ord_q(a_k)$.*

In specializing those results for our case of second order recurrences $u(P, Q)$ we obtain:

Corollary 3.4: *Let $u(P, Q)$ with $u_0 = 0$ and $u_1 = 1$ be a second order recursion sequence with irreducible characteristic polynomial f over F_q. Then $e = 2$ iff $\nu_2(ord_q(Q)) < \nu_2(q-1)$, where $\nu_2(a)$ denotes the highest power of 2 in the prime decomposition of a, and $e = 1$ otherwise. Further,*

1. $\alpha = \dfrac{q+1}{h}$, *where* $gcd\left(\dfrac{h}{\bar{e}}, ord_q(Q)\right) = 1$.

2. *If Q is a fixed nonzero element of F_q, d is of the form $\dfrac{q+1}{h}$ with $h \neq q+1$ and $gcd\left(\dfrac{h}{\bar{e}}, ord_q(Q)\right) = 1$ then there exists a unit sequence with $\alpha = d$.*

3.

$$
\begin{cases}
ord_q(Q) \mid \beta \mid 2 ord_q(Q), & \text{if } 2 \nmid ord_q(Q), \\
\beta = 2 ord_q(Q), & \text{if } (\nu_2(ord_q(Q)) < \nu_2(q-1), \text{ and } 2 \mid ord_q(Q)), \\
\beta = ord_q(Q), & \text{otherwise, i.e. when } \nu_2(ord_q(Q)) = \nu_2(q-1).
\end{cases}
$$

Proof: The assertions follow from Proposition 3.1, 3.2, 3.3 by specializing $k = 2$.   □

3.1.2: $\boxed{\text{Let } \left(\dfrac{D}{p}\right) = \left(\dfrac{P^2 - 4Q_0}{p}\right) = 1.}$

Lemma 3.5: *Suppose that the characteristic polynomial $x^2 - Px + Q$ splits over F_p with roots r_1 and r_2. Then*

- *For each root r_1 there exists a uniquely determined value G (mod p) such that for the conjugate root r_2, $r_2 \equiv Gr_1$ (mod p).*

- *For $G \equiv \dfrac{r_2}{r_1}$ (mod p) one has $\left(\dfrac{Q}{p}\right) = \left(\dfrac{G}{p}\right)$.*

- *If Q is fixed and P is being varied, then there are $\dfrac{p-1}{2}$ respectively $\dfrac{p-3}{2}$ such incongruent elements G modulo p, according as $\left(\dfrac{Q}{p}\right) = -1$ or 1.*

Proof: For each root $r_1 = \dfrac{P + \sqrt{D}}{2}$ (mod p) there exists a uniquely determined conjugate root $r_2 \equiv \dfrac{P - \sqrt{D}}{2}$ (mod p), where $D = D(P) = P^2 - 4Q_0$. By assumption, both r_1 and r_2 are elements of F_p with $r_1 r_2 = Q$ coprime to p so that the equation $r_2 \equiv Gr_1$ has a unique solution G modulo p.

Moreover, it follows from [18], p. 336 that $Q^{\frac{p-1}{2}} \equiv G^{\frac{p-1}{2}}$ (mod p).

Now assume that $Q = Q_0$ is kept fixed and P is being varied. According to Proposition 3.1 of [9] there are exactly $\frac{p-1}{2} - \frac{1}{2}\left(1 + \left(\frac{Q_0}{p}\right)\right)$ parameters $P \in \mathbf{Z}_p$ for which $\left(\frac{P^2 - 4Q_0}{p}\right) = \left(\frac{D}{p}\right) = 1$. Each of those P determines the roots $r_1 \equiv r_1(P)$ (mod p) and $r_2 \equiv r_2(P)$ (mod p).

We will show that the number of G's (mod p) equals the number of P's obtained (mod p) for a fixed $Q = Q_0$.

First, it follows that if $P \equiv 0$ (mod p), then exactly one G is obtained (mod p), namely $G \equiv -1$ (mod p), since $r_2(P) \equiv -r_1(P)$ (mod p) and $\frac{r_2(P)}{r_1(P)} \equiv \frac{r_1(P)}{r_2(P)} \equiv -1$ (mod p).

Now, if $P \not\equiv 0$ (mod p), then one obtains two distinct G's (mod p) for this value of P since the roots $r_1(P)$ and $r_2(P)$ can be listed in either order. One now notes that if $P \not\equiv 0$ (mod p), then $\frac{r_2(P)}{r_1(P)} \equiv G_1$ (mod p), while $\frac{r_1(P)}{r_2(P)} \equiv G_2$ (mod p), and $G_1 \not\equiv G_2$ (mod p). This follows since if $\frac{r_1(P)}{r_2(P)} \equiv \frac{r_2(P)}{r_1(P)}$ (mod p), then either $r_1(P) \equiv r_2(P)$ (mod p), which implies that $D \equiv 0$ (mod p), which is impossible, or $r_1(P) \equiv -r_2(P)$ (mod p), which implies that $P \equiv 0$ (mod p) since $P = r_1(P) + r_2(P)$, and this is also impossible. If $P' \equiv -P \not\equiv 0$ (mod p), then one also obtains the same two G's (mod p) for P' as for P, since $r_1(P') \equiv -r_2(P)$ (mod p) and $r_2(P') \equiv -r_1(P)$ (mod p).

It remains to show that if P' is different from P (mod p) and $P' \not\equiv -P$ (mod p), then the G's obtained for P' (mod p) are distinct from the G's obtained for P (mod p). Suppose there exists a P' (mod p) with $P' \not\equiv \pm P$ (mod p) such that the G's obtained form P and P' coincide (mod p). This means that one of the following four congruences holds: $G_1 \equiv \frac{r_2(P)}{r_1(P)} \equiv G_1' \equiv \frac{r_2(P')}{r_1(P')}$ (mod p), $G_2 \equiv \frac{r_1(P)}{r_2(P)} \equiv G_2' \equiv \frac{r_1(P')}{r_2(P')}$ (mod p), $G_1 \equiv G_2'$ (mod p), or $G_1' \equiv G_2$ (mod p). These congruences either imply $r_2(P)r_1(P') \equiv r_2(P')r_1(P)$ (mod p), or $r_1(P)r_1(P') \equiv r_2(P)r_2(P')$ (mod p).

If the former condition holds, then $r_2(P)r_1(P)\frac{r_1(P')}{r_1(P)} \equiv r_2(P')r_1(P')\frac{r_1(P)}{r_1(P')}$ (mod p), and, since $r_1(P)r_2(P) \equiv r_1(P')r_2(P') \equiv Q$ (mod p) by hypothesis, this gives $r_1(P) \equiv \pm r_1(P')$ (mod p). Similarly, $r_2(P) \equiv \pm r_2(P')$ (mod p). Since r_2 is the conjugate root of r_1, we only need to consider the cases $r_1(P) \equiv r_1(P')$, $r_2(P) \equiv r_2(P')$ (mod p), respectively $r_1(P) \equiv -r_1(P')$, $r_2(P) \equiv -r_2(P')$ (mod p). The former case yields $P \equiv P'$ (mod p) which has been excluded, whereas by the latter $P \equiv -P'$ (mod p), which also cannot be.

Similarly, the case that $r_1(P)r_1(P') \equiv r_2(P)r_2(P') \pmod{p}$ implies $P \equiv \pm P' \pmod{p}$. This follows since the congruence either implies $r_1(P) \equiv r_2(P') \pmod{p}$ or $r_1(P) \equiv -r_2(P') \pmod{p}$. By taking conjugates, the assertion now follows and we thus have shown that the number of G's $\pmod{p}$ equals the number of P's $\pmod{p}$ for a fixed $Q = Q_0$ when $\left(\frac{D}{P}\right) = 1$.

By using the notation of Lemma 3.5, we have $s \equiv r_1^\alpha \equiv r_2^\alpha \equiv G^\alpha r_1^\alpha \pmod{p}$, and therefore (cf. [18]).

Corollary 3.6: *If the characteristic polynomial $f(x) = x^2 - Px + Q$ splits over F_p, then for the value G of Lemma 3.5 one has: $\rho(p, P, Q_0) = \mathrm{ord}_p(G)$.*

Now we are in the position to establish the number of P with the same rank of appearance modulo a prime p.

3.2 Counting the number of parameters.

3.2.1: $\boxed{\text{The case } \left(\frac{Q_0}{p}\right) = -1.}$

As, according to [11], $v_{(p-(D/p))/2}(P,Q) \equiv 0 \pmod{p}$ and therefore $u_{(p-(D/p))/2}(P,Q) \not\equiv 0 \pmod{p}$, whenever $\left(\frac{Q}{p}\right) = -1$, it suffices to count the parameters with rank equal to d, where d is a divisor of $p - \left(\frac{D}{P}\right)$ with $\nu_2(d) = \nu_2\left(p - \left(\frac{D}{P}\right)\right)$.

Consider firstly the case of irreducible characteristic polynomials.

By hypothesis, we have $\nu_2(\mathrm{ord}_p(Q_0)) = \nu_2(p-1)$, and it follows from Corollary 3.4 that every $\alpha = \rho$ has the form $\frac{p+1}{t}$, where t is any integer relatively prime to $\mathrm{ord}_p(Q)$ such that $t \mid (p+1)$.

Now, let $d = \frac{p+1}{t}$ arbitrarily. We wish to evaluate the number of sequences $u = u(P)$ with $\rho(p,P) = d$. By Corollary 3.4, $\beta = \mathrm{ord}_p(Q)$ because of $\nu_2(\mathrm{ord}_p(Q)) = \nu_2(p-1)$. Therefore, as $Q = Q_0$ is fixed, we are by property (2.7) counting the sequences with $d' := \mu = \mathrm{ord}_p(Q)\frac{p+1}{t}$. However, the number of sequences with irreducible characteristic polynomial and fixed Q_0 such that $\mu = d'$, is according to Theorem 2 of [16] equal to $\frac{\phi(d')}{2\phi(\mathrm{ord}_p(Q_0))}$. By extracting the powers of 2 and applying the multiplicative property of the Euler ϕ-function, this quantity becomes $\phi\left(\frac{p+1}{t}\right) = \phi(d)$. We thus have shown

Theorem 3.7: *Let d be a divisor of $p+1 = p - \left(\frac{D}{P}\right)$ with $\nu_2(d) = \nu_2(p+1)$. Then there are $\phi(d)$ parameters $P \in \mathbb{Z}$ for which $\rho(p, P, Q_0) = d$, where $Q = Q_0$ is kept fixed such that $\left(\frac{Q_0}{P}\right) = -1$.*

Remark: Observe that the condition $\nu_2(\mathrm{ord}_p(Q_0)) = \nu_2(p-1)$ is essential for allowing us to apply Corollary 3.4.

The corresponding result is also true when the characteristic polynomial splits modulo p.

Theorem 3.8:   Let d be a divisor of $p-1 = p - \left(\dfrac{D}{p}\right)$ with $\nu_2(d) = \nu_2(p-1)$. Then there are $\phi(d)$ parameters $P \in \mathbf{Z}$ for which $\rho(p, P, Q_0) = d$, when $Q = Q_0$ is kept fixed such that $\left(\dfrac{Q_0}{p}\right) = -1$.

Proof: The desired number can by Lemma 3.5 and Corollary 3.6 be determined as the number of G for which $\mathrm{ord}_p(G) = d$ and $\left(\dfrac{G}{p}\right) = -1$. Obviously, the number of all quadratic nonresidues a with $\mathrm{ord}_p(a) = d$ equals the number of all $a \in \mathbf{Z}_p$ with $\mathrm{ord}_p(a) = d$, when $\nu_2(d) = \nu_2(p-1)$. Now Theorem 1.2 establishes the result. $\Box$

3.2.2: $\boxed{\text{The case } \left(\dfrac{Q_0}{p}\right) = 1.}$

It turns out that the methods developed above do not carry over to the case that $\left(\dfrac{Q_0}{p}\right) = 1$. We therefore have to introduce some new concepts. Below, we let $\left(\dfrac{Q_0}{p}\right) = \pm 1$ unless it is explicitly stated that $\left(\dfrac{Q_0}{p}\right) = 1$.

Definition 3.9:   For any fixed integer Q_0 we define the following polynomials, where $\gamma \equiv k+1$ (mod 2) and $\gamma \subset \{0,1\}$:

$$S_k(P) = \sum_{r=0}^{[k/2]} \binom{k}{2r} P^{k-2r} (P^2 - 4Q_0)^r \tag{3.1}$$

$$T_k(P) = \sum_{r=0}^{[k/2]-\gamma} \binom{k}{2r+1} P^{k-2r-1} (P^2 - 4Q_0)^r \tag{3.2}$$

It can easily be seen that $T_k(P)$ is a polynomial of degree $k-1$ with leading coefficient $\sum_{i=0}^{[k/2]-\gamma} \binom{k}{2i+1}$, while $S_k(P)$ is of degree k whose leading coefficient is $\sum_{i=0}^{[k/2]} \binom{k}{2i}$.

By means of property (IV.8) of [11] we obtain

Corollary 3.10:

$$2^{k-1} u_k(P, Q_0) = T_k(P)$$

$$2^{k-1} v_k(P, Q_0) = S_k(P).$$

Now, $T_{k+m}(P) = 2^{k+m-1} u_{k+m}(P, Q_0)$. By using the formula (IV.3) of [11]

$$u_{k+m}(P, Q) = u_k(P, Q) v_m(P, Q) - Q^m u_{k-m}(P, Q),$$

the latter equation becomes

$$T_{k+m}(P) = 2S_m(P)T_k(P) - (4Q_0)^m T_{k-m}(P).$$ (3.3)

In particular, if $k = m$ then we have $T_{2m}(P) = 2S_m(P)T_m(P)$. Further, if $k = tm$, then by induction,

$$T_k(P) = T_{m+(t-1)m}(P) = T_m(P)R_{k,m}(P),$$ (3.4)

where $R_{k,m}(P)$ is a polynomial with integer coefficients and $deg(R_{k,m}) = deg(S_{k-m}) = k - m$.

To make notation simpler we introduce the following concept.

Definition 3.11: Let $Q_0 \in \mathbb{Z}_p^*$, $\varepsilon \in \{-1,1\}$ be fixed, and $d > 1$ be any divisor of $p - \varepsilon$. The function $\bar{\psi}(d, Q_0, \varepsilon)$ is defined to be the number of distinct values of P modulo p for which $\rho(p, P, Q_0) = d$.

Remark: Note that $\bar{\psi}(d, Q_0, \varepsilon)$ counts the values of P with both $\varepsilon(p) = \left(\dfrac{P^2 - 4Q_0}{p}\right) = \varepsilon$ and $\rho(p) = d \mid (p - \varepsilon)$ if $d > 2$.

Now, if P is any integer, then

$$T_{p-\varepsilon(p)}(P, Q_0) \equiv 0 \pmod{p},$$

where p is any odd prime coprime to $P^2 - 4Q_0$. This identity follow easily from the definition of $T_{p-\varepsilon(p)}$ and the fact that $u_{p-\varepsilon(p)}(P, Q_0) \equiv 0 \pmod{p}$.

Moreover, if $\left(\dfrac{Q_0}{p}\right) = 1$ then

$$T_{p-\varepsilon(p)}(P, Q_0) \equiv 0 \pmod{p} \text{ if and only if } T_{(p-\varepsilon(p))/2}(P, Q_0) \equiv 0 \pmod{p}.$$ (3.5)

Firstly, suppose that $d > 2$ and $\left(\dfrac{Q_0}{p}\right) = 1$. Suppose $\varepsilon(p) > 0$ and put $\delta = \dfrac{p - \varepsilon(p)}{2}$. Assume that the polynomial congruence

$$T_d(P) \equiv 0 \pmod{p}$$ (3.6)

has j solutions. If $d = \delta$, then (3.5) states that every integer P with $\left(\dfrac{P^2 - 4Q_0}{p}\right) = \varepsilon(p)$ is a solution. This number is according to Proposition 3.1 of [9] equal to

$$\frac{p - \varepsilon(p)}{2} - \frac{1}{2}\left(1 + \left(\frac{Q_0}{p}\right)\right) = \frac{p - \varepsilon(p)}{2} - 1 = d - 1.$$ (3.7)

However, for $d < \delta$ the equality

$$T_\delta(P) = T_d(P)R_{\delta,d}(P).$$ (3.8)

shows that

$$R_{\delta,d}(P) \equiv 0 \pmod{p}$$ (3.9)

has $\delta - 1 - j$ solutions since $T_\delta(P)$ has $\dfrac{p - \varepsilon(p)}{2} - 1$ solutions. As we have seen above, $R_{\delta, d}(P)$ is of degree $\delta - d$ with leading coefficient relatively prime to p. Thus (3.9) cannot have more than $\delta - d$ solutions. We note that $j \leq d - 1$ since $T_d(P)$ has degree $k - 1$ with leading coefficient relatively prime to p. Suppose $j < d - 1$. Then $\delta - 1 - d > \delta - d$, a contradiction, and therefore, $j = d - 1$.

Referring to the definition of $\bar\psi(d, Q_0, \varepsilon)$ and the preceding deduction we see that

$$\sum_{h \mid d, h \neq 1} \bar\psi(h, Q_0, \varepsilon) = j = d - 1. \tag{3.10}$$

Putting $\chi(h) = \bar\psi(h, Q_0, \varepsilon)$ for $h \neq 1$ and $\chi(1) = 1$ we obtain

$$\sum_{h \mid d} \chi(h) = d,$$

and by Möbius' inversion

$$\chi(d) = \sum_{h \mid d} \mu(h) \frac{d}{h} = d \sum_{h \mid d} \frac{\mu(h)}{h} = \phi(d). \tag{3.11}$$

We thus have the following result.

Theorem 3.12: Let $Q = Q_0 \in \mathbf{Z}_n^*$ be fixed such that $\left(\dfrac{Q_0}{p}\right) = 1$ and let $D(P) = P^2 - 4Q_0$ be coprime to p. If d is a divisor of $p - \left(\dfrac{D(P)}{p}\right)$ with $\nu_2(d) < \nu_2\left(p - \left(\dfrac{D(P)}{p}\right)\right)$, then there are $\phi(d)$ parameters $P \in \mathbf{Z}$ for which $\rho(p, P, Q_0) = d$.

In summarizing the above results and noting that $u_2(P, Q_0) \equiv 0 \pmod{p}$ exactly for $P \equiv 0 \pmod{p}$, we therefore obtain the main result in terms of the $\bar\psi(d, Q_o, \varepsilon)$-function as follows.

Corollary 3.13: Let $d \geq 2$ and suppose d is a divisor of $p - \varepsilon$ where $\varepsilon \in \{-1, 1\}$.

1. If

$$\left(\frac{Q_0}{p}\right) = -1 \quad then \quad \bar\psi(d, Q_0, \varepsilon) = \begin{cases} \phi(d), & if\ v_2(d) = v_2(p - \varepsilon) \\[2ex] 0, & otherwise. \end{cases}$$

2. Further if

$$\left(\frac{Q_0}{p}\right) = 1 \quad then \quad \bar\psi(d, Q_0, \varepsilon) = \begin{cases} \phi(d), & if\ v_2(d) < v_2(p - \varepsilon) \\[2ex] 0, & otherwise. \end{cases}$$

Remark: Observe that in Theorem 12 on pages 325-326 of [14], Somer showed that $\bar\psi(d, Q_0, \varepsilon) \geq 1$ for all values of d such that $\bar\psi(d, Q_0, \varepsilon) = \phi(d)$ in cases 1 and 2 of Corollary 3.13.

4. KEEPING $P = P_0$ FIXED AND VARYING Q.

The above method developed for $\left(\dfrac{Q_0}{P}\right) = 1$ carries over to the case where the roles of P and Q are being reversed. The analogous definition to the above case is

<u>Definition 4.1:</u> For any fixed integer P_0 we define

$$\tilde{S}_k(Q) = \sum_{r=0}^{[k/2]} \binom{k}{2r} P_0^{k-2r}(P_0^2 - 4Q)^r \tag{4.1}$$

$$\tilde{T}_k(Q) = \sum_{r=0}^{[k/2]-\gamma} \binom{k}{2r+1} P_0^{k-2r-1}(P_0^2 - 4Q)^r, \tag{4.2}$$

where $\gamma \equiv k+1 \pmod 2$ and $\gamma \in \{0,1\}$.

In this case $\tilde{T}_k(Q)$ has degree $\lfloor \frac{k-1}{2} \rfloor$ (where $\lfloor x \rfloor$ is the greatest integer function or floor function), which is $\frac{k-1}{2}$, respectively $\frac{k-2}{2}$, according as k is odd or even. The leading coefficient is $(-4)^{(k-1)/2}$, respectively $(-4)^{(k-2)/2}$. Moreover, the polynomial $\tilde{S}_k(Q)$ has degree $\lfloor k/2 \rfloor$ with leading coefficient $(-4)^{[k/2]}$. Using the above arguments we see that

$$\tilde{T}_{k+m}(Q) = 2\tilde{S}_m(Q)\tilde{T}_k(Q) - (4Q)^m\tilde{T}_{k-m}(Q), \tag{4.3}$$

and again by induction, for $m \mid k$

$$\tilde{T}_k(Q) = \tilde{T}_m(Q)\tilde{R}_{k,m}(Q).$$

From the induction it can easily be seen that $deg(\tilde{R}_{k,m})$ is the integral value $\lfloor \frac{k-m}{2} \rfloor$ with leading coefficient a power of 4.

As above, we give the following

<u>Definition 4.2:</u> Let $P_0 \in \mathbf{Z}_p^*$, $\varepsilon \in \{-1,1\}$ be fixed integers and $d > 1$ be any divisor of $p - \varepsilon$. Denote by $\psi(d, P_0, \varepsilon)$ the number of distinct values of $Q \in \mathbf{Z}_p^*$ for which $\rho(p, P_0, Q) = d$.

<u>Remark:</u> Obviously we now count the number of Q's such that $\varepsilon(p) = \left(\dfrac{P_0^2 - 4Q}{p}\right) = \varepsilon$ and $\rho(p) = d \mid (p - \varepsilon)$.

Now we are in the position to prove the main result of this section.

<u>Theorem 4.3:</u> Let $d > 2$, $d \mid (p - \varepsilon)$, with $|\varepsilon| = 1$. Then $\psi(d, P_0, \varepsilon) = \dfrac{\phi(d)}{2}$.

<u>Proof:</u> We follow the line of thought of the proof to Theorem 3.13. Again, let $\tilde{T}_d(Q) \equiv 0 \pmod p$ have j solutions. If, firstly $d = \delta(p) = \delta$, where $\delta(p) = \delta = p - \varepsilon(p)$, then

$$\tilde{T}_{p-\varepsilon(p)}(Q) \equiv 0 \pmod p$$

has according to Proposition 3.1 of [9] exactly

$$\frac{p-2-\varepsilon}{2} = \frac{d-2}{2}$$

solutions $Q \in \mathbf{Z}_p^*$.

For $d < \delta$ the congruence

$$\tilde{T}_\delta(Q) = \tilde{T}_d(Q)\tilde{R}_{\delta,d}(Q) \tag{4.4}$$

shows that

$$\tilde{R}_{\delta,d}(Q) \equiv 0 \pmod{p} \tag{4.5}$$

has $\frac{\delta-2}{2} - j$ solutions. On the other hand we know from the degree of $\tilde{R}_{\delta,d}(Q)$ that the number of solutions of (4.5) is at most $\frac{\delta-d}{2}$ for d even, respectively $\frac{\delta-d-1}{2}$ for d odd. This gives $j \leq \frac{d-1}{2}$ in the first case, and $j \leq \frac{d-2}{2}$ in the latter case. If we thus assume $j < \frac{d-2}{2}$ for $2 \mid d$ then (4.5) has $\frac{\delta-2}{2} - j > \frac{\delta-2}{2} - \frac{d-2}{2} = \frac{\delta-d}{2}$ solutions which cannot occur. In the same manner we conclude that the assumption $j < \frac{d-1}{2}$ when d is odd, leads to a contradiction.

By introducing the analogous function $\chi(h)$ we therefore can argue as above and obtain

$$\sum_{h \mid d} \chi(h) = \begin{cases} \frac{d}{2}, & \text{for } 2 \mid d \\ \frac{d-1}{2}, & \text{for } 2 \nmid d. \end{cases} \tag{4.6}$$

If d is odd, the Möbius inversion formula gives

$$\chi(d) = \sum_{h \mid d} \mu(h)\frac{d-h}{2h} = \frac{d}{2}\sum_{h \mid d}\frac{\mu(h)}{h} - \frac{1}{2}\sum_{h \mid d}\mu(h) = \frac{\phi(d)}{2},$$

since for $d > 1$ the right sum reduces to zero.

From this also the result for d even immediately follows, which concludes the proof. $\square$

Remark: It should be noted that in Theorem 14 on pages 331-332 of [14], Somer proved that $\underline{\psi}(d, P_0, \varepsilon) \geq 1$ for all values of d in Theorem 4.3 for the case in which $\varepsilon = 1$.

SUMMARY

In regarding the Lucas sequence $u(P,Q)$ as a polynomial in one indeterminate by keeping either P or Q fixed and varying the other, we have shown that the rank of appearance of (u) fulfills properties very similar to the order of a group element. In particular, this can be seen by the number of parameters modulo a prime p with the same rank of appearance. For establishing this formula, we have employed very particular properties of the period of second-order recurrence sequences, as well as the number of zeros of the Lucas functions.

ACKNOWLEDGEMENTS

I am deeply grateful to Lawrence Somer for carefully reading the manuscript and for providing references [13]-[19] and [22]. His many insightful comments, corrections, and remarks have tremendously enhanced the quality of the paper. Also, I am very thankful to Winfried B. Müller for numerous helpful discussions, and for his patience.

REFERENCES

[1]   Baillie, R., Wagstaff, S., Jr. "Lucas pseudoprimes." *Math. Comp.*, Vol. *35* (1980): pp. 1391-1417.

[2]   Carmichael, R.D. "On the numerical factors of the arithmetic forms $a^n \pm b^n$." *Ann. of Math.*, Vol. *15* (1913): pp. 30-70.

[3]   Dresel, L.A.G. "On Pseudoprimes Related to Generalized Lucas Sequences." *The Fibonacci Quarterly*, Vol. *35.1* (1997): pp. 35-42.

[4]   Guillaume, D. and Morain, F. "Building pseudoprimes with a large number of prime factors." *AAECC*, Vol. *7.4* (1996): pp. 263-277.

[5]   Kowol, G. "On strong Dickson pseudoprimes." *AAECC*, Vol. *3* (1992): pp. 129-138.

[6]   Lidl, R., Mullen, G.L. and Turnwald, G. "Dickson Polynomials." Pitman Monographs and Surveys in Pure and Applied Mathematics, Vol. *65*. Longman, London, 1993.

[7]   Lidl, R. and Niederreiter, H. "Finite Fields." Encyclopedia of Mathematics and its Applications, Volume 20. Addison-Wesley, Reading, Massachusetts, 1983.

[8]   Müller, S. "On Strong Lucas Pseudoprimes." Combination to General Algebra, Volume 10. Edited by D. Dorninger, G. Eigenthaler, H.K. Kaiser, H. Kautschitsch, W. More and W.B. Müller. Proceedings of the Klagenfurt Conference. Klagenfurt: Johannes Heyn, 1998: pp. 237-249.

[9]   Müller, S. "Carmichael Numbers and Lucas Tests." *Contemporary Mathematics*, Volume *225*. Edited by R.C. Mullin and G. Mullen. Proceedings of the Fourth International Conference on Finite Fields and Applications. University of Waterloo, Ontario, Canada (1997): pp. 193-202.

[10]   Pinch, R.G.E. "The Carmichael numbers up to 10^{15}." *Math. Comp.*, Vol. *61* (1993): pp. 381-391.

[11]   Ribenboim, P. The Book of Prime Number Records. Springer Verlag, Berlin, 1988.

[12]   Riesel, H. Prime Numbers and Computer Methods for Factorization. Birkhäuser, Boston, Basel, Stuttgart, 1985.

[13]   Somer, L. "The Divisibility Properties of Primary Lucas Recurrences with Respect to Primes." *The Fibonacci Quarterly*, Vol. *18* (1980): pp. 316-334.

[14]   Somer, L. "Possible Periods of Primary Fibonacci-Like Sequences with Respect to a Fixed Odd Prime." *The Fibonacci Quarterly*, Vol. *20* (1982): pp. 311-333.

[15]   Somer, L. "The Divisibility and Modular Properties of kth - Order linear Recurrences Over the Ring of Integers of an Algebraic Number Field with Respect to Prime Ideals." Ph.D. Dissertation, The University of Illinois at Urbana-Champaign (1985).

[16]   Somer, L. "Periodicity Properties of kth Order Linear Recurrences with Irreducible Characteristic Polynomial Over a Finite Field." Finite Fields, Coding Theory and Advances in Communications and Computing. Edited by Gary L. Mullen and Peter Jau-Shyong Shiue. Marcel Dekker Inc., 1993: pp. 195-207.

[17]   Somer, L. "Periodicity Properties of kth Order Linear Recurrences Whose Characteristic Polynomial Splits completely Over a Finite Field I." *Contemporary Mathematics*, Vol. *168* (1994): pp. 327-339.

[18]   Somer, L. "Periodicity Properties of kth Order Linear Recurrences Whose Characteristic Polynomial Splits completely Over a Finite Field II." Finite Fields and their Applications. Edited by S. Cohen, H. Niederreiter. Cambridge University Press, 1996: pp. 333-347.

[19]   Somer, L. "On Lucas d-Pseudoprimes. Applications of Fibonacci Numbers, Volume 7. Edited by G.E. Bergum, A.F. Horadam and A.N. Philippou. Kluwer Academic Publishers, Dordrecht, The Netherlands, 1998: 369-375.

[20]   Vinson, J. "The Relation of the Period Modulo m to the Rank of Apparition of m in the Fibonacci sequence." *The Fibonacci Quarterly*, Vol. *1* (1963): pp. 37-45.

[21]   Williams, H.C. "On numbers analogous to the Carmichael - numbers." *Canad. Math. Bull.*, Vol. *20* (1977): pp. 133-143.

[22]   Wyler, O. "On Second-Order Recurrences." *Amer. Math. Monthly*, Vol. *72* (1965): pp. 500-506.

AMS Classification Numbers: 11B37, 11B39

ALGORITHMIC SIMPLIFICATION OF RECIPROCAL SUMS

Stanley Rabinowitz

1. INTRODUCTION.

Algorithms for evaluating or simplifying sums of the form

$$\sum_{n=1}^{N} \frac{1}{F_{n+a_1} F_{n+a_2} \cdots F_{n+a_r}}$$

where the F_i are Fibonacci numbers and the a_i are integers have been discussed in [13]. It is the goal of this paper to generalize these results to arbitrary second-order linear recurrences.

Consider the second order linear recurrences defined by

$$u_{n+2} = Pu_{n+1} - Qu_n, \quad u_0 = 0, \quad u_1 = 1, \tag{1}$$

$$v_{n+2} = Pv_{n+1} - Qv_n, \quad v_0 = 2, \quad v_1 = P, \tag{2}$$

and

$$w_{n+2} = Pw_{n+1} - Qw_n, \quad w_0, \quad w_1 \text{ arbitrary.} \tag{3}$$

Let r_1 and r_2 denote the roots of the characteristic equation $x^2 - Px + Q = 0$. Let

$$D = P^2 - 4Q \quad \text{and} \quad e = w_0 w_2 - w_1^2. \tag{4}$$

Throughout this paper, we shall assume that $D \neq 0$, $e \neq 0$, $Q \neq 0$, and $w_n \neq 0$ for $n > 0$.

In the case of the Fibonacci sequence, we showed [13] that all reciprocal sums can be expressed in closed form in terms of

This paper is in final form and no version of it will be submitted for publication elsewhere.

277

$$F_N = \sum_{n=1}^{N} \frac{1}{F_n}, \quad G_N = \sum_{n=1}^{N} \frac{(-1)^n}{F_n}, \quad \text{and } K_N = \sum_{n=1}^{N} \frac{1}{F_n F_{n+1}}. \tag{5}$$

It is our intent to generalize these results to apply to the sequence $\langle w_n \rangle$.

For the following definitions, let r be a positive integer and let $a_1, a_2, \cdots, a_r$ be distinct nonnegative integers.

Definition 1 (Unit Reciprocal Sum): A unit reciprocal sum of order r is a sum of the form

$$\sum_{n=1}^{N} \frac{1}{w_{n+a_1} w_{n+a_2} \cdots w_{n+a_r}}.$$

Definition 2 (Q- Reciprocal Sum): A Q-Reciprocal sum of order r is a sum of the form

$$\sum_{n=1}^{N} \frac{Q^{kn}}{w_{n+a_1} w_{n+a_2} \cdots w_{n+a_r}}$$

where $k = \lfloor r/2 \rfloor$.

Definition 3 (Reciprocal Sum): A Reciprocal sum is a unit reciprocal sum or a Q-reciprocal sum.

Definition 4 (Rational Sum): A rational sum of order r is a sum of the form

$$\sum_{n=1}^{N} \frac{f(x_1, x_2, \cdots, x_s)}{w_{n+a_1} w_{n+a_2} \cdots w_{n+a_r}}$$

where $f(x_1, x_2, \cdots, x_s)$ is a polynomial with each of the variables x_i being of the form w_{n+c_i} or Q^n.

In this paper, we will show that all reciprocal sums of order 1 and 2 can be expressed in closed form in terms of

$$X_N = \sum_{n=1}^{N} \frac{1}{w_n}, \quad Y_N = \sum_{n=1}^{N} \frac{Q^n}{w_n}, \quad \text{and } W_N = \sum_{n=1}^{N} \frac{1}{w_n w_{n+1}}. \tag{6}$$

We will also show that all Q-reciprocal sums (of any order) can be expressed in closed form in terms of W_N, X_N, and Y_N.

Finally, we shall show that if $Q = \pm 1$, then all rational sums can be expressed in terms of U_N, V_N, W_N, X_N, and Y_N, where

$$U_N = \sum_{n=1}^{N} \frac{w_{n+1}}{w_n} \quad \text{and } V_N = \sum_{n=1}^{N} \frac{Q^n w_{n+1}}{w_n}.$$

We shall also present algorithms for finding these closed forms.

We need the following results.

Theorem 1 (The Representation Theorem): If a, b, and c are integers and $u_{a-b} \neq 0$, then

$$w_{n+c} = \frac{u_{c-b}}{u_{a-b}} w_{n+a} + \frac{u_{c-a}}{u_{b-a}} w_{n+b}. \tag{7}$$

(This expresses w_{n+c} in terms of w_{n+a} and w_{n+b}.)

Proof: The identity can be mechanically verified by using algorithm LucasSimplify from [11]. □

Theorem 1 can be put into a more symmetrical form:

Theorem 2: For all integers a, b, and c,

$$Q^c u_{b-c} w_{n+a} + Q^a u_{c-a} w_{n+b} + Q^b u_{a-b} w_{n+c} = 0. \tag{8}$$

(This gives a symmetric connection between w_{n+a}, w_{n+b}, and w_{n+c}.)

Proof: This follows from the Representation Theorem by making use of the well-known Negation Formula [11]: $u_{-n} = -u_n Q^{-n}$. □

Theorem 3: If a, b, and c are integers and $u_{a-b} \neq 0$, then

$$\frac{1}{w_{n+a} w_{n+b}} = \frac{A}{w_{n+c} w_{n+a}} + \frac{B}{w_{n+c} w_{n+b}} \tag{9}$$

where

$$A = \frac{u_{c-a}}{u_{b-a}} \quad \text{and} \quad B = \frac{u_{c-b}}{u_{a-b}}.$$

(This allows one to convert reciprocal sums of order 2 to those in which w_{n+c} occurs as a factor of the denominator.)

Proof: This is an immediate consequence of the Representation Theorem. □

2. RECIPROCAL SUMS OF ORDER 1.

There are no known elementary forms for the reciprocal sums of order 1, so we shall give them names:

$$X_N = \sum_{n=1}^{N} \frac{1}{w_n} \quad \text{and} \quad Y_N = \sum_{n=1}^{N} \frac{Q^n}{w_n}. \tag{10}$$

Strictly speaking, we should write these as $X_N(w_0, w_1, P, Q)$ and $Y_N(w_0, w_1, P, Q)$; but we simply write X_N and Y_N when w_0, w_1, P, and Q are fixed.

If $a > 0$, we have

$$\sum_{n=1}^{N} \frac{1}{w_{n+a}} = X_{N+a} - X_a \tag{11}$$

and

$$\sum_{n=1}^{N} \frac{Q^n}{w_{n+a}} = Y_{N+a} - Y_a. \tag{12}$$

Thus all reciprocal sums of order 1 can be expressed in terms of X_N and Y_N.

3. Q-RECIPROCAL SUMS OF ORDER 2.

Theorem 4: If $a > 0$, then

$$u_a \sum_{n=1}^{N} \frac{Q^n}{w_n w_{n+a}} = \frac{Q}{e} \left[\sum_{n=1}^{a} \frac{w_{n-1}}{w_n} - \sum_{n=1}^{a} \frac{w_{N+n-1}}{w_{N+n}} \right]. \tag{13}$$

Proof: We begin with the identity

$$w_{n+a} w_{n-1} - w_n w_{n+a-1} = Q^{n-1} e u_a \tag{14}$$

which comes from d'Ocagne's Identity (see [12]). Thus, we have

$$\frac{w_{n-1}}{w_n} - \frac{w_{n+a-1}}{w_{n+a}} = \frac{Q^{n-1} e u_a}{w_n w_{n+a}}.$$

Summing from 1 to N yields

$$u_a \sum_{n=1}^{N} \frac{Q^{n-1}}{w_n w_{n+a}} = \frac{1}{e} \left[\sum_{n=1}^{N} \frac{w_{n-1}}{w_n} - \sum_{n=1}^{N} \frac{w_{n+a-1}}{w_{n+a}} \right] = \frac{1}{e} \left[\sum_{n=1}^{a} \frac{w_{n-1}}{w_n} - \sum_{n=1}^{a} \frac{w_{N+n-1}}{w_{N+n}} \right]$$

which is the desired result.   □

This can be put into a more symmetrical form. The following theorem is a generalization of a result by Good [4] and was proven by André-Jeannin [1].

Theorem 5 (symmetry Property for Reciprocal Sums): If $a > 0$, then

$$u_a \sum_{n=1}^{N} \frac{Q^n}{w_n w_{n+a}} = u_N \sum_{n=1}^{a} \frac{Q^n}{w_n w_{n+N}}. \tag{15}$$

Proof: Again we use d'Ocagne's identity. Putting $a = N$ in formula (14) gives

$$w_{n-1} w_{N+n} - w_n w_{N+n-1} = Q^{n-1} e u_N.$$

Combining the two sums on the right-hand side of Theorem 4 and applying this identity yields the desired result.   □

Corollary (letting $a = 1$):

$$\sum_{n=1}^{N} \frac{Q^n}{w_n w_{n+1}} = \frac{Q u_N}{w_1 w_{N+1}} = \frac{Q}{e}\left[\frac{w_0}{w_1} - \frac{w_N}{w_{N+1}}\right]. \tag{16}$$

4. UNIT RECIPROCAL SUMS OF ORDER 2.

For non-alternating reciprocal Fibonacci sums, we had to introduce (in [13]) the symbol

$$\mathsf{K}_N = \sum_{n=1}^{N} \frac{1}{F_n F_{n+1}} \tag{17}$$

for a sum with no known simple closed form. In a similar manner, we need to do the same thing for unit reciprocal sums for the sequence $\langle w_n \rangle$.

Let

$$\mathsf{W}_N = \sum_{n=1}^{N} \frac{1}{w_n w_{n+1}} \tag{18}$$

with the understanding that $\mathsf{W}_0 = 0$. Again, we should really write this as $\mathsf{W}_N(w_0, w_1, P, Q)$ but if $\langle w_n \rangle$ is a fixed sequence, we will simply write this as W_N.

Theorem 6: If $c > 0$ and $u_c \neq 0$, then

$$\sum_{n=1}^{N} \frac{1}{w_n w_{n+c}} = \frac{1}{u_c} \sum_{i=0}^{c-1} Q^i(\mathsf{W}_{N+i} - \mathsf{W}_i). \tag{19}$$

Proof: Letting $a = i$, $b = i+1$, and $c = 0$ in Theorem 3, we get

$$\frac{u_{i+1}}{w_n w_{n+i+1}} - \frac{u_i}{w_n w_{n+i}} = \frac{Q^i}{w_{n+i} w_{n+i+1}}$$

using the Negation Formula $u_{-n} = -u_n Q^{-n}$. Summing as n goes for 1 to N yields

$$u_{i+1}\mathsf{W}_N(i+1) - u_i\mathsf{W}_N(i) = Q^i(\mathsf{W}_{N+i} - \mathsf{W}_i)$$

where

$$\mathsf{W}_N(a) = \sum_{n=1}^{N} \frac{1}{w_n w_{n+a}}.$$

Now sum as i goes from 0 to $c-1$. The left side telescopes and we get

$$u_c\mathsf{W}_N(c) = \sum_{i=0}^{c-1} Q^i(\mathsf{W}_{N+i} - \mathsf{W}_i)$$

which gives our desired result.   □

Thus, all reciprocal sums of order 2 can be expressed in terms of W_N.

5. THE REDUCTION PROCESS.

We now show how to simplify certain reciprocal sums with three or more factors in the denominator.

Theorem 7 (The Partial Fraction Decomposition Formula for w):

For all n,

$$\frac{Q^n}{w_{n+a}w_{n+b}w_{n+c}} = \frac{A}{w_{n+a}} + \frac{B}{w_{n+b}} + \frac{C}{w_{n+c}} \tag{20}$$

where

$$A = \frac{-Q^{-a}}{eu_{b-a}u_{c-a}}, \quad B = \frac{-Q^{-b}}{eu_{c-b}u_{a-b}}, \quad \text{and} \quad C = \frac{-Q^{-c}}{eu_{a-c}u_{b-c}}. \tag{21}$$

Proof: This result can be mechanically proven using algorithm LucasSimplify from [11]. □

Theorem 8 (The Reduction Theorem for w): If $r > 2$, then any Q-reciprocal sum of order r can be expressed in terms of Q-reciprocal sums of order $r-2$.

Proof: By Theorem 7, we have

$$\frac{Q^{kn}}{w_{n+a}w_{n+b}w_{n+c}} = \frac{AQ^{(k-1)n}}{w_{n+a}} + \frac{BQ^{(k-1)n}}{w_{n+b}} + \frac{CQ^{(k-1)n}}{w_{n+c}} \tag{22}$$

with A, B, and C as given in display (21). If $r > 2$ and $k = \lfloor r/2 \rfloor$, then we can take the last three factors in the denominator and apply Theorem 7. This breaks the given sum up into sums with $r-2$ factors in the denominator. The numerators have terms that are constant multiples of $Q^{(k-1)n}$ where $k-1 = \lfloor (r-2)/2 \rfloor$, thus making these sums multiples of Q-reciprocal sums of order $r-2$.

Corollary: Any Q-reciprocal sum of order r can be expressed in terms of reciprocal sums of order 1 or 2. If $Q = \pm 1$, then any reciprocal sum of order r can be expressed in terms of reciprocal sums of order 1 or 2.

Proof: Apply Theorem 8 repeatedly, until the order of the Q-reciprocal sum becomes 1 or 2. If $Q = \pm 1$, then formula (20) can be written in the form

$$\frac{1}{w_{n+a}w_{n+b}w_{n+c}} = \frac{A(-1)^n}{w_{n+a}} + \frac{B(-1)^n}{w_{n+b}} + \frac{C(-1)^n}{w_{n+c}}. \tag{23}$$

Applying this repeatedly reduces the order of the reciprocal sum to 1 or 2.   □

By induction, we can state a more general form of the Partial Fraction Decomposition Theorem.

Theorem 9 (The Generalized Partial Fraction Decomposition Formula): If r is a positive integer, then

$$\frac{1}{w_{n_1} w_{n_2} w_{n_3} \cdots w_{n_{2r+1}}} = \sum_{i=1}^{2r+1} \frac{A_i}{w_{n_i}}$$ (24)

where $A_i^{-1} = (-eQ^{n_i})^r \prod_{j \neq i} u_{n_j - n_i}$.

6. THE SIMPLIFICATION ALGORITHM IN TERMS OF U_N, V_N, W_N, X_N, and Y_N.

We can also handle sums similar to reciprocal sums, but in which the numerators are polynomials in the w's. These are called rational sums.

We need to add in two new primitives:

$$U_N = \sum_{n=1}^{N} \frac{w_{n+1}}{w_n} \quad \text{and} \quad V_N = \sum_{n=1}^{N} \frac{Q^n w_{n+1}}{w_n}.$$ (25)

Once again, these would more properly be written as $U_N(w_0, w_1, P, Q)$ and $V_N(w_0, w_1, P, Q)$; but U_N and V_N will suffice when the sequence is fixed.

We now show how to evaluate a wide class of reciprocal and rational sums in closed form in terms of the quantities U_N, V_N, W_N, X_N, and Y_N.

Definition: A w-polynomial in the variable n is any polynomial $f(x_1, x_2, \cdots, x_r)$ with constant coefficients where each x_i is of the form w_x or Q^x, with each x of the form $n + c_j$, where the c_j are integer constants. For purposes of this definition, the quantities P, Q, w_0, w_1, and e are to be considered constants.

Theorem 10 (The Simplification Theorem for $Q = \pm 1$): Suppose that P, Q, w_0, and w_1 are fixed constants, thereby determining the sequence $\langle w_n \rangle$. Let $f(n)$ be any w-polynomial in the variable n. For r a positive integer, let c_j, $j = 1, 2, \cdots, r$ be distinct integers. Assume that $w_n > 0$ and $u_n > 0$ for all $n > 0$. Furthermore, if $Q = \pm 1$, then we can find

$$\sum_{n=1}^{N} \frac{f(n)}{w_{n+c_1} w_{n+c_2} \cdots w_{n+c_r}}$$

in closed form in terms of U_N, V_N, W_N, X_N, and Y_N.

Proof: As proof, we give the following algorithm.

Algorithm WReciprocalSum to evaluate certain reciprocal sums in closed form:

INPUT: A rational sum meeting the conditions of Theorem 9.

OUTPUT: A closed form for the sum expressed in terms of the quantities U_N, V_N, W_N, X_N, and Y_N.

STEP 1: [Reduce the order.] If the denominator consists of three or more terms of the form w_x, choose any three of them, say w_{n+a}, w_{n+b}, and w_{n+c}, and make the following substitution:

$$\frac{1}{w_{n+a}w_{n+b}w_{n+c}} = \frac{AQ^{-n}}{w_{n+a}} + \frac{BQ^{-n}}{w_{n+b}} + \frac{CQ^{-n}}{w_{n+c}} \tag{26}$$

where A, B, and C are given by formula (21). Expand out and make the transformation

$$\sum_{n=1}^{N}[f(n)+g(n)] = \sum_{n=1}^{N} f(n) + \sum_{n=1}^{N} g(n) \tag{27}$$

summing each term on the right by this algorithm.

STEP 2: [Normalize subscripts in denominator.] If the denominator is of the form w_{n+a} or of the form $w_{n+a}w_{n+b}$ with $a \neq 0$ and $a < b$, then apply one of the following transformations:

$$\sum_{n=1}^{N}\frac{f(n)}{w_{n+a}w_{n+b}} = \sum_{n=a+1}^{a+N}\frac{f(n-a)}{w_n w_{n+b-a}} \tag{28}$$

$$\sum_{n=1}^{N}\frac{f(n)}{w_{n+a}} = \sum_{n=a+1}^{a+N}\frac{f(n-a)}{w_n}. \tag{29}$$

STEP3: [Normalize index of summation.] If the index of summation does not start at 1, then add or subtract a finite number of terms to make the index start at 1. Specifically, if n_0 is a constant and $n_0 \neq 1$, then apply the transformation

$$\sum_{n=n_0}^{N} f(n) = \begin{cases} \displaystyle\sum_{n=1}^{N} f(n) - \sum_{n=1}^{n_0-1} f(n), & \text{if } n_0 > 1, \\[3mm] \displaystyle\sum_{n=1}^{N} f(n) + \sum_{n=n_0}^{0} f(n), & \text{if } n_0 \leq 0. \end{cases} \tag{30}$$

STEP 4: [Break up sums.] Expand out the numerator. If the numerator consists of a sum of terms, then sum each term individually. That is, apply the transformation

$$\sum_{n=1}^{N} \frac{f(n)+g(n)}{d} = \sum_{n=1}^{N} \frac{f(n)}{d} + \sum_{n=1}^{N} \frac{g(n)}{d}. \tag{31}$$

In each fraction, cancel any factors of the form w_{n+c} common to the numerator and denominator. Then evaluate each sum recursively using this algorithm. Return the sum of the results so obtained.

STEP 5: [Normalize numerator.] If the denominator is of the form $w_n w_{n+a}$ with $a > 0$ and if the numerator contains a subexpression of the form w_{n+c} where $c \neq 0$ and $c \neq a$, then express this subexpression in terms of w_n and w_{n+a} by using the Representation Theorem. Specifically, make the substitution

$$w_{n+c} = \frac{Q^a u_{c-a}}{u_a} w_n + \frac{u_{c-a}}{u_a} w_{n+a}. \tag{32}$$

If the numerator contains a subexpression of the form Q^n, then express this subexpression in terms of w_n and w_{n+a} by using the formula

$$Q^n = \frac{v_a w_n w_{n+a} - Q^a w_n^2 - w_{n+a}^2}{e u_a^2}. \tag{33}$$

Go back to step 4.

STEP 6: [Normalize numerator (continued).] If the denominator is of the form w_n, and if the numerator contains a subexpression of the form w_{n+c} where $c \neq 0$ and $c \neq 1$, then express this subexpression in terms of w_n and w_{n+1} by using the Representation Theorem. Specifically, make the substitution

$$w_{n+c} = u_c w_{n+1} - Q u_{c-1} w_n. \tag{34}$$

Go back to step 4.

STEP 7: [Evaluate polynomial sums.] If the summand is a w-polynomial in the variable n, evaluate the sum by using algorithm LucasSum from [11]. Exit.

STEP 8: [Reduce numerator.] If the denominator is of the form w_n and if the numerator contains a subexpression of the form w_{n+1}^r with $r > 1$, then write w_{n+1}^r as $w_{n+1}^{r-2} w_{n+1}^2$ and reduce the exponent by 1 by applying the substitution

$$w_{n+1}^2 = P w_n w_{n+1} - Q w_n^2 - e Q^n. \tag{35}$$

Expand the numerator. If $Q = -1$, replace any terms of the form Q^{rn+d} by $Q^d(Q^r)^n$. Repeat this step as long as possible, then go back to step 4.

STEP 9: [Pull out constants.] Replace any expressions of the form Q^{n+b} where b is a constant by $Q^b Q^n$. If the numerator is of the form c, cQ^n, $cQ^n w_x$, or cw_x^r, where c is a constant ($c \neq 0$ and $c \neq 1$), then apply the transformation

$$\sum_{n=1}^{N} cf(n) = c \sum_{n=1}^{N} f(n). \tag{36}$$

STEP 10: [Handle sums of order 2.] If the denominator is of the form $w_n w_{n+a}$, then evaluate the sum by using one of the following formulas:

$$\sum_{n=1}^{N} \frac{Q^n}{w_n w_{n+a}} = \frac{u_N}{u_a} \sum_{n=1}^{a} \frac{Q^n}{w_n w_{n+N}}; \tag{37}$$

$$\sum_{n=1}^{N} \frac{1}{w_n w_{n+a}} = \frac{1}{u_a} \sum_{i=0}^{a-1} Q^i (W_{N+i} - W_i). \tag{38}$$

STEP 11: [Handle basic sums.] If the summand is of one of the following forms, make the substitution shown.

$$\sum_{n=1}^{N} \frac{w_{n+1}}{w_n} = U_N$$

$$\sum_{n=1}^{N} \frac{Q^n w_{n+1}}{w_n} = V_N$$

$$\sum_{n=1}^{N} \frac{1}{w_n w_{n+1}} = W_N \tag{39}$$

$$\sum_{n=1}^{N} \frac{1}{w_n} = X_N$$

$$\sum_{n=1}^{N} \frac{Q^n}{w_n} = Y_N.$$

Proof of Correctness:

Step 1 reduces the order to 1 or 2. This step introduces terms of the form Q^{-n}. If $Q = \pm 1$, then $Q^{-n} = Q^n$. Thus, the numerator will remain a w-polynomial if $Q = \pm 1$.

Step 2 guarantees that if there is a denominator, then its first factor will be w_n.

Step 3 ensures that the index of summation begins with 1. The upper limit can be any expression, since N need not be just a variable, but may be any expression.

Step 4 guarantees that there will be no sums (or differences) in the numerator.

Step 5 is justified by the Representation Theorem. Formula (33) comes from [12]. At the end of step 5, there will be no terms of the form w_x in the numerator of any reciprocal sum of order 2.

Step 6 is justified by the Representation Theorem. After step 4, the numerator consists only of a product of terms.

Steps 5 and 6 ensure that these terms only involve w's that cancel with w's in the denominator or are of the form w_{n+1}. Thus, by the time we get to step 7, the only w's left in the numerator are those of the form w_{n+1}. Of course, if the denominator went away, then we are left with a w-polynomial and it is easily summed in step 7.

Step 8 reduces the degree of the variable w_{n+1} to 0 or 1.

Step 9 removed any constants from the numerator.

Step 10 is justified by Theorems 5 and 6. None of the previous steps introduce terms of the form Q^{rn} in the numerator (for $r > 1$). Thus steps 10 and 11 handle all the remaining cases. $\qquad\qquad\qquad\qquad\qquad\qquad\qquad\qquad\qquad\qquad\qquad\qquad\qquad\qquad\qquad$ $\square$

Note. It should be noted that Algorithm WReciprocalSum also works in the cases where $r < 3$ and deg $f(n) < 2$ or for any r if $f(n) = Q^{kn}$ where $k = \lfloor r/2 \rfloor$. In step 1, if $f(n) = Q^{kn}$, the Q^{-n} term introduced changes Q^{kn} into $Q^{(k-1)n}$ and the order of the sum decrements by 1 until it reaches 1 or 2. The degree of Q^n will increase only if the degree of $f(n)$ was larger than 1, so if deg $f(n) < 2$, no terms of the form Q^{cn} are introduced with $c > 1$.

This gives us the following two theorems.

Theorem 11 (The Simplification Theorem for Q-reciprocal sums): Let r be a positive integer and let $k = \lfloor r/2 \rfloor$. Let c_j, $j = 1, 2, \cdots, r$ be distinct integers. Then we can find

$$\sum_{n=1}^{N} \frac{Q^{kn}}{w_{n+c_1} w_{n+c_2} \cdots w_{n+c_r}}$$

in closed form in terms of U_N, V_N, W_N, X_N, and Y_N.

Theorem 12 (The Simplification Theorem for Low-Order Reciprocal sums): Let $f(n)$ be any w-polynomial in the variable n with deg $f(n) < 2$. Let a and b be distinct integers. Then we can find

$$\sum_{n=1}^{N} \frac{f(n)}{w_{n+a}} \text{ and } \sum_{n=1}^{N} \frac{f(n)}{w_{n+a} w_{n+b}}$$

in closed form in terms of U_N, V_N, W_N, X_N, and Y_N.

7. SOME GENERAL FORMULAS.

We have given an algorithm for evaluating certain reciprocal sums. However, in some special cases, simple explicit formulas can be given.

We can take a formula, such as that given by Theorem 4, which involves expressions of the form w_{n+a} and turn it into a valid formula involving expressions of the form $w_{k(n+a)}$ where k is a fixed positive integer. We do this by applying the Dilation Theorem (see [12]) which says we can transform an identity into another identity by replacing all occurrences of w_x by w_{kx} provided that we also replace u_x by u_{kx}/u_k, v_x by v_{kx}, Q by Q^k, P by v_k, and e by eu_k.

Applying the Dilation Theorem to Theorem 4 gives us the following theorem:

Theorem 13: If $a > 0$, $k > 0$, $u_k \neq 0$, and $u_{ka} \neq 0$, then

$$\sum_{n=1}^{N} \frac{Q^{kn}}{w_{kn} w_{k(n+a)}} = \frac{Q^k}{eu_k u_{ka}} \left[\sum_{n=1}^{a} \frac{w_{k(n-1)}}{w_{kn}} - \sum_{n=1}^{a} \frac{w_{k(N+n-1)}}{w_{k(N+n)}} \right]$$

$$= \frac{1}{eu_k u_{ka}} \left[\sum_{n=1}^{a} \frac{w_{k(N+n+1)}}{w_{k(N+n)}} - \sum_{n=1}^{a} \frac{w_{k(n+1)}}{w_{kn}} \right]. \tag{40}$$

The last equality comes from the identity

$$\frac{w_{n+1}}{w_n} = \frac{Pw_n - Qw_{n-1}}{w_n} = P - Q\frac{w_{n-1}}{w_n} \tag{41}$$

which when dilated by k gives

$$\frac{w_{k(n+1)}}{w_{kn}} = v_k - Q^k \frac{w_{k(n-1)}}{w_{kn}}. \tag{42}$$

Corollary (letting $a = 1$): If $k > 0$ and $u_k \neq 0$, then

$$\sum_{n=1}^{N} \frac{Q^{kn}}{w_{kn} w_{k(n+1)}} = \frac{Q^k}{eu_k^2} \left[\frac{w_0}{w_k} - \frac{w_{kN}}{w_{k(N+1)}} \right] = \frac{1}{eu_k^2} \left[\frac{w_{k(N+2)}}{w_{k(N+1)}} - \frac{w_{2k}}{w_k} \right]. \tag{43}$$

This agrees with the result given by Lucas [7] in 1878 for the sequence $\langle u_n \rangle$ and $\langle v_n \rangle$. Furthermore, when $k = 1$, we get the result found by Kappus [6] which generalized the result of Hillman in [6]. When $w_n = F_n$, this reduces to the results found by Swamy in [14]. When w_n is either the Pell polynomials or the Pell-Lucas polynomials, formula (43) is equivalent to results found by Mahon and Horadam in [8].

In a similar manner, applying the Dilation Theorem to Theorem 5 yields Theorem 14:

Theorem 14: If $a > 0$ and $k > 0$, then

$$u_{ka} \sum_{n=1}^{N} \frac{Q^{kn}}{w_{kn} w_{k(n+a)}} = u_{kN} \sum_{n=1}^{a} \frac{Q^{kn}}{w_{kn} w_{k(n+N)}}. \tag{44}$$

<u>Theorem 15:</u> If $a > 0$, $b > 0$, $k > 0$, and $u_{ka} \neq 0$, then

$$\sum_{n=1}^{N} \frac{Q^{kn}}{w_{kn+b} w_{k(n+a)+b}} = \frac{u_{kN}}{u_{ka}} \sum_{n=1}^{a} \frac{Q^{kn}}{w_{kn+b} w_{k(n+N)+b}}. \tag{45}$$

<u>Proof:</u> Apply the Translation Theorem (see [12]) to Theorem 6 to convert the sequence $\langle w_n \rangle$ into the sequence $\langle w_{n+b} \rangle$. □

If $P = x$ and $Q = -1$, Theorems 5, 6, and 14 give results about partial sums of Fibonacci polynomials that were found by Bergum and Hoggatt [2].

<u>Corollary (letting $a = 1$):</u> If $b > 0$, $k > 0$, and $u_k \neq 0$, then

$$\sum_{n=1}^{N} \frac{Q^{k(n-1)}}{w_{kn+b} w_{k(n+1)+b}} = \frac{u_{kN}}{u_k w_k + b w_{k(N+1)+b}}. \tag{46}$$

This is equivalent (with $b = a - k$) to the results found by Popov [10] for the sequences $\langle u_n \rangle$ and $\langle v_n \rangle$.

<u>Theorem 16:</u> If $a < b$, $k > 0$, and $u_{k(b-a)} \neq 0$, then

$$\sum_{n=1}^{N} \frac{Q^{kn}}{w_{k(n+a)} w_{k(n+b)}} = \frac{u_{kN}}{u_{k(b-a)}} \sum_{n=1}^{b-a} \frac{Q^{kn}}{w_{k(n+a)} w_{k(n+N+a)}}. \tag{47}$$

<u>Proof:</u> Apply the Translation Theorem (see [12]) to change the sequence $\langle w_m \rangle$ in Theorem 6 into the sequence $\langle w_{m+ka} \rangle$. Then let $c = b - a$. □

<u>Theorem 17:</u> If $k > 0$, $c > 0$, and $u_{kc} \neq 0$, then

$$\sum_{n=1}^{N} \frac{1}{w_{kn} w_{k(n+c)}} = \frac{u_k}{u_{kc}} \sum_{i=0}^{c-1} Q^{ki} (W_{k,N+i} - W_{k,i}) \tag{48}$$

where

$$W_{k,N} = \sum_{n=1}^{N} \frac{1}{w_{kn} w_{k(n+1)}}.$$

<u>Proof:</u> Apply the Dilation Theorem to Theorem 5. □

Applying the Translation Theorem to Theorem 15 gives us the following result.

<u>Theorem 18:</u> If $a < b$ and $u_{k(b-a)} \neq 0$, then

$$\sum_{n=1}^{N} \frac{1}{w_{k(n+a)}w_{k(n+b)}} = \frac{u_k}{u_{k(b-a)}} \sum_{i=0}^{b-a-1} Q^{ki}(W_{k,a+N+i} - W_{k,a+i}). \qquad (49)$$

8. SPECIAL CASES.

Although unit reciprocal sums of order 2 cannot in general be evaluated in closed form (without involving terms of the form U_N, V_N, W_N, X_N, or Y_N), a closed form can be found for some important special cases (such as when $Q = \pm 1$).

<u>Theorem 19:</u> If $Q = -1$, then

$$\sum_{n=1}^{N} \frac{1}{w_{n+a}w_{n+a+2}} = \frac{1}{P}\left[\frac{1}{w_{a+1}w_{a+2}} - \frac{1}{w_{N+a+1}w_{N+a+2}}\right]. \qquad (50)$$

<u>Proof:</u> Since $Q = -1$, we have $w_{m+2} = Pw_{m+1} + w_m$. Thus,

$$\frac{P}{w_{n+a}w_{n+a+2}} = \frac{Pw_{n+a+1}}{w_{n+a}w_{n+a+1}w_{n+a+2}} = \frac{w_{n+a+2} - w_{n+a}}{w_{n+a}w_{n+a+1}w_{n+a+2}}$$

$$= \frac{1}{w_{n+a}w_{n+a+1}} - \frac{1}{w_{n+a+1}w_{n+a+2}}.$$

Summing from 1 to N gives the desired result since the right-hand side telescopes. $\qquad\square$

<u>Lemma:</u> If $Q = 1$, then

$$c_k = \frac{r_1^k w_{k(n+1)} - w_{kn}}{r_1^{k(n+1)}} \qquad (51)$$

is independent of n. In particular, $c_k = (w_1 - w_0)r_2 u_k$.

<u>Proof:</u> Since $Q = 1$, we have $r_1 r_2 = 1$. The Binet form for w_n is known to be

$$w_n = Ar_1^n + Br_2^n$$

where $A = \frac{w_1 - w_0 r_2}{r_1 - r_2}$ and $B = \frac{w_0 r_1 - w_1}{r_1 - r_2}$. Then

$$c_k r_1^{k(n+1)} = r_1^k w_{k(n+1)} - w_{kn} = r_1^k\left[Ar_1^{k(n+1)} + Br_2^{k(n+1)}\right] - \left[Ar_1^{kn} + Br_2^{kn}\right]$$

$$= Ar_1^{kn+2k} - Ar_1^{kn}$$

$$= Ar_1^{kn}\left[r_1^{2k} - 1\right]$$

$$= Ar_1^{kn}\left[r_1^{2k} - (r_1 r_2)^k\right]$$

$$= A r_1^{k(n+1)} \left[r_1^k - r_2^k \right]$$

$$= A r_1^{k(n+1)} (r_1 - r_2) u_k.$$

Therefore $\quad c_k = A(r_1 - r_2) u_k = (w_1 - w_0 r_2) u_k.$ $\qquad\qquad$ □

Theorem 20: If $Q = 1$, $k \neq 0$, and $u_k \neq 0$, then

$$\sum_{n=1}^{N} \frac{1}{w_{kn} w_{k(n+1)}} = \frac{1}{(w_1 - w_0 r_2) r_1^k u_k} \left[\frac{1}{w_k} - \frac{1}{r_1^{kN} w_{k(N+1)}} \right]. \qquad (52)$$

Proof: Using the Lemma, we have

$$\frac{1}{r_1^{kn} w_{kn}} - \frac{1}{r_1^{k(n+1)} w_{k(n+1)}} = \frac{r_1^k w_{k(n+1)} - w_{kn}}{r_1^{k(n+1)} w_{kn} w_{k(n+1)}} = \frac{(w_1 - w_0 r_2) u_k}{w_{kn} w_{k(n+1)}}.$$

Summing as n goes from 1 to N, we find that the left side telescopes and we reach the stated result. $\qquad\qquad$ □

This theorem generalizes the results found by Melham and Shannon [9]. The idea for the proof comes from that paper. Alternatively, we could let $Q = 1$ in formula (43).

Corollary (letting $k = 1$): If $Q = 1$, then

$$\sum_{n=1}^{N} \frac{1}{w_n w_{n+1}} = \frac{1}{(w_1 - w_0 r_2) r_1} \left[\frac{1}{w_1} - \frac{1}{r_1^N w_{N+1}} \right]. \qquad (53)$$

9. OPEN PROBLEMS.

Query 1: Is there a simple closed form for any of the quantities U_N, V_N, W_N, X_N, or Y_N?

Query 2: Is there a simple algebraic relationship between any of the quantities U_N, V_N, W_N, X_N, and Y_N?

Query 3: Can $\displaystyle\sum_{n=1}^{N} \frac{1}{w_{2n} w_{2n+1}}$ be expressed in terms of U_N, V_N, W_N, X_N, and Y_N?

Query 4: Can $\displaystyle\sum_{n=1}^{N} \frac{1}{w_n^2}$ be expressed in terms of U_N, V_N, W_N, X_N, and Y_N?

Query 5: Can $\displaystyle\sum_{n=1}^{N} \frac{1}{w_n w_{n+1} w_{n+2}}$ be expressed in terms of U_N, V_N, W_N, X_N, and Y_N?

Query 6: Can $\displaystyle\sum_{n=1}^{N} \frac{1}{w_{n+a} w_{n+b} w_{n+c}}$ be expressed in terms of U_N, V_N, W_N, X_N, Y_N, a, b, and c?

REFERENCES

[1]   André-Jeannin, R. "Summation of Reciprocals in Certain Second-Order Recurring
       Sequences." *The Fibonacci Quarterly*, Vol. *35.1* (1997): pp. 68-74.

[2]   Bergum, G.E. and Hoggatt Jr., V.E. "Infinite Series with Fibonacci and Lucas
       Polynomials." *The Fibonacci Quarterly*, Vol. *17.2* (1979): pp. 147-151.

[3]   Brousseau, Brother Alfred. "Fibonacci-Lucas Infinite Series - Research Topic." *The
       Fibonacci Quarterly*, Vol. *7.2* (1969): pp. 211-217.

[4]   Good, I.J. "A Symmetry Property of alternating sums of Products of Reciprocals." *The
       Fibonacci Quarterly*, Vol. *32.3* (1994): pp. 284-287.

[5]   Hillman, A.P. "Solution to Problem B-697: A Sum of Quotients." *The Fibonacci
       Quarterly*, Vol. *30.3* (1992): pp. 280-281.

[6]   Kappus, Hans. "Solution to Problem B-747: Great Sums from Partial Sums." *The
       Fibonacci Quarterly*, Vol. *33.1* (1995): pp. 87-88.

[7]   Lucas, Edouard. "Théorie des Fonctions Numériques Simplement Périodiques."
       American Journal of Mathematics, Vol. *1* (1878): pp. 184-240, 289-321.

[8]   Mahon, Br. J.M. and Horadam, A.F. "Infinite Series Summation Involving Reciprocals
       of Pell Polynomials." Applications of Fibonacci Numbers, Volume 1. Edited by G.E.
       Bergum, A.F. Horadam and A.N. Philippou. D. Reidel Publishing Company, Dordrecht,
       The Netherlands, 1986: pp. 163-180.

[9]   Melham, R.S. and Shannon, A.G. "On Reciprocal Sums of Chebyshev Related
       Sequences." *The Fibonacci Quarterly*, Vol. *33.3* (1995): pp. 194-202.

[10]   Popov, Blagoj S. "Summation of Reciprocal Series of Numerical Functions of Second
       Order." *The Fibonacci Quarterly*, Vol. *24.1* (1986): pp. 17-21.

[11]   Rabinowitz, Stanley. "Algorithmic Manipulation of Fibonacci Identities." Applications
       of Fibonacci Numbers, Volume 6. Edited by G.E. Bergum, A.F. Horadam and
       A.N. Philippou. Kluwer Academic Publishers, Dordrecht, The Netherlands, 1996: pp.
       389-408.

[12]   Rabinowitz, Stanley. "Algorithmic Manipulation of Second-Order Linear Recurrences."
       The Fibonacci Quarterly, Vol. *37.2* (1999): pp. 162-177.

[13]   Rabinowitz, Stanley. "Algorithmic Summation of Reciprocals of Products of Fibonacci
       Numbers." *The Fibonacci Quarterly*, Vol. *37.2* (1999): pp. 122-127.

[14]   Swamy, M.N.S. "More Fibonacci Identities." *The Fibonacci Quarterly*, Vol. *4.4* (1966):
       pp. 369-372.

AMS Classification Numbers: 11B37, 11Y16, 11B39

SOLVED AND UNSOLVED PROBLEMS ON PSEUDOPRIME NUMBERS AND THEIR GENERALIZATIONS

Andrzej Rotkiewicz

1. PROBLEMS ON PERIODIC SEQUENCES OF PSEUDOPRIMES CONNECTED WITH CARMICHAEL NUMBERS.

Following recent papers [1], [5], [22], [23] a composite n is called a pseudoprime to base b if $b^{n-1} \equiv 1 \mod n$.

This definition does not coincide with the definition given in my book [35] where I defined a pseudoprime as a composite number dividing $2^n - 2$, "pseudoprimes with respect to b" as composite numbers n dividing $b^n - b$ and composite number n that divides $b^n - b$ for every integer b was called on absolute pseudoprime (see also [48]).

Leibniz in September 1680 and December 1681 gave incorrect proof that the number $2^n - 2$ is not divisible by n unless n is a prime [6], [15]. The first pseudoprime to base 2 was not discovered until 1819 when Sarrus noted that the number 341 is a pseudoprime to base 2. Since then many writers have given further pseudoprimes.

Following recent papers a composite number n is called a Carmichael number if $a^n \equiv a \mod a$ for every integer $a \geq 1$. The smallest Carmichael number is $561 = 3 \cdot 11 \cdot 17$. By Korselt's criterion [13] n is a Carmichael number if and only if n is squarefree and $p - 1$ divides $n - 1$ for all primes dividing n. The set of Carmichael numbers coincides with the set of composite n for which $a^{n-1} \equiv 1 \mod n$ ([24] pp. 118-119, [48] p. 217).

This paper is in final form and no version of it will be submitted for publication elsewhere.

In 1994 Alford, Granville and Pomerance [1] proved that there exist infinitely many Carmichael numbers and there are more than $x^{2/7}$ Carmichael numbers up to x for sufficiently large x. Recently, Conway, Guy, Schneeberger, Sloane [5] introduced the following

Definition 1: Any composite number q such that $b^q \equiv b \bmod q$, is called a *prime pretender* to base b.

Definition 2: By q_b we denote the least prime pretender q to base b and call such q *primary pretender*.

In the paper [43] I proved that for every $b > 1$ there exist infinitely many prime pretenders which are not pseudoprime to base b.

Schinzel [44] already in 1958 proved that $q_x(x = 1, 2, \cdots)$ is a periodic function of x with period 561! and that there exists b such that $q_b = 561$.

Schinzel [44] proved that $q_b \neq 4$, 6 if and only if $b \equiv 2$, 11 mod 12 and put forward the following problem:

Find all distinct primary pretenders [47].

Conway, Guy, Schneeberger, Sloane [5] proved that there are only 132 distinct primary pretenders and the least period of the function $q_x(x = 1, 2, \cdots)$ is the 122-digit number.

19 5685843334 6007258724 5340037736 2789820172 1382933760 4336734362 —

2947386477 7739548319 6097971852 9992599213 2923650684 2360439300

Let l_b denote the lease pseudoprime to base b. By theorem of Cipolla [4] the number $\dfrac{(n!)^{2p} - 1}{(n!)^2 - 1}$, where p is any odd prime such that p does not divide $(n!)^2 - 1$ is a pseudoprime to base $n!$. If k is a pseudoprime to base $n!$, then $(n!)^{k-1} \equiv 1 \bmod k$, hence $(k, n!) = 1$ and $k \geq l_{n!} > n$. Thus the number of distinct values of l_b is not bounded, since $l_{n!} > n$ and l_b is not a periodic function of b.

We shall introduce the following definition of the function l_x^C

Definition 3: Let C be a given Carmichael number. Then

$$l_x^C = \begin{cases} l_x & \text{if } (x, C) = 1 \\ 1 & \text{if } (x, c) \neq 1. \end{cases}$$

We have

$$l_1^{561} = l_1 = 4, \; l_2^{561} = l_2 = 341, \; l_3^{561} = 1, \; l_4^{561} = l_4 = 15, \; l_5^{561} = l_5 = 4, \; l_6^{561} = 1,$$

$$l_7^{561} = l_7 = 6, \; l_8^{561} = l_8 = 9, \; l_9^{561} = 1, \; l_{10}^{561} = l_{10} = 9.$$

We have $a^{C-1} \equiv 1 \mod C$ for every a coprime to C. Let $b \equiv a \mod C!$, then $b^{h-1} - 1 \equiv a^{h-1} - 1 \mod C!$, hence, for every $h \leq C$, $a^{h-1} \equiv 1 \mod h$ if and only if $b^{h-1} \equiv 1 \mod h$, hence $l_a^C = l_b^C$ for $(a, C) = 1$ and $b \equiv a \mod C!$. Thus in the sequence $\{l_a^C\}_{a=1}^{\infty}$, the numbers greater than 1 appear with period $C!$. On the other hand, in the sequence $\{l_a^C\}_{a=1}^{\infty}$ numbers equal 1 appear with period C. Since l.c.m. $(C!, C) = C!$, the sequence $\{l^C\}_{x=1}^{\infty}$ is a periodic sequence with period $C!$ and the function l_x^C has period $C!$. The following problems arise.

Problem 1: Given a Carmichael number C. Find the least period of the function l_x^C.

Problem 2: Given a Carmichael number C. Find all composite numbers n which are values of the function l_x^C.

Definition 4: The Carmichael number C has *property D* if there exist a natural base a, coprime to C such that $l_a^C = C$.

Definition 5: The Carmichael number C has *property A* if there exist a Carmichael number $C_1 < C$ such that $C_1 \mid C$.

Definition 6: The Carmichael number C has *property B* if there does not exist a Carmichael number $C_1 < C$ such that $C_1 \mid C$.

Denote by C_n the n-th Carmichael number. Among first 55 Carmichael numbers 7 have property A. These are: $C_{15} = 7 \cdot 13 \cdot 19 \cdot 37$, $C_{19} = 7 \cdot 13 \cdot 19 \cdot 73$, $C_{21} = 7 \cdot 13 \cdot 31 \cdot 61$, $C_{22} = 7 \cdot 13 \cdot 19 \cdot 109$, $C_{24} = 5 \cdot 17 \cdot 29 \cdot 113$, $C_{39} = 7 \cdot 13 \cdot 19 \cdot 433$, $C_{43} = 7 \cdot 13 \cdot 19 \cdot 577$. Five numbers: C_{15}, C_{19}, C_{22}, C_{39}, C_{43} are divisible by $C_3 = 7 \cdot 13 \cdot 19$ and $5 \cdot 17 \cdot 29 = C_4 \mid C_{24}$, $7 \cdot 17 \cdot 31 = C_5 \mid C_{24}$, $7 \cdot 13 \cdot 31 = C_5 \mid C_{21}$. The other 48 Carmichael numbers have property B.

The following theorem holds

Theorem 1 [43]: *A Carmichael number C has property D if an only if it has property B.*

I raised the question: Do there exist infinitely many Carmichael numbers with property D?

A. Schinzel proved that answer to this question is in the affirmative and the following theorem holds.

Theorem 2 [43]: *There exist infinitely many Carmichael numbers with property D. There exist infinitely many Carmichael numbers with property A.*

In the paper [43] I solved Problem 1.

Let $p!_k = p_1 p_2 \cdots p_k$ denote the product of the first k primes. Let ρ denote the least period of the function $l_x^C (x = 1, 2, \cdots)$ and $[a_1, a_2, \cdots, a_n]$ denote the least common multiple of the integer $a_1, a_2, \cdots, a_n$.

The following theorem holds

Theorem 3 [43]: *If a Carmichael number C has property D then the function $l_x^C (x = 1, 2, \cdots)$ has the period $C!$ and the least period of the function l_x^C is $\rho = p!_m p!_r$, where p_m is the largest prime such that $2p_m < C$ and p_r is the largest prime such that $(p_r)^2 < C$.*

If a Carmichael number C does not have property D, let C_1 denote the least Carmichael number such that $C_1 \mid C$, then the function $l_x^C (x = 1, 2, \cdots)$ has a period equal to $[C_1!, C]$ and the least period of the function l_x^C is equal to $[p!_{\overline{m}} p!_{\overline{r}}, C]$ where $p_{\overline{m}}$ denote the largest prime such that $2p_{\overline{m}} < C_1$ and $(p_{\overline{r}})^2 < C_1$.

2. PROBLEM OF POMERANCE.

Let α and β be algebraic numbers such that $\alpha + \beta = P$ and $\alpha\beta = Q$ are rational integers.

$$L(P, Q; n) = (\alpha^n - \beta^n)/(\alpha - \beta) \text{ for } n \geq 0 \qquad (1)$$

is called a Lucas sequence with parameters P and Q (α and β are distinct roots of the trinomial $z^2 - Pz + Q$; its discriminant is $D = P^2 - 4Q$).

If, instead of supposing that $\alpha + \beta \in \mathbb{Z}$, we only suppose that $(\alpha + \beta)^2 = L$ is a nonzero rational integer (that is α and β are roots of the trinomial $x^2 - \sqrt{L}x + Q$), then we define the Lehmer numbers $P(L, Q; n)$ by

$$P(L, Q; n) = \begin{cases} (\alpha^n - \beta^n)/(\alpha - \beta) & \text{if } n \text{ is odd} \\[2mm] (\alpha^n - \beta^n)/(\alpha^2 - \beta^2) & \text{if } n \text{ is even.} \end{cases} \qquad (2)$$

To eliminate trivial cases we assume that $\alpha\beta \neq 0$, and α/β is not a root of umty.

Definition 7: A composite n is called a Lucas pseudoprime with parameters P and Q if $(n, 2QD) = 1$ and

$$L(P, Q; n - (D/n)) \equiv 0 \bmod n, \qquad (3)$$

where (D/n) is the Jacobi symbol.

The above definition is a generalization of a pseudoprime to any base $b > 1$. Let $\Theta_b(x)$ denote the number of pseudoprimes to base b not exceeding x. Pomerance [22] showed that, for sufficiently large x, $\Theta_b(x) \geq \exp\{(\log x)^{5/14}\}$.

The Lucas pseudoprimes with $P = 1$, $D = 5$ are called Fibonacci pseudoprimes. From the paper of E. Lehmer [14] it follows that there exist infinitely many Fibonacci pseudoprimes n with $(5/n) = -1$.

Baillie and Wagstaff [2] suggest some primality tests which combine Lucas pseudoprimes with parameters P and Q so that $(D/n) = -1$. There exist very few n will be both Lucas pseudoprimes with parameters P and Q satisfying $(D/n) = -1$ and at the same time pseudoprime to the fixed base.

In the paper of Erdös, Kiss and Sárközy [7] it is shown that there are $> e^{(\log x)^c}$ Lucas pseudoprimes up to x with coprime parameters P, Q, where $c > 0$ does not depend on P, Q and in his recent dissertation J. Grantham [9] has shown that there are $> x^c$ Lucas pseudoprimes up to x. In both cases, the Jacobi symbol is $+1$, not -1. In this connection Pomerance put forward the following problems.

Given $x^2 - Px + Q$ with $D = P^2 - 4Q$ not a square

1. Does there exist at least one Lucas pseudoprime with parameters P and Q satisfying $(D/n) = -1$?

2. Do there exist infinitely many Lucas pseudoprimes n with parameters P and Q satisfying $(D/n) = -1$?

I proved [42] that the answer to the above problems is in the affirmative and even it is in the affirmative when Lucas pseudoprimes are replaces by the strong Lucas pseudoprimes in arithmetic progression.

The Lucas number V_k is defined by $V_k = \alpha^k + \beta^k$, where α and β are distinct roots of trinomial $x^2 - Px + Q$.

Definition 8: Let $n - (D/n) = 2^r s$, where s is odd. An odd composite n is a strong Lucas pseudoprime with parameters P, Q if $(n, 2QD) = 1$ and $L_s = (\alpha^s - \beta^s)/(\alpha - \beta) \equiv 0 \mod n$ or $V_{2^t s} = \alpha^{2^t s} + \beta^{2^t s} \equiv 0 \mod n$ for some t, $0 \leq t < r$, where α and β are roots of the trinomial $x^2 - Px + Q$.

Definition 9: A composite number n is called a Lehmer pseudoprime with parameters L and Q if $(n, 2LQD) = 1$ and $P(L, Q; (n - (DL/n))) \equiv 0 \mod n$.

The following theorems hold [42]

Theorem 4: *Given $x^2 - Px + Q$ with $D = P^2 - 4Q$ not a square. There exist infinitely many Lucas pseudoprimes n with parameters P, Q and $(D/n) = -1$.*

Theorem 5: *If D is not a square, then every arithmetic progression $ax + b$, where $(a,b) = 1$, which contains prime number p with $(D/p) = -1$ contains also an infinite number of odd strong Lucas pseudoprimes with parameters P and Q and $(D/n) = -1$.*

Theorem 6: *Given $x^2 - Px + Q$ with $D = P^2 - 4Q$ not a square. Let $\overline{R}(x)$ denote the number of Lucas pseudoprimes n with parameters P, Q and $(D/n) = -1$ not exceeding x. Then*

$$\overline{R}(x) \gg c(P,Q)\frac{\ln x}{\ln \ln x},$$

where $c(P,Q)$ is a positive constant depending on P and Q.

Williams [49] introduced the Lucas-Carmichael numbers.

Definition 10: (see [49]). An odd composite integer n is called a Carmichael-Lucas number associated to D if for a given value of D, has the following property (4)

$$
\begin{cases}
\text{For all integers } P, Q \text{ such that} \\
(P,Q) = 1, \ P^2 - 4Q = D, \ (n, QD) = 1 \\
\\
\text{we have} \\
L_{n - (D/n)}(P,Q) \equiv 0 \bmod n.
\end{cases}
\tag{4}
$$

If $D = 1$, then n is a Carmichael number.

The above follows from the following theorem of Williams [49].

If, for a fixed D, n is a Carmichael-Lucas number, then n is the product of k distinct primes $p_1, p_2, \cdots, p_k$ and

$$p_i - (D/p_i) \mid n - (D/p_i) \quad (i = 1, 2, \cdots, k).$$

Note that $323 = 17 \cdot 19$ is a Carmichael-Lucas numbers with $D = 5$; but it cannot be a Carmichael number, because it is the product of only two distinct primes.

Williams [49] asked whether there exists a Lucas-Carmichael number n with $(D/n) = -1$.

If such numbers D and n exist, n must be the product of an odd number of distinct primes $p_1, p_2, \cdots, p_k$, $(D/p_i) = -1 (i = 1, 2, 3, \cdots, k)$, and $p_i + 1 \mid n + 1$, $p_i - 1 \mid n - 1$ for

$i = 1, 2, 3, \cdots, k$, and k must be ≥ 5 (see Williams [49]).

It is not known whether any such number exist.

3. PROBLEMS OF SIERPIŃSKI.

Prime pretenders to base $2''$, we shall call "pseudoprime numbers" such terminology is usually used.

The most important theorems on pseudoprimes which answer to questions raised by Sierpiński are:

1. *Every arithmetical progression* $ax + b(x = 1, 2, \cdots)$, *where* $(a, b) = 1$ *contains an infinite number of pseudoprimes* [28], [32].

2. *There exist infinitly many squarefree pseudoprimes divisible by an arbitrary given prime* $p > 2$ [27].

3. *There exist infinitely many arithmetic progressions formed of four pseudoprimes* [36].

4. *There exist infinitely many pseudoprimes which are at the same time triangular* [30].

5. *There exist infinitely many pseudoprimes which are at the same time pentagonal* [31].

<u>Definition 11:</u> An odd composite n is an Euler pseudoprime to base a if $(a, n) = 1$ and $a^{(n-1)/2} \equiv (a/n) \bmod n$, where (a/n) is the Jacobi symbol.

In 1980 [37] I proved that for every odd $a \geq 3$ there exist infinitely many arithmetic progressions formed of three Euler pseudoprimes to base a.

Let $F_n = (\alpha^n - \beta^n)/(\alpha - \beta)$, where $\alpha = (1 + \sqrt{5})/2$ and $\beta = (1 - \sqrt{5})/2$ denote the n-th Fibonacci number. A composite n is called Fibonacci pseudoprime if $F_{n-(5/n)} \equiv 0 \bmod n$, where $(5/n)$ is the Jacobi symbol. The smallest Fibonacci pseudoprime is $323 = 17 \cdot 19$. The following three Fibonacci pseudoprimes form an arithmetical progression:

$$F_{73}, F_{74}, L_{73} = \alpha^{73} + \beta^{73} \text{ with the difference } F_{72}.$$

In 1995 I proved [40] that there are infinitely many arithmetical progressions formed by three different Fibonacci pseudoprimes.

Later [41] I generalized above theorem to Lucas pseudoprimes with the discriminant $D > 0$, such that $\overline{D} \equiv 1 \bmod 4$, where $\overline{D}$ denote the square-free kernel of D.

THE PRIME k-TUPLES CONJECTURE.

If a_i, b_i are integers > 0 for $i = 1, 2, \cdots, k$ and if the number of solution of

$$(a_1 x + b_1)(a_2 x + b_2) \cdots (a_k x + b_k) \equiv 0 \bmod p$$

is less than p for every p, then there are infinitely many integers m such that $a_1 m + b_1$, $a_2 m + b_2, \cdots, a_k m + b_k$ are prime numbers.

It is special case of conjecture H. of A. Schinzel [45]. Let s be a natural number and let $f_1(x), \cdots, f_s(x)$ be irreducible polynomials with integral coefficients such that the leading coefficient is positive. Then if there is no natural number > 1 which is a divisor of the product $f_1(x) \cdots f_s(x)$ for every integer x, then there exist infinitely many natural numbers x for which each of the numbers ia a prime number.

Sierpiński raised the question whether it follows from hypothesis H that there exist infinitely many arbitrarily long arithmetical progressions whose terms are pseudoprime numbers.

In 1969 [34] I proved that the answer to this question is in the affirmative. (Even if we instead of conjecture H, use the prime k-tuples conjecture.)

Later I generalized above theorem to Lehmer pseudoprimes [39].

In 1972 I proposed [35] (problem #29) whether it follows from hypothesis H. of A. Schinzel that there exist arbitrarily long arithmetic progressions of Carmichael numbers. Recently (letter of 21st of December 1995) Granville [10] proved the following theorem.

The prime k-tuples conjecture implies that there are arbitrarily long arithmetic progressions of Carmichael numbers.

The problem #28 of [35]:

Does there exist an infinity of arithmetic progressions consisting of five pseudoprimes? is still open. One such progression is: 172081, 285541, 393001, 512461, 625921.

Problem:

Does there exist for every even a an infinity of arithmetic progressions consisting of three pseudoprimes to base a?
is still open.

A composite number n is called super pseudoprime to base a if each divisor of n is a prime or pseudoprime to base a.

In 1968 [33] I proved that there are infinitely many super pseudoprimes which are products of exactly three distinct primes.

In 1983 J. Fehér and P. Kiss [8] proved the following theorems: *Let a be an integer with conditions $a > 1$ and $4 \nmid a$. Then there exist infinitely many super pseudoprime numbers to* *base a which are products of exactly three distinct primes.*

The problem of K. Szymiczek ([35] problem 31):

Do there exist infinitely many arithmetic progressions formed by three super pseudoprime numbers?

is still open.

4. PROBLEMS ON THE CONGRUENCES $a^{n-k} \equiv b^{n-k} \bmod n$, $a^{f(n)} \equiv b^{f(n)} \bmod n$ (cf. Ribenboim [24]).

The congruence $a^{n-k} \equiv 1 \bmod n$.

I asked in my book [35] (problem 18, p. 138) the following question:

Let $a, k > 1$ be fixed positive integers. Do there exist infinitely many composite integers n such that $n \mid (a^{n-k} - 1)$?

First we shall give a sketch of the proof of the theorem [38]: *There exist infinitely many positive integers n such that $2^{n-2} \equiv 1 \bmod n$.*

It all begins with D.H. and Emma Lehmer ([11], p. 250, F10) who found the number $n = 4700063497 = 19 \cdot 47 \cdot 5263229$, which, at present, is the only known integer such that $2^m \equiv 3 \bmod m$. Now we remark that if $2^m \equiv 3 \bmod m$ then $n = 2^m - 1$ satisfies the congruence $2^{n-2} \equiv 1 \bmod n$. Next, if p is a primitive prime factor of $2^{n-2} - 1$, then $n_1 = np$ also satisfies $2^{n_1 - 2} \equiv 1 \bmod n_1$.

This result was extended by Shen [46], who proved the following Theorem.

Each of the congruences $2^{n-k_i} \equiv 1 \bmod n(i = 0, 1, 2, \cdots)$, where $k_0 = 2$, $k_{i+1} = 2^{k_i} - 1$, has infinitely many solutions n. He found the following solutions of $2^{n-2} \equiv 1 \bmod n$: $20737 = 89 \cdot 233$, $93527 = 7 \cdot 31 \cdot 431$, $228727 = 127 \cdot 1801$, $373457 = 7 \cdot 31 \cdot 1721$, $540857 = 31 \cdot 73 \cdot 239$.

Benkoski observes, in his review [3] that Shen's five solutions are each congruent to 7 modulo 10 and asked if there was solution to $2^{n-2} \equiv 1 \bmod n$ which didn't end in 7 when written in decimal notation. Zhang [50] gave the solutions where $n \equiv 1$ or $3 \bmod 10$ and asked if there were any solution with $n \equiv 9 \bmod 10$.

We note here that Wayne McDaniel [18] answers Problem #22 of my book [35]:

Does there exist a pseudoprime of the form $2^n - 2$ McDaniel [18] shown, that for $N = 465794$, $2^N - 2$ is pseudoprime. McDaniel observes that if $2^N \equiv 3 \bmod N - 1$ then

$$(N-1) \mid (2^N - 3) \Rightarrow (2^{N-1} - 1) \mid (2^{2^N - 3} - 1) \Rightarrow (2^N - 2) \mid (2^{2^N - 2} - 2).$$

Numbers $n > 3$ for which $n \mid a^{n-3} - 1$ holds for $(a, n) = 1$ have been considered by D.C. Morrow [20], who has called them D numbers.

It is easy to see that every numbers of the form $n = 3p$, where p is a prime ≥ 3, is a D number.

A. Mąkowski [16] remarked that if $a^n \equiv k$ mod n, then $a^{s - (k-1)} \equiv 1$ mod s for $s = a^n - 1$.

A. Mąkowski [16] has proved that for any number $k \geq 2$ there exist infinitely many composite natural numbers n such that the relation $n \mid a^{n-k} - 1$ holds for any integer a with $(a, n) = 1$ (see also Sierpiński [48], p. 271).

In the Mąkowski proof $(k, a) = 1$, and so the question remained unanswered if a and k are fixed and $(k, a) > 1$.

In 1987 Kiss and Phong [12] gave a general solution by proving the following theorem: *Let $a \geq 2$ and k be fixed positive integers. Then there are infinitely many composite integers n such that $a^{n-k} \equiv 1$ mod n.*

THE CONGRUENCE $a^{n-k} \equiv b^{n-k}$ mod n.

In 1959 I proved [25] that there are infinitely many solutions of the above congruence for $k = 1$ and in 1972 [35] that there are infinitely many solutions for $k = 3$.

In 1987 Phong [21] proved the following theorem (T): *Let a and b be positive. Then for each integer $k \geq 2$ there are infinitively many composite n, which satisfy the congruence $a^{n-k} \equiv b^{n-k}$ mod n except possibly in the following cases (a, b, k): $(2^s + 1, 2^s - 1, 3)$ for $s \geq 2$, $(5 \cdot 2^t + 1, 5 \cdot 2^5 - 1, 3)$ for $t \geq 0$, $(c + 1, c, 2)$, $(c + 3, c, 2)$ for $c \geq 2$.*

THE CONGRUENCE $a^{f(n)} \equiv b^{f(n)}$ mod n.

Independently McDaniel [17] proved the theorem (T) and generalized it for the congruence $a^{f(n)} \equiv b^{f(n)}$ mod n [19], where f is a polynomial with integer coefficients which has a rational zero $r = k/c$ and some other conditions are satisfied.

Phong [21] in 1987 generalized problem for the Lehmer numbers and proved the following theorem: *Let $U(L, M)$ be the nondegenerate Lehmer sequence. Then there is an positive integer $k_0 = k_0(L, M)$ such that for each integer $k \geq k_0$ there are infinitely many composite numbers n, which satisfy the congruence*

$$U_{n-k(LK/n)} \equiv 0 \bmod n.$$

Moreover, if $k \geq 2$ and $(k, M) = 1$, then the congruence

$$U_{n-k} \equiv 0 \bmod n$$

has infinitely many composite solutions n.

In 1962 I considered [26] the congruence $a^{f(n)} \equiv b^{f(n)} \bmod n$, where $a > b \geq 1$, $(a, b) = 1$, $f(n)$ is a function of n such that $(p-1, f(n)) \mid f(np)$ for every prime p not dividing n.

In 1972 I proved [35] the following theorem: *Let f be a function such that if $p \nmid n$ and $p \equiv 1 \bmod f(n)$, then $f(n) \mid f(np)$. If $f(n)$ and $f(np)$ are divisible by same power of 2 and n_0 is a solution of*

$$a^{f(n)} + b^{f(n)} \equiv 0 \bmod n, \tag{5}$$

with the properties that $(a, b, f(n_0), n_0) \neq (2, 1, 3, 3)$, $1 < f(n_0) \geq n_0/2$ and $f(n) > n/4$ for $n > n_0$, then the congruence (5) has infinitely many composite solutions.

The problem 19 [35]: Let $f(n)$ be a function of n such that $(p-1, f(n)) \mid f(np)$ for every prime $p \nmid n$ and let $f(n) > \frac{n}{10}$ for $n > n_0$. Do there exist infinitely many positive integers n such that $n \mid 2^{f(n)} - 1$?

and problem 25 [35]: Do there exist infinitely many composite integers n such that $n \mid a^{\sigma(n)} - b^{\sigma(n)}$, where $\sigma(n)$ denote the sum of the natural divisors of a natural number n, for every n such that $(ab, n) = 1$ are still open. There are 7 such numbers of the form pq, where p and q are distinct primes. These are $2 \cdot 3$, $2 \cdot 7$, $3 \cdot 5$, $5 \cdot 7$, $5 \cdot 13$, $7 \cdot 13$, $7 \cdot 17$ and $13 \cdot 29$.

REFERENCES

[1]   Alford, W.R., Granville, A. and Pomerance C. "There are infinitely many Carmichael numbers." *Ann. of Math.*, Vol. *140* (1994): pp. 703-722.

[2]   Baillie R. and Wagstaff Samuel S. "Lucas pseudoprimes." *Math. Comp.*, Vol. *35* (1980): pp. 1391-1417.

[3]   Benkoski, S.J. Review of On the congruence $2^{n-k} \equiv \pmod{n}$. MR87e: 11005.

[4]   Cipolla, M. "Sui numeri composti P che verificaro la congruenza di Fermat $a^{P-1} \equiv 1 \pmod{P}$." *Annali di Matematica (3)*, Vol. *9* (1904): pp. 139-160.

[5]   Conway, J.H., Guy, R.K., Schneeberger, W.A. and Sloane, N.J.A. "The primary pretenders." *Acta Arith.*, Vol. *78* (1997): pp. 307-313.

[6]   Dickson, L.E. History of the Theory of Numbers, Carnegie Institute, Washington, 1919, 1920, 1923; reprinted Stechert, New York, 1934; Chelsea, New York, 1952, 1960, Vol. I, Chap. III.

[7]   Erdős, P., Kiss, P. and Sárközy, A. "Lower bound for the counting function." *Math. Comp.*, Vol. *51* (1988): pp. 315-323.

[8]   Fehér, J. and Kiss, P. "Note on super pseudoprime numbers." *Ann. Univ. Sci. Budapest, Eötvös Sect. Math.*, Vol. *26* (1983): pp. 157-159.

[9]   Grantham, J.F. Frobenius pseudoprimes. A dissertation submitted to the Graduate Faculty of the University of Georgia, Athens GA 1997.

[10]   Granville, A. *The prime k-tuplets conjecture implies that there are arbitrarily long arithmetic progressions of Carmichael numbers*, (written communication).

[11]   Guy, R.K. Unsolved Problems in Number Theory, Springer-Verlag, New York - Heidelberg, XVI + 285, p.p. 1994.

[12]   Kiss, P. and Phong, B.M. "On a problem of A. Rotkiewicz." *Math. Comp.*, Vol. *48* (1987): pp. 751-755.

[13]   Korselt, A. "Problème chinois." *L'intermédaire des mathématiciens*, Vol. *6* (1899): pp. 142-143.

[14]   Lehmer E. "On the infinitude of Fibonacci pseudoprimes." *The Fibonacci Quarterly*, Vol. *2* (1964): pp. 229-230.

[15]   Mahnke, D. "Leibniz and der Suche nach einer allgemeinem Primzahlgleichung." *Bibliotheca Math.*, Vol. *13* (1913): pp. 29-61.

[16]   Mąkowski, A. "Generalization of Morrow's D numbers." *Simon Stevin*, Vol. *36* (1962): p. 71.

[17]   McDaniel, W.L. "The generalized pseudoprime congruence $a^{n-k} \equiv b^{n-k} \pmod{n}$." *C. R. Math. Rep. Acad. Sci. Canada*, Vol. *9* (1987): pp. 143-148.

[18]   McDaniel, W.L. "Some pseudoprimes and related numbers having special forms." *Math. Comp.*, Vol. *53* (1989): pp. 407-409.

[19]   McDaniel, W.L. "The existence of solutions of the generalized pseudoprime congruence $a^{f(n)} \equiv b^{f(n)} \pmod{n}$." *Colloq. Math.*, Vol. *49* (1990): pp. 177-190.

[20]   Morrow, D.C. "Some properties of D numbers." *Amer. Math. Monthly*, Vol. *58* (1951): pp. 329-330.

[21]   Phong, B.M. "Generalized solutions of Rotkiewicz's problem." *Matematikai Lapok*, Vol. *34 (1-3)* (1987): pp. 109-119.

[22]   Pomerance, C. "A new lower bound for the pseudoprimes counting function." *Illinois J. Math.*, Vol. *26* (1982): pp. 4-9; MR 83h: 1012.

[23]   Pomerance, C., Selfridge, I.L. and Wagstaff, Samuel S. "The pseudoprimes to $25 \cdot 10^9$." *Math. Comp.*, Vol. *35* (1980): pp. 1003-1026.

[24]   Ribenboim, P. The New Book of Prime Number Records, Springer-Verlag, New York - Berlin - Heidelberg, 1995.

[25]   Rotkiewicz, A. "Sur les nombres composés n qui divisent $a^{n-1} - b^{n-1}$." *Rend. Circ. Mat. Palermo (2)*, Vol. *8* (1959): pp. 115-116; MR. 23#A1579.

[26]   Rotkiewicz, A. "Sur quelques généralisations des nombres pseudopremiers." *Colloq. Math.*, Vol. *9* (1962): pp. 109-113.

[27]   Rotkiewicz, A. "Sur les nombres premiers p et q tels que $pq \mid 2^{pq} - 2$." *Rend Circ. Mat. Palermo*, Vol. *11* (1962): pp. 280-282.

[28]   Rotkiewicz, A. "Sur les nombres pseudopremiers de la forme $ax + b$." *C. R. Acad. Sci. Paris*, Vol. *257* (1963): pp. 2601-2604.

[29]   Rotkiewicz, A. "Sur progressions arithmétiques et gométriques formés de trois nombres pseudopremiers distincts." *Acta Arith.*, Vol. *10* (1964): pp. 325-328.

[30]   Rotkiewicz, A. "Sur les nombres pseudopremiers triangulaires." *Elem. Math.*, Vol. *19* (1964): pp. 82-83.

[31]   Rotkiewicz, A. "Sur les nombres pseudopremiers pentagonaux." *Bull. Soc. Roy. Sci. Liège*, Vol. *33* (1964): pp. 261-263.

[32]   Rotkiewicz, A. "On the pseudoprimes of the form $ax + b$." *Proc. Combridge Phil. Soc.*, Vol. *63* (1963): pp. 389-392.

[33]   Rotkiewicz, A. "On the prime factors of the number $2^{p-1} - 1$." *Glasgow Math. J.*, Vol. *9* (1968): pp. 83-86.

[34]   Rotkiewicz, A. "On arithmetical progressions formed by k different pseudoprimes." *Journal of Math. Sciences*, Vol. *4* (1969): pp. 5-10.

[35]   Rotkiewicz, A. "Pseudoprime Numbers and Their Generalizations." Student Association of Faculty of Sciences, Univ. of Novi Sad, 1972, i + 169, pp.

[36]   Rotkiewicz, A. "The solution of W. Sierpiński's problem." *Rend. Circ. Mat. Palermo*, Vol. *28* (1979): pp. 62-64.

[37]   Rotkiewicz, A. "Arithmetical progressions formed from three different Euler pseudoprimes for the odd base a." *Rend. Circ. Mat. Palermo*, Vol. *29* (1980): pp. 420-426.

[38] Rotkiewicz, A. "On the congruence $2^{n-2} \equiv 1 \pmod n$." *Math. Comp.* , Vol. *43* (1984): pp. 271-272.

[39] Rotkiewicz, A. "Arithmetical progressions formed by k different pseudoprimes." *Rend. Circ. Mat. Palermo*, Vol. *69* (1994).

[40] Rotkiewicz, A. "There are infinitely many arithmetical progressions formed by three different Fibonacci pseudoprimes." Applications of Fibonacci Numbers, Volume 7. Edited by G.E. Bergum, A.F. Horadam and A.N. Philippou. Kluwer Academic Publishers, Dordrecht, The Netherlands, 1998: pp. 327-332.

[41] Rotkiewicz, A. "Arithmetical progressions formed by Lucas pseudoprimes." Number Theory, Proceedings Eger Conference. Edited by K. Györy, A. Pethő, V.T. Sós, Walter de Gruyter. Berlin-New York 1998: pp. 465-472.

[42] Rotkiewicz, A. "Lucas and Lehmer pseudoprimes with negative Jacobi symbol." (to appear in *Acta Arith.*).

[43] Rotkiewicz, A. "Periodic sequences of pseudoprimes connected with Carmichael numbers and the least period of the function l_x^C." (to appear in *Acta Arith.*).

[44] Schinzel, A. "Sur les nombres composés n qui divisent $a^n - a$." *Rend. Circ. Mat. Palermo (2)*, Vol. *7* (1958): pp. 37-41.

[45] Schinzel, A. and Sierpiński, W. "Sur certaines hypothèses concernnt les nombres premiers." *Acta Arith.*, Vol. *4* (1958): pp. 185-208, Correction ibid. Vol. *5* (1958): p. 259.

[46] Shen, Mok-King. "On the congruence $2^{n-k} \equiv 1 \pmod n$." *Math. Comp.*, Vol. *46* (1986): pp. 715-716.

[47] Sierpiński, W. "Remark on composite numbers m dividing $a^m - a$ (in Polish)." *Wiadom. Mat.*, Vol. *4* (1961): pp. 183-184.

[48] Sierpiński, W. "Elementary Theory of Numbers." *Monografie Matematyczne*, Vol. *42* PWN, Warsaw (1964), (second edition, North-Holland, Amsterdam, 1987).

[49] Williams, H.C. "On numbers analogous to the Carmichael numbers." *Canad. Math. Bull.*, Vol. *20.1* (1977): pp. 133-143.

[50] Zhang, Ming-Zhi. "A note on the congruence $2^{n-2} \equiv 1 \bmod n$." (Chinese and English summary), *Sichuan Daxue Xuebao*, Vol. *27* (1990): pp. 130-131.

AMS Classification Numbers: 11A07, 11B39

SOME RELATIONSHIPS AMONG VIETA, MORGAN-VOYCE AND JACOBSTHAL POLYNOMIALS

A.G. Shannon and A.F. Horadam

1. INTRODUCTION

We define below the Morgan-Voyce polynomials $B_n(x)$ and $b_n(x)$ [21], and the 'companion' Morgan-Voyce polynomials $C_n(x)$ and $c_n(x)$ which were found in embryonic form in [26] and formalized, particularly in terms of notation, by Horadam [12, 16]. Horadam [11, 13] has also developed properties for the Jacobsthal polynomials, $J_n(x)$. It is the purpose of this paper to reveal some new inter-relationships among these polynomials and between them and the Fibonacci and Lucas numbers.

We shall utilize the notation of Horadam [9] wherein the general second-order linear recursive sequence $[w_n] \equiv \{w_n(a, b; p, q)\}$ is defined by the homogeneous linear recurrence relation

$$w_n = pw_{n-1} - qw_{n-2}, n \geq 2, \tag{1.1}$$

with initial conditions $w_0 = a$, $w_1 = b$, and where $a, b, p, q \in \mathbb{Z}$. Accordingly we utilise this notation to define the polynomials under discussion in Table 1, where we also include the well-known sequences of Fibonacci and Lucas for reference purposes.

This paper is in final form and no version of it will be submitted for publication elsewhere.

a	b	p	q	w_n	Sequence
0	1	1	-1	F_n	Fibonacci
2	1	1	-1	L_n	Lucas
0	1	$x+2$	1	$B_n(x)$	Morgan-Voyce Even Fibonacci
1	1	$x+2$	1	$b_n(x)$	Morgan-Voyce Odd Fibonacci
2	$x+2$	$x+2$	1	$C_n(x)$	Morgan-Voyce Even Lucas
-1	1	$x+2$	1	$c_n(x)$	Morgan-Voyce Odd Lucas
0	1	1	$-x$	$J_n(x)$	Jacobsthal-Fibonacci
2	1	1	$-x$	$j_n(x)$	Jacobsthal-Lucas
0	1	x	1	$V_n(x)$	Vieta-Fibonacci
2	x	x	1	$v_n(x)$	Vieta-Lucas

Table 1. Integer and Polynomial Sequences $\{w_n(a,b;p,q)\}$.

The names assigned to the various sequences require some explanation. For example, it will be shown in the next section in Table 3(s) that the sums of the coefficients of $B_n(x)$ are even-suffixed Fibonacci numbers; similarly for the other Morgan-Voyce polynomials. The Jacobsthal-Fibonacci and Jacobsthal-Lucas numbers not only have the sums of their coefficients equal to Fibonacci and Lucas numbers respectively, but also, from Table 1,

$$\{J_n(1)\} \equiv \{F_n\} \text{ and } \{j_n(1)\} \equiv \{L_n\}.$$

The Vieta-Fibonacci polynomials, which have been studied as the Vieta Polynomials by Robbins [22] and which have links with the Chebychev polynomials [10], are mentioned in the next section. (We have changed their symbol to lower-case for consistency.) From Table 1 we can establish that

$$\{V_n(3) \equiv \{F_{2n}\} \equiv \{B_n(1)\}$$

and

$$\{v_n(3) \equiv \{L_{2n}\} \equiv \{C_n(1)\}.$$

so that they are related to the Morgan-Voyce polynomials and to one another as the Fibonacci are to the Lucas numbers.

2. SUMS OF COEFFICIENTS

The general terms of the polynomials are

$$B_n(x) = \sum_{k=0}^{n-1} \binom{n+k}{2k+1} x^k, \tag{2.1}$$

$$C_n(x) = \sum_{k=0}^{n-1} \frac{2n}{n-k} \binom{n+k-1}{n-k-1} x^k + x^n, \tag{2.2}$$

$$J_n(x) = \sum_{k=0}^{\lfloor (n-1)/2 \rfloor} \binom{n-k-1}{k} x^k \tag{2.3}$$

which suggest that some links may be found by investigating the coefficients of the various polynomials. These are set out in Tables 2(a) and (b) for the Morgan-voyce Fibonacci polynomials, Tables 3(a) and (b) for the Morgan-Voyce Lucas polynomials, and Tables 4(a) and (b) for the Jacobsthal polynomials. Included in the tables are the associated diagonal sums, row sums, and partial column sums where various patterns also emerge.

Before setting out these tables it is convenient at this point to list the general terms for $b_n(x)$, $c_n(x)$ and $j_n(x)$. These are developed analogously from (2.1), (2.2) and (2.3) from the general term for generalized Fibonacci numbers in Filipponi [5] and the general term for generalized Lucas numbers in Carlitz [2] which have been collated in a current paper by Melham [20].

$$b_n(x) = \sum_{k=0}^{\lfloor n/2 \rfloor} \frac{n-k}{n+k} \binom{n+k}{2k} x^k, \tag{2.4}$$

$$c_n(x) = \sum_{k=0}^{n-1} \frac{2n+1}{n-k-1} \binom{n+k-1}{n-k-2} x^k + x^n, \tag{2.5}$$

$$j_n(x) = \sum_{k=0}^{\lfloor n/2 \rfloor} \frac{n}{n-k} \binom{n-k}{k} x^k. \tag{2.6}$$

n	Diagonal Sums	Row	Partial
	0 1 2 4 8 16 32 64 128	Sums	Column Sums
0	0	0	0
1	1	1	1
2	2 1	3	3 1
3	3 4 1	8	6 5 1
4	4 10 6 1	21	10 15 7 1
5	5 20 21 8 1	55	15 35 28 9 1
6	6 35 56 36 10 1	144	21 70 84 45 11 1
7	7 56 126 120 55 12 1	377	28 126 210 165 66 13 1

Table 2(a). Coefficients and Sums Associated with
Morgan-Voyce Even Fibonacci Polynomials, $B_n(x)$.

n	Diagonal Sums	Row	Partial
	1 1 1 2 4 8 16 32 64 128	Sums	Column Sums
0	1	1	1
1	1	1	2
2	1 1	2	3 1
3	1 3 1	5	4 4 1
4	1 6 5 1	13	5 10 6 1
5	1 10 15 7 1	34	6 20 21 8 1
6	1 15 35 28 9 1	89	7 35 56 36 10 1
7	1 21 70 84 45 11 1	233	8 56 126 120 55 12 1

Table 2(b). Coefficients and Sums Associated with
Morgan-Voyce Odd Fibonacci Polynomials, $b_n(x)$.

n	Diagonal Sums	Row	Partial
	2 2 3 6 12 24 48 96	Sums	Column Sums
0	2	2	2
1	2 1	3	4 1
2	2 4 1	7	6 5 1
3	2 9 6 1	18	8 14 7 1
4	2 16 20 8 1	47	10 30 27 9 1
5	2 25 50 35 10 1	123	12 55 77 44 11 1
6	2 36 105 112 54 12 1	322	14 91 182 156 65 13 1
7	2 49 196 294 210 77 14 1	843	16 140 378 450 275 90 15 1

Table 3(a). Coefficients and Sums Associated with
Morgan-Voyce Even Lucas Polynomials, $C_n(x)$.

n	Diagonal Sums	Row	Partial
	− 1 1 3 6 1 2 24 48 96	Sums	Column Sums
0	−1	−1	−1
1	1	1	0
2	3 1	4	3 1
3	5 5 1	11	8 6 1
4	7 14 7 1	29	15 20 8 1
5	9 30 27 9 1	76	24 50 35 10 1
6	11 55 77 44 11 1	199	35 105 112 54 12 1
7	13 91 182 156 65 13 1	521	48 196 294 210 77 14 1

Table 3(b). Coefficients and Sums Associated with
Morgan-voyce Odd Lucas Polynomials, $c_n(x)$.

n	Diagonal Sums	Row	Partial
	0 1 1 1 2 3 4 6 9 13 19	Sums	Column Sums
0	0	0	0
1	1	1	1
2	1	1	2
3	1 1	2	3 1
4	1 2	3	4 3
5	1 3 1	5	5 6 1
6	1 4 3	8	6 10 4
7	1 5 6 1	13	7 15 10 1

Table 4(a). Coefficients and Sums Associated with
Jacobsthal-Fibonacci Polynomials, $J_n(x)$.

n	Diagonal Sums	Row	Partial
	2 1 1 3 4 5 8 12 17 25	Sums	Column Sums
0	2	2	2
1	1	1	3
2	1 2	3	4 2
3	1 3	4	5 5
4	1 4 2	7	6 9 2
5	1 5 5	11	7 14 7
6	1 6 9 2	18	8 20 16 2
7	1 7 14 7	29	9 27 30 9

Table 4(b). Coefficients and Sums Associated with
Jacobsthal-Lucas Polynomials, $j_n(x)$.

The coefficients of the Jacobsthal-Lucas polynomial for $n > 0$ form Vieta's array [22], and

$$v_n(x) = x^n j_n(-1/x^2). \tag{2.7}$$

From Equations (2.1) and (2.3) we also obtain

$$B_n(x) = x^{n-1} J_{2n}(1/x). \tag{2.8}$$

which we can also observe from the coefficients in Tables 3(a) and 5(a). Since we use Equation (2.5) in the next section we outline its proof, namely:

$$x^{n-1} J_{2n}(1/x) = \sum_{k=0}^{n-1} \binom{2n-k-1}{k} x^{n-k-1}$$

$$= \sum_{k=0}^{n-1} \binom{n+k}{n-k-1} x^k$$

$$= \sum_{k=0}^{n-1} \binom{n+k}{2k+1} x^k$$

$$= B_n(x). \qquad \qquad \square$$

In Table 5 we observe various sequences among the diagonal, row and partial column sums. We shall then explore them further to come up with a Fibonacci triangle.

Sequence	Diagonals*	Rows	Partial Columns
$B_n(x)$	$a_n = 2a_{n-1}, a_0 = 1$	F_{2n}	Coeffts of $b_n(x)$***
$b_n(x)$	$a_n = 2a_{n-1}, a_0 = 1$	F_{2n-1}	Coeffts of $B_n(x)$
$C_n(x)$	$a_n = 2a_{n-1}, a_0 = 3$	L_{2n}	Coeffts of $c_n(x)$
$c_n(x)$	$a_n = 2a_{n-1}, a_0 = 3$	L_{2n-1}	Coeffts of $C_n(x)$
$J_n(x)$	$a_n = a_{n-1} + a_{n-3}, a_0 = 0$**	F_n	Coeffts of $J_n(x)$
$j_n(x)$	$a_n = a_{n-1} + a_{n-3}, a_0 = 2$	L_n	Coeffts of $j_n(x)$

Table 5. Patterns among Row, Diagonal and Column Sums;

*some initial terms are ignored; *$a_1 = a_2 = 1$; Sequence M0571 of [23];

***In some of these, the constant's column does not occur: e.g., $b_n(x)$

and $B_n(x)$, but we are concerned with an overall perspective.

The choice of nomenclature for the various sequences now appears reasonably logical. We can, however, learn much more from these patterns. For example, the coefficients of $J_n(x)$ and $j_n(x)$ satisfy the partial recurrence relation

$$a_{n,m} = a_{n-1,m} + a_{n-2,m-1}, \quad n \geq 1, m \geq 1,$$

with boundary conditions $a_{n,1} = 1 - a_{n,n}, a_{n,m} = 0, m > n$, which suggests that they could be studied in terms of lattice permutations along the lines of Carlitz and Riordan [3].

We not look at the coefficients in Tables 3(a) and (b) for the Morgan-Voyce Lucas polynomials. If we specify the symbolism

$$a_{n,m} \epsilon C_n(x) \text{ and } a'_{n,m} \epsilon c_n(x).$$

then we find that the coefficients are inter-related by the partial recurrence relations

$$a_{n,m} = a'_{n,m-1} + a_{n-1,m},$$

and

$$a'_{n,m} = a_{n-1,m} + a'_{n-1,m}.$$

The significance of these is that it means that the coefficients of the Morgan-Voyce Lucas polynomials are the elements of the Lucas triangle defined by Feinberg [4] (except for $n = 0$) and displayed in Table 6. The elements are generated by the coefficients in the expansion of

$$(a-b)^{n-1}(a+2b), \quad n \geq 1.$$

$n \backslash m$	0 1 2 3 4 5 6 7 8
0	2
1	1 2
2	1 3 2
3	1 4 5 2
4	1 5 9 7 2
5	1 6 14 16 9 2
6	1 7 20 30 25 11 2
7	1 8 27 50 55 36 13 2
8	1 9 35 77 105 91 49 15 2

Table 6. Lucas Triangle.

The elements of this triangle were shown by Gould and Greig [6] to satisfy the partial recurrence relation

$$a_{n,m} = a_{n-1,m} + a_{n-1,m-1}. \tag{2.9}$$

with boundary conditions $a_{1,0} = 1$, $a_{1,1} = 2$ and $a_{n,m} = 0$ for $m < 0$ or $m > n$. If we reverse the procedure we used in Tables 3(a) and (b), then we can use the coefficients of the Morgan-Voyce Fibonacci polynomials in Tables 2(a) and (b) to generate a Fibonacci triangle which corresponds to the Lucas triangle. This is a feature which is missing from Bondarenko's compendium of such triangles [1]. In order to achieve this we use the same partial difference equation (2.5) with $m > 1$, but with the boundary conditions:

$$a_{2n,0} = 0, \ a_{2n+1,0} = 1, \ a_{n,1} = 1, \ a_{n,n} = n, \ a_{n,m} = 0 \text{ if } n < m.$$

We then get the Fibonacci Triangle displayed in Table 7. The elements are generated by the coefficients in the expansion of

$$(a-b)^{n-1}(a+2b), \quad n \geq 1.$$

$n\backslash m$	0 1 2 3 4 5 6 7 8
0	0
1	1 1
2	0 1 2
3	1 1 3 3
4	0 1 4 6 4
5	1 1 5 10 10 5
6	0 1 6 15 20 15 6
7	1 1 7 21 35 35 21 7
8	0 1 8 28 56 70 56 28 8

Table 7. Fibonacci Triangle.

Observe that this contains the coefficients of the Morgan-Voyce Fibonacci polynomials in the same way as the Lucas triangle contains the coefficients of the Morgan-Voyce Lucas polynomials, and that the diagonal sums yield the Fibonacci numbers, just as the diagonal sums of the Lucas triangle yield the Lucas numbers. It is of interest to note that Fibonacci triangles are appearing in the emerging theory of vines [19].

3. DIFFERENCES AMONG POLYNOMIALS

We define difference operators δ_1 and δ_2 for a general recursive polynomial, $w_n(x)$ [cf. 25], by

$$\delta_1 w_n(x) = w_{n+1}(x) - w_n(x), \quad n \geq 0.$$

$$\delta_2 w_n(x) = w_n(x+1) - w_n(x), \quad \text{for any } x.$$

Immediately from Tables 2, 3 and 4, we observe that

$$\delta_1 b_n(x) = x B_n(x), \tag{3.1a}$$

$$\delta_1 c_n(x) = C_n(x), \tag{3.1b}$$

$$\delta_1 j_n(x) = x j_{n-1}(x). \tag{3.1c}$$

Less obviously we can obtain

$$\delta_1^m b_n(x) = x^{m+n-1} J_{m=2n-1}(1/x), \tag{3.2a}$$

$$\delta_1^m c_n(x) = x^{m+n-1} j_{m+2n-1}(1/x), \tag{3.2b}$$

$$\delta_1^m j_n(x) = x^m j_{n-m}(x). \tag{3.2c}$$

The result (3.2c) follows by induction from repeated application of (3.1c). We give some examples of (3.2a) and (3.2b).

From the theory of difference operators [7] it can be readily established that

$$\delta_1^r w_n(x) = \sum_{k=0}^{r} (-1)^k \binom{r}{k} w_{n+r-k}(x),$$

$$\delta_2^r w_n(x) = \sum_{k=0}^{r} (-1)^k \binom{r}{k} w_n(x+r-k).$$

Thus,

$$\delta_1^3 b_2(x) = b_5(x) - 3b_4(x) + 3b_3(x) - b_2(x)$$

$$= x^4 + 4x^3 + 3x^2$$

$$= x^4\left(1 + \frac{4}{x} + \frac{3}{x^2}\right)$$

$$= x^4 J_6(1/x),$$

and

$$\delta_1^3 c_3(x) = c_6(x) - 3c_5(x) + 3c_4(x) - c_3(x)$$

$$= 2x + 16x^2 + 20x^3 + 8x^4 + x^5$$

$$= x^5\left(1 + \frac{8}{x} + \frac{20}{x^2} + \frac{16}{x^3} + \frac{2}{x^4}\right)$$

$$= x^5 j_8(1/x).$$

To prove (3.2 a or b) we need first to establish by induction on m that, for instance,

$$\delta_1^m b_n(x) = \begin{cases} x^{m*+1} B_{n+m*}(x), & m \text{ odd}, \tag{3.3a} \\ \\ x^{m*} b_{n+m*}(x), & m \text{ even}, \tag{3.3b} \end{cases}$$

in which $m* = \lfloor m/2 \rfloor$. Examples of this are seen later. The inductive step can be illustrated as follows:

$$\delta_1^m b_n(x) = \delta_1(\delta_1^{2k} b_n(x)) \text{ if } m = 2k+1 \text{ say,}$$

$$= \delta_1 x^k b_{n+k}(x),$$

$$= x^{k+1} B_{n+k}(x).$$

We then utilize

$$\delta_1^k b_n(x) = \delta_1(\delta_1^{k-1} b_n(x))$$

$$= \delta_1 x^{(k-1)/2} b_{n+(k-1)/2}(x) \qquad \text{(if } k \text{ is odd, by (3.3b))}$$

$$= x^{(k+1)/2} B_{n+(k-1)/2}(x) \qquad \text{(by (3.1a))}$$

$$= x^{k+n-1} J_{2n+k-1}(1/x) \qquad \text{(by (2.4))}.$$

We can also obtain neat results by successive application of the operator δ_2:

$$\delta_2^m b_m(x) = 0, \qquad \qquad (3.4a)$$

$$\delta_2^m c_m(x) = 0, \qquad \qquad (3.4b)$$

$$\delta_2^m j_m(x) = 0. \qquad \qquad (3.4c)$$

For example

$$\delta_2^3 c_3(x) = c_3(x+3) - 3c_3(x+2) + 3c_3(x+1) - 3c_3(x)$$

$$= (x^2 + 11x + 29) - 3(x^2 + 9x + 19) + 3(x^2 + 7x + 11) - (x^2 + 5x + 5)$$

$$= 0.$$

The proof for (3.4a) is typical for all three results in (3.3). We use induction on n. Initially, for $n = 1$,

$$\delta_2^1 b_1(x) = b_1(x+1) - b_1(x) = 0.$$

Suppose the result is true for $n = 2, 3, \cdots, m-1$. Then

$$\delta_2^m b_m(x) = \delta_2(\delta_2^{m-1} b_m(x)), \text{ and from the definition of } b_n(x),$$

$$= \delta_2(\delta_2^{m-1}(x+2)b_{m-1}(x) - \delta_2^{m-1} b_{m-2}(x))$$

$$= (x+2)\delta_2(\delta_2^{m-1} b_{m-1}(x)) - \delta_2^2(\delta_2^{m-2} b_{m-2}(x))$$

$$= 0 \qquad \qquad \text{by the inductive hypothesis,}$$

and so Equation (3.3a) follows by the Principle of Mathematical Induction.

In Table 8, values of $b_n(x)$ are displayed for various values of x and n. Some of the pertinent sequences from Sloane and Plouffe [23] are also noted. In particular, the bisections of the Fibonacci and Lucas sequences are generated from the following recurrence relations and initial conditions

$$b_n(1) = 3b_{n-1}(1) - b_{n-2}(1), n \geq 2, b_0(1) = b_1(1) = 1,$$

and

$$b_n(-5) = -3b_{n-1}(-5) - b_{n-2}(-5), n \geq 2, b_0(-5) = b_1(-5) = 1.$$

$x \backslash n$	0	1	2	3	4	5	Sloane
-6	1	1	-5	19	-71	265	$M3890$
-5	1	1	-4	11	-29	76	$M4320$
-4	1	1	-3	5	-7	9	$M2400$
-3	1	1	-2	1	1	-2	R_n
-2	1	1	-1	-1	1	1	S_n
-1	1	1	0	-1	-1	0	T_n
0	1	1	1	1	1	1	$M0003$
1	1	1	2	5	13	34	$M1439$
2	1	1	3	11	41	153	$M2894$
3	1	1	4	19	91	436	$M3553$
4	1	1	5	29	169	985	$M3955$

Table 8. $b_n(x)$.

We observe in passing the periodicity in Table 8 of the sequences:

$$R_n = R_{n+3}, \quad S_n = S_{n+4}, \quad T_n = T_{n+6}.$$

For those who wish to pursue these further, we observe that from Simson's identity we have another periodic sequence:

$$Q_n = Q_{n+2},$$

where

$$Q_n = F_n^2 - F_{n+1}F_{n-1}.$$

Tables 9(a) and (b) illustrate the application of the difference operators to the elements of Table 8.

$x \backslash n$	0	1	2	3	4	5
-6	0	0	1	-10	68	-401
-5	0	0	1	-8	42	-189
-4	0	0	1	-6	22	-67
-3	0	0	1	-4	8	-11
-2	0	0	1	-2	0	3
-1	0	0	1	0	-2	-1
0	0	0	1	2	2	1
1	0	0	1	4	12	33
2	0	0	1	6	28	119

Table 9(a). $\delta_2 b_n(x)$.

$x \backslash n$	1	2	3	4	5	6
-6	0	-6	24	-90	336	-1254
-5	0	-5	15	-40	105	-275
-4	0	-4	8	-12	16	-20
-3	0	-3	3	0	-3	3
-2	0	-2	0	2	0	-2
-1	0	-1	-1	0	1	1
0	0	0	0	0	0	0
1	0	1	3	8	21	55
2	0	2	8	30	112	418

Table 9(b). $\delta_1 b_n(x)$.

Table 10 shows the effects of Equation (3.2a). For example,

$$\delta_1^1 b_1(x) = x;$$

$$\delta_1^2 b_2(x) = x^2 b_3(x);$$

$$\delta_1^4 b_4(x) = x^4 b_6(x).$$

More specifically

$$\delta_1^2 b_2(3) = 3^3 J_5(1/3)$$

$$= (27 \times 19)/9$$

$$= 57;$$

$$\delta_1^3 b_3(3) = 3^5 J_8(1/3)$$

$$= (3^5 \times 115)/27$$

$$= 1035;$$

$$\delta_1^4 b_4(2) = 2^7 J_{11}(1/2)$$

$$= (2^7 \times 571)/32$$

$$= 2284.$$

$x = \delta_1^1 b_1(x)$	$\delta_1^2 b_2(x)$	$\delta_1^3 b_3(x)$	$\delta_1^4 b_4(x)$
1	5	21	89
2	22	224	2284
3	57	1035	18801
4	116	3264	91856
5	205	8225	330025
6	330	17856	966204
7	497	34839	2442209
8	712	62720	55225056

Table 10. $\delta_1^r b_r(x)$.

Other difference properties can be developed. For instance,

$$x B_n(x) = b_k(x) b_{n-k+2}(x) - b_{k-1}(x) b_{n-k-1}(x). \qquad (3.4)$$

Proof: $\quad x B_n(x) = \delta_1 b_n(x) \qquad\qquad\qquad\qquad$ by (3.1a)

$$= b_{n+1}(x) - b_n(x) \qquad\qquad\qquad \text{by definition of } \delta_1$$

$$= (x+1) b_n(x) - b_{n-1}(x) \qquad\qquad \text{by definition of } b_n(x)$$

$$= ((x+1)(x+2) - 1) b_{n-1}(x) - (x+1) b_{n-2}(x)$$

$$= (x^2 + 3x + 1) b_{n-1}(x) - (x+1) b_{n-2}(x)$$

$$= (x^3 + 5x^2 + 6x + 1) b_{n-2}(x) - (x^2 + 3x + 1) b_{n-3}(x)$$

$$= b_4(x) b_{n-2}(x) - b_3(x) b_{n-3}(x)$$

$$= \cdots$$

$$= b_k(x) b_{n-k+2}(x) - b_{k-1}(x) b_{n-k+1}(x). \qquad\qquad\qquad \square$$

The analogous result for the Jacobsthal polynomials is found similarly by successively applying the recurrence relation for $j_n(x)$, so that we obtain

$$j_{n+1}(x) + x j_n(x) = (1+x) j_n(x) + x j_{n-1}(x)$$

$$= J_3(x) j_n(x) + x J_2(x) j_{n-1}(x),$$

and eventually

$$j_{n+r}(x) + x j_{n+r-1}(x) = J_{r+2}(x) j_n(x) + x J_{r+1}(x) j_{n-1}(x). \qquad (3.5)$$

When $x = 1$, this becomes

$$L_{n+r-1} = F_{r+2}L_n + F_{r+1}L_{n-1}$$

which is a particular case of Equation (3.14) of Horadam [8].

4. CONCLUSION

There are other ways in which the Morgan-Voyce and Jacobsthal polynomials can be related. One such is with the Vieta-Fibonacci polynomial, $V_n(x)$, defined by

$$V_n(x) = xV_{n-1}(x) - V_{n-2}(x), \quad n \geq 2, \tag{4.1}$$

with $V_0(x) = 0$, $V_1(x) = 1$, which are part of a fourth order sequence [17].

The next few polynomials in this sequence are

$$V_2(x) = x,$$
$$V_3(x) = x^2 - 1,$$
$$V_4(x) = x^3 - 2x.$$

The two types of Vieta polynomials are related by

$$v_n(x) = V_{n+1}(x) - V_{n-1}(x), \quad n \geq 1. \tag{4.2}$$

When $x = 3$, this becomes (see the Introduction)

$$L_{2n} = F_{2n+2} \quad F_{2n-2},$$

which, when $n = 2$ for example, becomes

$$L_4 = 7 = 8 - 1 = F_6 - F_2.$$

It is left to the reader to establish that

$$V_{2n+1}(x) = (-1)^{n+1}(xB_n(-x^2) - b_n(-x^2)),$$
$$V_{2n}(x) = (-1)^{n+1}(b_n(-x^2) - xB_{n-1}(-x^2)),$$
$$V_n(x) = x^{n-2}J_{n-1}(-1/x^2) - x^{n-3}J_{n-3}(-1/x^2).$$

Various elegant identities for $V_n(x)$, and hence for the Morgan-Voyce and Jacobsthal polynomials, can be produced by including a negative argument. For instance,

$$V_n(x)V_{n-1}(-x) + V_n(-x)V_{n-1}(x) = 0, \quad n \geq 2. \tag{4.3}$$

Proof: We use induction on n. When $n = 2$,

$$V_2(x)V_1(-x) + V_2(-x)V_1(x) = 0.$$

Assume the result is true for $n = 3, 4, \cdots, m-1$. Then

$$V_m(x)V_{m-1}(-x) + V_m(-x)V_{m-1}(x)$$

$$= xV_{m-1}(x)V_{m-1}(-x) - V_{m-2}(x)V_{m-1}(-1)$$

$$- xV_{m-1}(x)V_{m-1}(-x) - V_{m-2}(-x)V_{m-1}(x)$$

$$= -(V_{m-1}(x)V_{m-2}(-x) + V_{m-1}(-x)V_{m-2}(x))$$

$$= 0 \quad \text{(from the inductive hypothesis)}. \qquad \qquad \square$$

Similar development of further results is left to the interested reader who might note recent developments in the theory of Jacobsthal-type polynomials [14,15].

Gratitude is expressed to an anonymous referee for suggestions to improve the presentation of this paper and for some corrections.

REFERENCES

[1]   Bondarenko, B.A. Generalized Pascal Triangles and Pyramids, (tr. from the Russian by R.C. Bollinger), Fibonacci Association, Santa Clara, 1993.

[2]   Carlitz, L. "A Fibonacci Array." *The Fibonacci Quarterly*, Vol. *1.2* (1963): pp. 17-27.

[3]   Carlitz, L. and Riordan, J. "Two element lattice permutations and their *q*-generalizations." *Duke Mathematical Journal*, Vol. *31.4* (1964): pp. 371-388.

[4]   Feinberg, M.A. "A Lucas triangle." *The Fibonacci Quarterly*, Vol. *5.5* (1967): pp. 486-490.

[5]   Filipponi, P. "Some binomial Fibonacci identities." *The Fibonacci Quarterly*, Vol. *33.3* (1995): pp. 251-257.

[6]   Gould, H.W. and Greig, W.E. "A Lucas triangle primality criterion dual to that of Mann-Shanks." *The Fibonacci Quarterly*, Vol. *23.1* (1985): pp. 66-69.

[7]   Hartree, D. R. Numerical Analysis, Clarendon Press, Oxford, Second Edition, 1958, Ch. 4.

[8]   Horadam, A.F. "Basic properties of a certain generalized sequence of numbers." *The Fibonacci Quarterly*, Vol. *3.3* (1965): pp. 161-176.

[9]   Horadam, A.F. "Generating functions for powers of certain generalized sequence of numbers." *Duke Mathematical Journal*, Vol. *32.3* (1965): pp. 437-446.

[10]   Horadam, A.F. "Tschebyscheff and other functions associated with the sequence $\{w_n(a,b;p,q)\}$." *The Fibonacci Quarterly*, Vol. *7.1* (1969): pp. 14-22.

[11]   Horadam, A.F. "Jacobsthal representation numbers." *The Fibonacci Quarterly*, Vol. *34.1* (1996): pp. 40-54.

[12]   Horadam, A.F. "A composite of Morgan-Voyce generalizations." *The Fibonacci Quarterly*, Vol. *35.3* (1997): pp. 233-239.

[13]   Horadam, A.F. "Negative subscript Jacobsthal numbers." *Notes on Number Theory & Discrete Mathematics*, Vol. *3.1* (1997): pp. 9-22.

[14]   Horadam, A.F. "Rodriques' formulas for Jacobsthal-type polynomials." *The Fibonacci Quarterly*, Vol. *35.4* (1997): pp. 361-370.

[15]   Horadam, A.F. and Filipponi, P. "Derivative sequences of Jacobsthal and Jacobsthal-Lucas polynomials." *The Fibonacci Quarterly*, Vol. *35.4* (1997): pp. 352-357.

[16]   Horadam, A.F. "New aspects of Morgan-Voyce polynomials.", in press.

[17]   Horadam, A.F., Loh, R.P. and Shannon, A.G. "Divisibility properties of some Fibonacci-type sequences." Combinatorial Mathematics VI. Edited by A.F. Horadam and W.D. Wallis, Springer-Verlag, Berlin, 1979: pp. 55-64.

[18]   Horadam, A.F. and Mahon, J.M. "Pell and Pell-Lucas sequences." *The Fibonacci Quarterly*, Vol. *23.1* (1985): pp. 7-20.

[19]   Lavers, T.G. "The Fibonacci Pyramid." Applications of Fibonacci Numbers, Volume 7. Edited by G.E. Bergum, A.F. Horadam and A.N. Philippou. Kluwer Academic Publishers, Dordrecht, The Netherlands, 1998: pp 255-263.

[20]   Melham, R.S. "On certain polynomials of even subscripted Lucas numbers." Applications of Fibonacci Numbers, Volume 8. Edited by F.T. Howard. Kluwer Academic Publishers, Dordrecht, The Netherlands, 1999: pp. 251-258.

[21]   Morgan-Voyce, A.M. "Ladder networks analysis using Fibonacci numbers." *IRE Transactions on Circuit Theory*, Vol. *6.3* (1959): pp. 321-322.

[22]   Robbins, N. "Vieta's triangular array and a related family of polynomials." *Journal of Mathematics and Mathematical Sciences*, Vol. *14.2* (1991): pp. 239-244.

[23]   Sloane, N.J.A. and Plouffe, S. The Encyclopedia of Integer Sequences. Academic Press, San Diego, 1995.

[24]   Swamy, M.N.S. "Properties of the polynomials defined by Morgan-Voyce." *The Fibonacci Quarterly*, Vol. *4.1* (1966): pp. 73-81.

[25]   Swamy, M.N.S. "On a class of generalized polynomials." *The Fibonacci Quarterly*, Vol. *35.2* (1997): pp. 329-334.

[26]   Swamy, M.N.S. and Bhattacharya, B.B. "A study of recurrent ladders using the polynomials defined by Morgan-Voyce." *IEEE Transactions on Circuit Theory*, Vol. *14* (1967): pp. 260-264.

AMS Classification Numbers: 11B37, 11B39

SPECIAL MULTIPLIERS OF LUCAS SEQUENCES MODULO p^r

Lawrence Somer

1. INTRODUCTION

Throughout this paper let p be an odd prime and $r \geq 1$. We will introduce the concept of special multipliers modulo p^r for second-order linear recurrences. In particular, we will specialize this concept to Lucas sequences and show that in these cases special multipliers modulo p^r obey certain symmetry properties. We will then apply these symmetry properties to obtain results on the stability of certain Lucas sequences modulo p. This paper relies heavily on the results of the earlier work [16] and is primarily an application of the tools develped there.

Let $w(a,b) = (w)$ be a second-roder linear recurrence satisfying the relation

$$w_{n+2} = aw_{n+1} - bw_n, \tag{1.1}$$

where the parameters a and b and the initial terms w_0, w_1 are all integers. Let $D = a^2 - 4b$ be the discriminant of $w(a,b)$. We will distinguish two particular recurrences satisfying (1.1), $u(a,b) = (u)$, called the Lucas sequence of the first kind and having initial terms $u_0 = 0$, $u_1 = 1$, and $v(a,b) = (v)$, called the Lucas sequence of the second kind and having initial terms $v_0 = 2$, $v_1 = a$.

The recurrence $w(a,b)$ is called *regular modulo p*, or *p-regular* for short if

$$\det{}_w(a,b) = \begin{vmatrix} w_0 & w_1 \\ w_1 & w_2 \end{vmatrix} = w_0 w_2 - w_1^2 \not\equiv 0 \pmod{p}. \tag{1.2}$$

This paper is in final form and no version of it will be submitted for publication elsewhere.

We note that the recurrence $w(a,b)$ is p-regular if and only if $w(a,b)$ satisfies no recursion relation of order less than two modulo p. Lemma 1.1 below tells when $u(a,b)$ and $v(a,b)$ are p-regular.

Lemma 1.1: *The Lucas sequence $u(a,b)$ is always p-regular, while $v(a,b)$ is p-regular if and only if $p \nmid D$.*

Proof: For $u(a,b)$, using (1.2), we see that, for all p,

$$\det_u(a,b) = 0 \cdot a - 1^2 = -1 \not\equiv 0 \pmod{p}.$$

For $v(a,b)$, again using (1.2), we obtain

$$\det_v(a,b) = 2(a^2 + 2b) - a^2 = a^2 + 4b = D.$$

The result now follows from the definition of p-regularity.   □

Throughout this paper we will always assume that the recurrence $w(a,b)$ is p-regular.

2. PERIODS AND RESTRICTED PERIODS

The following natational conventions and well-known results are taken from [16]. The *period* of $w(a,b)$ modulo p^r, denoted by $\lambda_w(p^r) = \lambda(p^r)$, is the least positive integer t such that $w_{n+t} \equiv w_n \pmod{p^r}$ for all $n \geq 0$. We note that if $p \nmid b$, then $w(a,b)$ is purely periodic modulo p^r (see [9, pp. 344-45]). Accordingly, we assume throughout this paper that $p \nmid b$. The *restricted period* of $w(a,b)$ modulo p^r, denoted by $h_w(p^r) = h(p^r)$ is the least positive integer t such that

$$w_{n+t} \equiv M w_n \pmod{p^r} \tag{2.1}$$

for all $n \geq 0$ and some integer M such that $p \nmid M$. The integer $M = M_w(p^r)$, defined up to congruence modulo p^r, is called the *multiplier* of $w(a,b)$ modulo p^r. As was pointed out in [16], $h_w(p^r) \mid \lambda_w(p^r)$ and $E_w(p^r) = \lambda_w(p^r)/h_w(p^r)$ is the multiplicative order in $(Z/p^rZ)^*$ of the multiplier $M_w(p^r)$. Moreover, if $h = h_w(p^r)$ and $M = M_w(p^r)$, then for all n,

$$w_{n+ih} \equiv M^i w_n \pmod{p^r}. \tag{2.2}$$

It was further shown in [16] that all p-regular recurrences $w(a,b)$ satisfying the relation (1.1) have the same period, restricted period, and multiplier modulo p^r. Accordingly, we will sometimes drop the subscript w and let $\lambda(p^r) = \lambda_w(p^r)$, $h(p^r) = h_w(p^r)$, $M(p^r) = M_w(p^r)$ and $E(p^r) = E_w(p^r)$. For the p-regular recurrence $w(a,b)$, we define e to be the largest integer, if it exists, such that $h(p^e) = h(p)$. It was noted in [16] that if e does not exist, then $u_{h(p)} = 0$. For the p-regular recurrence $w(a,b)$, we also define the parameter f to be the largest integer, if it exists, such that $\lambda(p^f) = \lambda(p)$. It was observed in [16] that if e exists, then f also exists and $f \leq e$. It was also shown in Seciton 2.5 of [16] that if f exists, then f is the largest integer such that $1 \leq f \leq e$ and $\mathrm{ord}_{p^f}(u_{h(p)+1}) \mid p - 1$. We will need the following lemma concerning $h(p^r)$.

Lemma 2.1: *Suppose that e exists and $r \geq e$. Then*

$$h(p^r) = p^{r-e}h(p^e). \tag{2.3}$$

Proof: This is proved in [8, p. 42]. □

If $p \nmid w_n$, we say that w_n is a *p-regular* element of (w). For p-regular w_n, define $\rho_w(n, m) = \rho(n, m)$ to be the ratio w_{n+m}/w_n. Note that the definition of $\rho(n, m)$ differs slightly from that given in [16]. In this case, if $h = h_w(p^r)$, then, by (2.2), w_{n+ih} is also p-regular for all i, and it follows that $\rho_w(n, h)$ is well-defined in $(\mathbf{Z}/p^r\mathbf{Z})$ and $\rho_w(n, h) \equiv M(p^r)$ (mod p^r).

3. SPECIAL RESTRICTED PERIODS AND SPECIAL MULTIPLIERS

The concepts regarding special restricted periods and sepcial multipliers were introduced in [16] and are key to the arguments that follow. The following discussion is taken from [16].

Restricted periods and multipliers may be viewed from another perspective. If $h = h_w(p^r)$ nd $M = M_w(p^r)$, then for every n the sequence (w^*) defined by $w_t^* = w_{n+th}$ satifies the first-order recurrence relation $w_{t+1}^* \equiv Mw_t^*$ (mod p^r).

The restricted period can be characterized as the smallest integer h such that for all n the subsequences (w_{n+th}) satisfy the same first-order recurrence relation. It may happen, however, that for a fixed n there exists an integer $m < h$ such that the subseuqence defined by $w_t^* = w_{n+tm}$ satifies a first-order recurrence relation $w_{t+1}^* \equiv M^*w_t^*$ (mod p^r). We will be interested in this phenomenon when $m = h_w(p^c)$ for some $c < r$. (In this case, m becomes a restricted period when (w) is reduced modulo p^c.) This motivates the following definition.

Definition 3.1: For fixed n and r, let $h_w^*(n, p^r)$ be the least integer m of the set $\{h_w(p^c) \mid 1 \leq c \leq r\}$ for which the sequence $w_t^* = w_{n+tm}$ satisfies a first-order recurrence relation $w_{t+1}^* \equiv M^*w_t^*$ (mod p^r) for some integer M^*. The integer $h^* = h_w^*(n, p^r)$ is called the *special restricted period* and $M^* = M_w^*(n, p^r)$ (defined up to congruence modulo p^r) the *special multiplier* with respect to w_n modulo p^r.

For example, consider the Fibonacci sequence $u(1, -1)$. Here $h_u(3^4) = 108$ and $M_u(3^4) \equiv 80$ (mod 3^4). However, if $u_i^* = u_{1+12i}$, then $u_{i+1}^* \equiv 71u_i^*$ (mod 3^4) for all i.

Thus for this sequence $h_u^*(1, 3^4) = 12$ and $M_u^*(1, 3^4) \equiv 71 \pmod{3^4}$.

We note that $M_u^*(n, p^r) = \rho_u(n, h_u^*(n, p^r))$. The following theorems present known results concerning special restricted periods and special multipliers modulo p^r. For the remainder of the paper we let $r^* = \max(\lceil r/2 \rceil, e)$ and $e^* = \min(r, e)$.

Theorem 3.2: *For the sequence $w(a, b)$, suppose that $p \nmid w_n$. Then $h_w^*(n, p^r) = h(p^{r^*})$. In particular, $h_w^*(n, p^r)$ is independent of n and $M_w^*(n, p^r) \equiv \rho_w(n, h(p^{r^*})) \equiv M_w(p^{r^*}) \pmod{p^r}$. Moreover, $M_w^*(n, p^r) \equiv M_w(p^{e^*}) \pmod{p^{e^*}}$.*

Proof: This is proved in Theorem 3.5 of [16]. $\quad\square$

Theorem 3.3: *For the sequence $w(a, b)$, suppose that w_i and w_j are p-regular elements such that $0 \leq i < j < h(p^e)$. Then $M_w^*(i, p^{2e}) \not\equiv M_w^*(j, p^{2e}) \pmod{p^{2e}}$.*

Proof: This is a special case of Corollary 3.6 of [16]. $\quad\square$

4. THE MAIN RESULTS

Theorems 4.1 and 4.3 below show that the special multipliers modulo p^r for the Lucas sequences $u(a, b)$ and $v(a, b)$ obey certain symmetry properties.

Theorem 4.1: *For the sequence $u(a, b)$, suppose that $1 \leq n \leq h^* - 1$ and $p \nmid u_n$. Then*

$$M_u^*(n, p^r) \cdot M_u^*(h^* - n, p^r) \equiv b^{h^*} \pmod{p^r}. \tag{4.1}$$

Moreover,

$$[M_u(p^{r^*})]^2 \equiv b^{h^*} \pmod{p^r}. \tag{4.2}$$

Example 4.2: Consider the Fibonacci sequence $u(1, -1)$. Let $p = 11$ and $r = 2$. Then $e = 1$, $r^* = 1$, and $h^* = 10$. Then

$M_u^*(1, 11^2) \equiv 89 \pmod{121}$, $M_u^*(2, 11^2) \equiv 23 \pmod{121}$,

$M_u^*(3, 11^2) \equiv 56 \pmod{121}$, $M_u^*(4, 11^2) \equiv 45 \pmod{121}$,

$M_u^*(5, 11^2) \equiv 1 \pmod{121}$, $M_u^*(6, 11^2) \equiv 78 \pmod{121}$,

$M_u^*(7, 11^2) \equiv 67 \pmod{121}$, $M_u^*(8, 11^2) \equiv 100 \pmod{121}$,

and $M_u^*(9, 11^2) \equiv 34 \pmod{121}$.

One sees by inspection that Theorem 4.1 holds. Note that

$$M_u^*(n, 11^2) \equiv M(11) \equiv 1 \pmod{11}$$

for $1 \leq n \leq 9$.

Theorem 4.3: *For the sequence $v(a, b)$ suppose that $p \nmid D$. We suppose further that $p \nmid v_n$ and $0 \leq n \leq h^*$. Then*

$$M_v^*(n, p^r) \cdot M_v^*(h^* - n, p^r) \equiv b^{h^*} \pmod{p^r}. \tag{4.3}$$

Moreover,

$$[M_v(p^{r^*})]^2 \equiv b^{h^*} \pmod{p^{r^*}}. \tag{4.4}$$

For the residue d modulo p^r, let $\nu_w(d, p^r) = \nu(d, p^r)$ denote the number of times that d appears in a full period of the recurrence $w(a, b)$ modulo p^r. Then $\nu(d, p^r)$ is the frequency distribution function of the sequence $w(a, b)$ modulo p^r. We let $\Omega_w(p^r) = \Omega(p^r)$ denote the set of values that $\nu_w(d, p^r)$ takes on. The recurrence $w(a, b)$ is called *stable* modulo p if there exists a positive integer t such that $\Omega(p^r) = \Omega(p^t)$ for all $r \geq t$. The concept of stability for second-order recurrences was introduced by Jacobson in [12] and [10]. Theorem 4.4 and Corollaries 4.7 and 4.8 below provide results concerning the stability of $u(a, b)$ and $v(a, b)$ when $b = \pm 1$. We note that Theorem 4.4 is also presented in [17], however a new proof of Theorem 4.4, based on the symmetry properties given in Theorems 4.1 and 4.3, is provided in this paper.

Theorem 4.4: *If $b = \pm 1$ and $p \nmid D$, then $v(a, b)$ is not stable modulo p. If $b = \pm 1$ and $h(p)$ is even, then $u(a, b)$ is not stable modulo p, while if $b = \pm 1$ and $h(p)$ is odd, then $u(a, b)$ is stable modulo p.*

Remark 4.5: We note that criteria based on the Legendre symbol for determining whether $h(p)$ is even or odd are given in Theorems 13-18 of [14] in all cases when $b = \pm 1$, except for some of the cases in which $\left(\frac{D}{p}\right) = \left(\frac{b}{p}\right) = 1$.

Remark 4.6: Further resutls on stability of second-order recurrences modulo odd primes p are given in [1], [2], [7], [10], [11], [17], and [18]. Results on stability of second-order recurrences modulo 2 are given in [3], [4], [5], [6], [12], and [13].

Corollary 4.7: *If $b = \pm 1$ and $h(p)$ is even, then $\nu_u(u_{h^*/2}, p^r) \geq p^{r-r^*}$.*

Corollary 4.8: *If $b = \pm 1$ and $p \nmid D$, then $\nu_v(2, p^r) = \nu_v(v_0, p^r) \geq p^{r-r^*}$.*

5. PROOFS OF THEOREMS 4.1 AND 4.3

The following lemma will be needed for the proofs of Theorems 4.1 and 4.3.

Lemma 5.1: *Let $u(a,b)$ and $v(a,b)$ be Lucas sequences with the parameters a and b. Then*

$$u_{n+k}u_{n-k} - u_n^2 = -b^{n-k}u_k^2 \text{ and } v_{n+k}v_{n-k} - Dv_n^2 = b^{n-k}v_k^2. \qquad (5.1)$$

Proof: The identities in (5.1) follow from the Binet formulas.   □

Lemma 5.2: *For the sequence $u(a,b)$, let $h = h(p^r)$. Then*

$$[M_u(p^r)]^2 \equiv b^h \pmod{p^r}.$$

Proof: This is a special case of the lemma in [19, p. 504].   □

The hypotheses for Theorems 4.1 and 4.3 require that $p \nmid u_n$ and $p \nmid v_n$. Before proving these theorems, we will describe conditions for this to occur.

We define $R_w(p^r)$ to be the rank of appearance of p^r in $w(a,b)$. Then $R_w(p^r)$ is the least positive integer t, if it exists, such that $w_t \equiv 0 \pmod{p^r}$. Since $u_0 = 0$, it is clear that if $h = h_u(p^r)$, then $R_u(p^r) = h$ and $M_u(p^r) \equiv u_{h+1} \pmod{p^r}$. It is shown in [8, p. 38] that $u_m \equiv 0 \pmod{p^r}$ if and only if $R_u(p^r) \mid m$. It was further shown in [8, pp. 42-47] that $R_v(p^r)$ exists if and only if $R_u(p^r)$ is even. Moreover, if $R_u(p^r)$ is even, then it follows by Theorem XI on page 42 of [8] and part (iv) of Proposition 2 on pages 475-476 of [15] that $R_v(p^r) = (1/2)R_u(p^r)$. Furthermore, it was demonstrated in Theorem XI of [8] that if $R_v(p^r)$ exisits, then $v_m \equiv 0 \pmod{p^r}$ if and only if $m = (2k+1)R_v(p^r)$ for some nonnegative integer k.

Proof of Theorem 4.1: Let $1 \leq n \leq h^* - 1$ and suppose that $p \nmid u_n$. Noting that $u_{ih^*} \equiv 0 \pmod{p^{r^*}}$ for all $i \geq 1$, that $r \leq 2r^*$, and that $u_{n+kh^*} \equiv [M_u^*(n,p^r)]^k u_n \pmod{p^r}$ by Theorem 3.2, we obtain from Lemma 5.1

$$u_{2h^*+n}u_{2h^*-n} - u_{2h^*}^2 \equiv [M_u^*(n,p^r)]^2 u_n M_u^*(h^*-n,p^r)u_{h^*-n} - 0 \pmod{p^r}$$
$$\equiv -b^{2h^*-n}u_n^2 \pmod{p^r} \qquad (5.2)$$

and

$$u_{h^*+n}u_{h^*-n} - u_{h^*}^2 \equiv M_u^*(n,p^r)u_n u_{h^*-n} - 0 \pmod{p^r}$$
$$\equiv -b^{h^*-n}u_n^2 \pmod{p^r} \qquad (5.3)$$

Dividing congruence (5.2) by congruence (5.3) modulo p^r, we obtain

$$M_u^*(n,p^r)M_u^*(h^*-n,p^r) \equiv b^{h^*} \pmod{p^r}, \qquad (5.4)$$

as desired. Congruence (4.2) follows from Lemma 5.2.   □

Proof of Theorem 4.3: We note that by Lemma 1.1, the condition that $p \nmid D$ forces $v(a,b)$ to be p-regular. Let $0 \leq n \leq h^*$ and suppose that $p \nmid v_n$. Then by Lemma 5.1, since $r \leq 2r^*$,

$$v_{2h^* + n} v_{2h^* - n} - Du_{2h^*}^2 \equiv [M_v^*(n, p^r)]^2 v_n M_v^*(h^* - n, p^r) v_{h^* - n} - 0 \pmod{p^r}$$

$$\equiv b^{2h^*} - n v_n^2 \pmod{p^r} \tag{5.5}$$

and

$$v_{h^* + n} v_{h^* - n} - Du_{h^*}^2 \equiv M_v^*(n, p^r) v_n v_{h^* - n} - 0 \pmod{p^r}$$

$$\equiv b^{h^*} - n v_n^2 \pmod{p^r}. \tag{5.6}$$

Dividing congruence (5.5) by congruence (5.6) modulo p^r, we have

$$M_v^*(n, p^r) M_v^*(h^* - n, p^r) \equiv b^{h^*} \pmod{p^r},$$

as desired. Congruence (4.4) follows from Lemma 5.2 upon noting that $M_v(p^{r^*}) \equiv M_u(p^{r^*})$ (mod p^{r^*}) since $v(a,b)$ is p-regular and all p-regular recurrences have the same multiplier modulo p^{r^*} by the remarks following Lemma 2.4 of [16].   □

6. NECESSARY LEMMAS

The following lemmas will be necessary for the proofs of Theorem 4.4 and Corollaries 4.7 and 4.8. Let $s = E(p) = \text{ord}_p(M(p))$, where $\text{ord}_p(M(p))$ is the multiplicative order of $M(p)$ modulo p.

Lemma 6.1, given below, was proved in [16].

Lemma 6.1: *Let d be an integer such that $p \nmid d$ and supppose that there exists a largest positive integer m such that $\text{ord}_{p^m}(d) = \text{ord}_p(d)$. If $r \geq m$, then $\text{ord}_{p^r}(d) = p^{r-m} \text{ord}_{p^m}(d)$.*

Remark 6.2: It follows by Lemma 6.1 that if $p \nmid \text{ord}_{p^r}(d)$, then $\text{ord}_{p^r}(d) = \text{ord}_{p^c}(d)$ for all c such that $1 \leq c < r$.

Lemma 6.3: *For the sequence $w(a,b)$, suppose that $b = \pm 1$. Then*

$$\text{ord}_{p^r}(M(p^r)) = \text{ord}_p(M(p)) = s \tag{6.1}$$

for all $r \geq 1$. In particular, $\text{ord}_{p^r}(M(p^r)) \mid p - 1$ for all $r \geq 1$.

Proof: By part (c) of Theorem 2.11 in [16], (6.1) will follow if we can show that $e = f$. Applying our discussion regarding the parameter f in Section 2 of this paper, we will obtain $e = f$ by showing that $\text{ord}_{p^e}(u_{h(p)+1}) \mid p - 1$. By Lemma 6.1, this occurs if and only if $p \nmid \text{ord}_{p^e}(u_{h(p)+1})$. Let $t = h(p) = h(p^e)$. Then, by (5.1), with $k = 1$, we obtain

$$\pm 1 = b^t = b^t u_1^2 = u_{t+1}^2 - u_t u_{t+2} \equiv u_{t+1}^2 - 0 \cdot u_{t+2} \equiv u_{t+1}^2 \pmod{p^e}.$$

Hence, $\text{ord}_{p^e}(u_{h(p)+1}) \mid 4$ and $p \nmid \text{ord}_{p^e}(u_{h(p)+1})$.   □

Lemma 6.4: Let $\nu_n^*(d, p^r)$ denote the number of terms w_m in a period of the p-regular recurrence $w(a, b)$ such that $w_m \equiv d \pmod{p^r}$ and $m \equiv n \pmod{h^*}$. Then

$$\nu(d, p^r) = \sum_{n=0}^{h^*-1} \nu_n^*(d, p^r). \tag{6.2}$$

In particular, $\nu(d, p^r) \geq \nu_n^*(d, p^r)$ for $0 \leq n \leq h^* - 1$. Moreover, if $p \nmid d$, $\nu_n^*(d, p^r) \geq 1$, and $b = \pm 1$, then

$$\nu_n^*(d, p^r) = \frac{p^{r-r^*} s}{\operatorname{ord}_{p^r} M_w^*(n, p^r)}. \tag{6.3}$$

Proof: It was shown in Section 4 of [16] that (6.2) is satisfied. It follows from Lemma 4.4 of [16] that (6.3) is satisifed when $e = f$. However, it was shown in the proof of Lemma 6.3 that, in fact, $e = f$ when $b = \pm 1$.   □

In the proof of Theorem 4.4, we will need to make use of Hensel's lemma. We use the version of Hensel's lemma given in [16].

Lemma 6.5 (Hensel's Lemma): Suppose that $f(x)$ is a polynomial with integral coefficients. If $f(m) \equiv 0 \pmod{p^i}$ and $f'(m) \not\equiv 0 \pmod{p}$, then there is a unique t modulo p such that $f(m + tp^i) \equiv 0 \pmod{p^{i+1}}$.

Corollary 6.6: Let $f(x) = x^{2^k} - 1$, where $k \geq 0$. If $p \nmid d$, $f(d) \equiv 0 \pmod{p^i}$, and $j > i$, then there is a unique residue d' modulo p^j such that $d' \equiv d \pmod{p^i}$ and $f(d') \equiv 0 \pmod{p^j}$.

Proof: This follows from Lemma 6.5 upon noting that $(2^k, p) = 1$.   □

Lemma 6.7: If $p \mid d$, then

$$\nu_u(d, p^r) = \begin{cases} 0 & \text{if } p^{e^*} \nmid d \\ \\ s & \text{if } p^{e^*} \mid d. \end{cases}$$

Proof: This follows from Theorems 6.4, 6.8, and 6.9 of [16].   □

Lemma 6.8: The sequence $u(a, b)$ is stable modulo p if $p \mid \operatorname{ord}_{p^{2e}}(M_u^*(n, p^{2e}))$ for all n such that $1 \leq n < h(p^e)$.

Proof: We first suppose that $p \mid u_n$ and that $r > e$. By Lemma 6.7, if $p^e \nmid u_n$, then

$$\nu_u(u_n, p^r) = 0, \tag{6.4}$$

while if $p^e \mid u_n$, then

$$\nu_u(u_n, p^r) = s. \tag{6.5}$$

Now suppose that $p \nmid u_n$. By the discussion in Section 2, $f \le e$. Let c be the largest integer such that $p \nmid \mathrm{ord}_{p^{2e}}(M_u^*(n, p^{2e}))$. Then $1 \le c \le 2e - 1$. We note that if $n > h(p^e)$, then $M_u^*(n, p^{2e}) \equiv M_u^*(n - h(p^e), p^{2e}) \pmod{p^{2e}}$ by Theorem 3.2. We also note that $u_n \not\equiv 0 \pmod{p}$ for $1 \le n \le h(p^e)$. We first consider the case in which $e < f$. Noting that $h(p) = h(p^e)$ and that, by Theorem 3.2,

$$u_{h(p)+1} \equiv M_u(p^e) \equiv M^*(n, p^{2e}) \pmod{p^e}$$

for all n such that $p \nmid u_n$, we see, by the discussion in Section 2 regarding the parameter f, that $c = f$. It then follows, by the proof of Theorem 6.2 in [16] that if $r > f$, then

$$\nu_u(u_n, p^r) = \nu_u(u_n, p^f). \tag{6.6}$$

We now consider the case in which $f = e$. Then $c \ge e$. By a similar proof to that given for Theorem 6.2 of [16], we obtain

$$\nu_u(u_n, p^r) = \nu_u(u_n, p^c), \tag{6.7}$$

when $r > c$. By (6.4), (6.5), (6.6), and (6.7), we now see that $u(a,b)$ is stable modulo p. $\square$

7. PROOFS OF THEOREM 4.4 AND COROLLARIES 4.7 AND 4.8

Proof of Theorem 4.4: First consider the LSSK $v(a,b)$, where $b = \pm 1$. By Theorem 3.2, $M_v^*(0, p^r) \equiv M_v^*(h^*, p^r) \pmod{p^r}$. Thus, by Theorem 4.3,

$$[M_v^*(0, p^r)]^2 \equiv b^{h^*} \equiv \pm 1 \pmod{p^r}. \tag{7.1}$$

by Theorem 3.2,

$$M_v^*(0, p^r) \equiv M_v(p^{r^*}) \pmod{p^{r^*}}. \tag{7.2}$$

By (7.1), $\mathrm{ord}_{p^r}(M_v^*(0, p^r)) = 1, 2,$ or 4. Since $(2, p) = 1$ and $\mathrm{ord}_{p^{r^*}}(M_v(p^{r^*})) \mid p - 1$ by Lemma 6.3, it follows by Lemma 6.3, Lemma 6.1, and Remark 6.2 that

$$\mathrm{ord}_{p^r}(M_v^*(0, p^r)) = \mathrm{ord}_{p^{r^*}}(M_v(p^{r^*})) = s. \tag{7.3}$$

Noting that $v_0 = 2$, we see by (7.3) and by Lemma 6.4 that

$$\nu(v_0, p^r) = \nu(2, p^r) \ge \nu_0^*(v_0, p^r) = \nu_0^*(2, p^r) = p^{r - r^*}, \tag{7.4}$$

Thus, $v(a,b)$ is not stable modulo p, since $r - r^* \ge \lfloor r/2 \rfloor$ when $r \ge 2e$.

We now consider the LSFK $u(a,b)$ and suppose that $h(p)$ is even. By Lemma 2.1, it follows that h^* is also even, since $h(p^e) = h(p)$. By Theorem 4.1,

$$[M_u^*(u_{h^*/2}, p^r)]^2 \equiv b^{h^*} \equiv 1 \pmod{p^r}. \tag{7.5}$$

Thus, $\mathrm{ord}_{p^r}(M_u^*(u_{h^*/2}, p^r)) = 1$ or 2. By Lemma 6.3, Lemma 6.1, and Remark 6.2,

$$\mathrm{ord}_{p^r}(M_u^*(u_{h^*/2}, p^r)) \equiv \mathrm{ord}_{p^{r^*}}(M_u(p^{r^*})) = s \tag{7.6}$$

and, by Lemma 6.4,

$$\nu(u_{h^*/2}, p^r) \geq \nu_{h^*/2}^*(u_{h^*/2}, p^r) = p^{r - r^*}. \tag{7.7}$$

Consequently, $u(a, b)$ is not stable modulo p.

We finally consider $u(a, b)$ in the case that $h_u(p^e) = h_u(p)$ is odd. Let $h = h_u(p^e)$. By Theorem 3.3, the special multipliers $M_u^*(n, h(p^{2e}))$ are distinct for $1 \leq n \leq h - 1$. By Theorem 4.1, with $r = 2e$,

$$[M_u(p^e)]^2 \equiv b^h \equiv \pm 1 \pmod{p^e}. \tag{7.8}$$

by (7.8) and (7.1), $\mathrm{ord}_{p^e}(M_u(p^e)) \mid 4$ and $\mathrm{ord}_{p^{2e}}(M_v^*(0, p^{2e})) \mid 4$. Since $M_v^*(0, p^{2e}) \equiv M_v(p^e)$ (mod p^e) by Theorem 3.2, and since $M_v(p^e) \equiv M_u(p^e)$ (mod p^e) because $v(a, b)$ is p-regular by Lemma 1.1, we see by Lemma 6.1 and Remark 6.2 that

$$\mathrm{ord}_{p^e}(M_u(p^e)) = \mathrm{ord}_{p^{2e}}(M_v^*(0, p^{2e})) = t \tag{7.9}$$

where $t \mid 4$. Applying Hensel's lemma, Lemma 6.5, to the polynomial $f(x) = x^t - 1$, we see by Corollary 6.6 that $M_v^*(0, p^{2e})$ is the unique residue d modulo p^{2e} such that $d \equiv M_u(p^e)$ (mod p^e) and $f(d) = d^t - 1 \equiv 0$ (mod p^{2e}). If $1 \leq n \leq h - 1$, then

$$M_u^*(n, p^{2e}) M_u^*(h - n, p^{2e}) \equiv b^h \pmod{p^{2e}}, \tag{7.10}$$

but

$$M_u^*(n, p^{2e}) \not\equiv M_u^*(h - n, p^{2e}) \pmod{p^{2e}} \tag{7.11}$$

by Theorems 4.1 and 3.3. Hence, from (7.10), (7.11), and (7.1), we see that $M_u^*(n, p^{2e}) \not\equiv M_v^*(0, p^{2e})$ (mod p^{2e}). Thus

$$\mathrm{ord}_{p^{2e}}(M_u^*(n, p^{2e})) \neq \mathrm{ord}_{p^{2e}}(M_u(p^e)), \tag{7.12}$$

but, by Theorem 3.2,

$$M_u^*(n, p^{2e}) \equiv M_u(p^e) \pmod{p^e} \tag{7.13}$$

for $1 \leq n \leq h - 1$. Therefore, by Lemma 6.1 and Remark 6.2, $p \mid \mathrm{ord}_{p^{2e}}(M_u^*(n, p^{2e}))$ for $1 \leq n \leq h - 1$. Consequently, $u(a, b)$ is stable modulo p, by Lemma 6.8. $\qquad\square$

<u>Proof of Corollaries 4.7 and 4.8:</u> Corollaries 4.7 and 4.8 follow from equations (7.7) and (7.4), respectively, in the proof of Theorem 4.4. $\qquad\square$

ACKNOWLEDGMENT

I wish to express my deep appreciation to the referee for his or her careful reading of the manuscirpt and many helpful suggestions that improved this paper. I also wish to thank Walter Carlip for his helpful comments concerning this paper.

REFERENCES

[1]   Bumby, R.T. "A distribution property for linear recurrence of the second order." *Proc. Amer. Math. Soc.*, Vol. *50* (1975): pp. 101-106.

[2]   Bundschuh, Peter and Shiu, Jau Shyong. "Solution of a problem on the uniform distribution of integers." *Atti Accad. Naz. Lincei Rend. Cl. Sci. Fis. Mat. Natur.*, Vol. *55.8* (1973)-(1974): pp. 172-177.

[3]   Carlip, Walter and Jacobson, Eliot. "A criterion for stability of two-term recurrence sequences modulo 2^k." *Finite Fields Appl.*, Vol. *2.4* (1996): pp. 369-406.

[4]   Carlip, Walter and Jacobson, Eliot. "On the stability of certain Lucas sequences modulo 2^k." *The Fibonacci Quarterly*, Vol. *34.4* (1996): pp. 298-305.

[5]   Carlip, Walter and Jacobson, Eliot. "Unbounded stability of two-term recurrence sequences modulo 2^k." *Acta Arith.*, Vol. *74.4* (1996): pp. 329-346.

[6]   Carlip, Walter and Jacobson, Eliot. "Stability of two-term recurrence sequences with even parameter." *Finite Fields Appl*, Vol. *3.1* (1997): pp. 70-83.

[7]   Carlip, Walter, Jacobson, Eliot and Somer, Lawrence. "A criterion for stability of two-term recurrence sequences modulo odd primes." Applications of Fibonacci Numbers, Volume 7. Edited by G.E. Bergum, A.F. Horadam and A.N. Philippou. Kluwer Academic Publishers, Dordrecht, (1997): pp. 49-60.

[8]   Carmichael, R.D. "On the numerical factors of the arithmetic forms $\alpha^n \pm \beta^n$." *Ann. of Math.*, Vol. *15.2* (1913): pp. 30-70.

[9]   Carmichael, R.D. "On sequences of integers defined by recurrence relations." *Quart. J. Pure Appl. Math.*, Vol. *48* (1920): pp. 343-372.

[10]   Carroll, Dana, Jacobson, Eliot and Somer, Lawrence. "Distribution of two-term recurrence sequences mod p^e." *The Fibonacci Quarterly*, Vol. *32.3* (1994): pp. 260-265.

[11]   Darvasi, Gyula and Nagy, Mihály. "On repetitions in frequency blocks of the generalized Fibonacci sequence $u(3,1)$ with $u_0 = u_1 = 1$." *The Fibonacci Quarterly*, Vol. *34.2* (1996): pp. 176-180.

[12]   Jacobson, Eliot T. "Distribution of the Fibonacci numbers mod 2^k." *The Fibonacci Quarterly*, Vol. *30.3* (1992): pp. 211-215.

[13]   Morgan, Mark D. "The distribution of second order linear recurrence sequences mod 2^m." *Acta Arith.*, Vol. *83.2* (1998): pp. 181-195.

[14]   Somer, Lawrence. "The divisibility properties of primary Lucas recurrences with respect to primes." *The Fibonacci Quarterly*, Vol. *18.4* (1980): pp. 316-334.

[15]   Somer, Lawrence. "Divisibility of terms in Lucas sequenecs of the second kind by thier subscripts." Applications of Fibonacci Numbers, Volume 6. Edited by G.E. Bergum, A.F. Horadam and A.N. Philippou. Kluwer Academic Publishers, Dordrecht, 1996: pp. 473-486.

[16]   Somer, Lawrence and Carlip, Walter. "Bounds for frequencies of residues of regular second order recurrences modulo p^r. Proceedings of the International Number Theory Conference Dedicated to Professor Andrzej Schinzel, (Zakopane-Koscielisko, Poland, 1997),

        Walter de Gruyter, Berlin, (to appear).

[17]   Somer, Lawrence and Carlip, Walter. "Stability of second order recurrences modulo p^{r}", (submitted).

[18]   Webb, William A. and Long, Calvin T. "Distribution modulo p^h of the general linear second order recurrence." *Atti Accad. Naz. Lincei Rend. Cl. Sci. Fis. Mat. Natur.*, Vol. *58.2* (1975): pp. 92-100.

[19]   Wyler, Oswald. "On second-order recurrences." *Amer. Math. Monthly*, Vol. *72* (1965): pp. 500-506.

AMS Classification Numbers: 11B37, 11B39, 11B50

DIGITAL HALFTONING USING ERROR DIFFUSION AND LINEAR PIXEL SHUFFLING

John Szybist and Peter G. Anderson

<u>INTRODUCTION</u>

Digital halftoning is used to render continuous gray-scale images to a device, such as a laser printer or video display, that is only capable of binary output (black and white). Other applications include image transmission and storage where the gray-scale to binary data reduction translates to reduced transmission and storage costs. The idea is that the local density of dots mimics the original gray-scale and appears nearly the same to a viewer. This method works because the human visual system acts like an averaging or low-pass filtering device when presented with the relatively high frequency dot pattern. The challenge is to optimize the appearance by minimizing side effects such as undesirable texture artifacts and false contours, that can be produced by the pattern. The general issues that must be addressed are those concerned with the original image relative to dynamic range and spatial resolution of the gray-scale and those concerned with the visibility of the halftone pattern.

Digital halftoning techniques receive as input an image represented as an $m \times n$ array A of multi-level values (e.g., 256 levels of gray), and produce an $m \times n$ array B of bi-level values (black and white) that modulate the bi-level imaging system. "Good" algorithms do this in such a fashion that the average gray level in the neighborhood of B_{ij} approximates the value of A_{ij} at a reasonable viewing distance (generally considered 18 inches for 300 dots per inch).

This paper is in final form and no version of it will be submitted for publication elsewhere.

Digital halftone algorithms can be categorized as either "point" or "neighborhood" [6]. A point algorithm renders B_{ij} using information from A at A_{ij} only. Neighborhood algorithms use information from A_{ij} and pixels in some neighborhood of A_{ij} to render B_{ij}. The selection between these two classes of algorithms will usually trade off rendering speed (point algorithms) and resulting image quality (neighborhood algorithms). Point algorithms quite often show contouring (especially if they support too few gray levels), degrade contrast, and have quite prominent texture artifacts.

ERROR DIFFUSION

Error diffusion is a neighborhood method where the value of each input pixel, A_{ij}, is compared to a (usually) fixed threshold, θ, to determine the value of the corresponding output pixel, B_{ij} and the difference ("error") between A_{ij} and B_{ij} is diffused to some as yet unprocessed pixels in a small neighborhood of A. We suppose that the input pixel values are integers in the range 0-255, expressing a range of 256 gray values from white to black, and that the quantized output pixel values are either 0 or 255 (white or black, respectively). In this case $\theta = 128$. The following pseudo code expresses the general error diffusion algorithm:

```
for every (i, j) pixel location in the image
        {Quantize pixel (i, j)}
        if A_ij ≥ θ then B_ij ←255
            else B_ij ←0

        {Determine and disperse the quantization error}
        E←B_ij − A_ij
        divide up E, and add its portions
        to unprocessed neighbors of A_ij
```

In most error diffusion algorithms, the order of pixel processing corresponding to the code step

```
for every (i, j) pixel location in the image
```

generally takes the form of the following raster ordering:

```
for i←0, step 1, while i < max.i
    for j←0, step 1, while j < max.j
        process pixed (i, j)
```

One common variation on the left-to-right, row-by-row ordering is *serpentine* or *bostrouphedon* (as the ox plows) order, in which the processing order for lines alternates between left-to-right and right-to-left.

The literature has many references to variations on error diffusion algorithms that analyze the size of the neighborhood and the amount of error diffused to each neighbor (denoted as the diffusion kernel or mask). The classic diffusion mask developed by Floyd and Steinberg [4] will be used to benchmark our development:

$$D = \frac{1}{16} \begin{bmatrix} & P & 7 \\ 3 & 5 & 1 \end{bmatrix}$$

P represents the pixel A_{ij} under consideration. This mask diffuses the error from P to pixels yet to be rendered when scanning in a top-to-bottom, left-to-right, raster fashion:

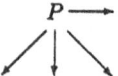

Specifically, the quantization error, $E = B_{ij} - A_{ij}$, is dispersed by adding $\frac{7}{16}E$ to $A_{i,j+1}$, $\frac{3}{16}E$ to $A_{i+1,j-1}$, $\frac{5}{16}E$ to $A_{i+1,j}$ and $\frac{1}{16}E$ to $A_{i+1,j+1}$. The pixel to the left of P, $A_{i,j-1}$ has already been quantized, as have all the pixels in the rows above P. It is possible, and not at all damaging, if the values of unquantized A_{ij}'s go outside the range 0-255. Thus, the error is pushed down and to the right—like snow before a plow, it eventually exceeds the capacity (i.e., θ), and becomes exposed in clumps. This results in the "worm" artifacts; see the hat in Fig. 1.

Error diffusion uses the error to bring the average gray level in the neighborhood of B_{ij} closer to the value of A_{ij}. Some problems evident in Figure 1 such as the "tearing" pattern at midtone and the "worm" effects at other tones will be addressed by our LPS error diffusion. The "tearing" evident down the center of Figure 1 is caused by the checkerboard pattern that Floyd-Steinberg generates at the neighborhood of 50% gray values and, to a lesser extent, around 25% and 75% gray.

LINEAR PIXEL SHUFFLING

Anderson [2, 3] developed a method, called *Linear Pixel Shuffling* (LPS), of visiting pixels in a smooth manner all over an image. This method extends to higher dimensions the one-dimensional permutation of the integers, $0, \cdots, F_n - 1$ given by the rule $v_k = kF_{n-1} \% F_n$ ("%" denoting remainder). For $n = 8$, this permutation of $0, \cdots, 20$ is

$$0, 13, 5, 18, 10, 2, 15, 7, 20, 12, 4, 17, 9, 1, 14, 6, 19, 11, 3, 16, 8.$$

The special property of this permutation—namely, that sequentially near values are numerically distant $\left(\,|p-q|\text{ is small}\Leftrightarrow|v_p-v_q|\text{ is large}\right)$ was exploited in [1]. For the two-dimensional generalization of this permutation, we utilize a Fibonacci-like sequence, G_n, third-order linear recurrence defined as follows:

$$G_0 = 0$$

$$G_1 = G_2 = 1$$

$$G_n = G_{n-1} + G_{n-3} \text{ for } n \geq 3.$$

This sequence begins:

$$0, 1, 1, 1, 2, 3, 4, 6, 9, 13, 19, 28, 41, 60, 88, 129, 189, 277, 406, 595, 872, 1278, 1873, 2745, \cdots$$

This sequence provides the parameters for LPS as follows. Suppose the image is contained within a $G_n \times G_n$ square. We introduce the LPS shuffling matrix

$$S = \begin{pmatrix} G_{n-2}^2 + G_{n-1}^2 & G_{n-1} \\ G_{n-1}^2 & G_n - G_{n-2} \end{pmatrix}$$

and specify the pixel visitation order for error diffusion using a change of basis:

> for $i \leftarrow 0$, step 1, while $i < G_n$
>> for $j \leftarrow 0$, step 1, while $j < G_n$
>>
>> {Perform matrix-vector multiplication}
>> process the pixel at location $S[i\ j]'$

For convenience, we describe the image as a $G_n \times G_n$ square. In practice, we must include the image in $G_n \times G_n$ square, and ignore the pixels in that square that are outside the image. This is reminiscent of the common computing practice of using a power of 2 for sizes of the data structures.

To describe this order of pixel processing, we introduce an auxiliary $G_n \times G_n$ array, T, of integers in the range $0, \cdots, G_n - 1$, with the rule

$$T_{ij} = (iG_{n-2} + jG_{n-1})\%G_n.$$

This is illustrated in Fig. 3 for $n = 10$, and $G_n = 19$. The specifically useful properties of T are:

- Numerically close values are geometrically dispersed; or, equivalently, neighboring values are numerically distant. The 0's in T are widely distributed; the 1's are also, and they are not near the 0's; and so on.

- T is specified by a linear rule, so that the neighborhoods of each entry are arithmetic translates (mod G_n) of neighborhoods of any other entry.

- T gives the LPS order of pixel visitation: the (i, j)-positions such that $T_{ij} = 0$ are visited first, then the positions with $T_{ij} = 1$, and so on. (The order in which the 0 positions are visited is irrelevant for the present application. This could be an opportunity to exploit parallelism, by processing all the 0 positions simultaneously, followed by the 1's, etc.)

The areas outlined in the center of Fig. 3 are discussed in the following section. That the matrix-vector multiplication indicated in the above pseudocode performs the desired transformations follows from these observations (developed in [3]):

- If $[p \ q]' = S[0 \ 1]'$, then $T_{pq} = 0$.

- If $[p \ q]' = S[1 \ 0]'$, then $T_{pq} = 1$.

- So, by linearity, if $[p \ q]' = S[i \ j]'$, then $T_{pq} = i$.

LPS ERROR DIFFUSION

The LPS halftoning algorithm scans the image using the LPS order described in the previous section, and diffuses error to unprocessed pixels in all directions using an algorithm similar to Floyd and Steinberg's. We use the following diffusion mask:

$$D = \frac{1}{32} \begin{bmatrix} & 1 & 1 & 1 & \\ 1 & 2 & 3 & 2 & 1 \\ 1 & 3 & P & 3 & 1 \\ 1 & 2 & 3 & 2 & 1 \\ & 1 & 1 & 1 & \end{bmatrix}$$

Our LPS error diffusion algorithm takes the form:

for $i \leftarrow 0$, step 1, while $i < G_n$

    Determine the unprocessed pixels at positions with label i

    for $j \leftarrow 0$, step 1, while $j < G_n$

        {Determine the next pixel to process}

        $[p \ q]' \leftarrow S[i \ j]'$

        {Note: $T_{pq} = i$}

        {Quantize and distribute the error}

        if $A_{pq} > \theta$ then $B_{pq} \leftarrow 255$

            else $B_{pq} \leftarrow 0$

        distribute the quantization error, $E \leftarrow B_{pq} - A_{pq}$

        over unprocessed pixels in the neighborhood of A_{pq}

The LPS scanning order allows us easily to determine which of the pixels in the 5×5 error diffusion kernel D have already been processed. Because the LPS rule is a simple linear rule, the neighborhood of every pixel labeled i is identical (see Fig. 3). The pixel labels in a kernel centered at $T_{pq} = i$ are a translation by $+i \pmod{G_n}$ of the labels in the kernels centered at $T_{00} = 0$. The following neighborhood diagram was excised from Fig. 3:

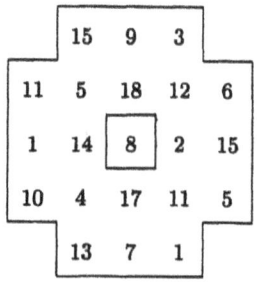

In the neighborhood of a $T_{pq} = 8$ we have 9 already processed pixels (those with labels less than eight, namely, 3, 5, 6, 1, 2, 4, 5, 7, and 1) and 11 unprocessed pixels (those with labels greater than 8, namely, 15, 9, 11, 18, 12, 14, 15, 10, 17, 11, and 13).

We easily keep the description of the error diffusion neighborhood up to date, for each new i in the program—this i is controlled by the outer loop, so there is minor impact on program performance. Donald Knuth employed a similar idea in his "dot diffusion" algorithm [5].

DISCUSSION

Figs. 1 and 2, *Woman With a French Horn*[1] is a scene with several natural and man-made objects and their textures. A principal advantage of Floyd-Steinberg error diffusion is its enhancement of edges and preservation of high frequency image information (here in the woman's hair and sweater). These images show that our LPS error diffusion appears to do admirably in this aspect as well.

Differences between the two error diffusion algorithms are also evident in two artificial images, *arctangent* (Figs. 4 and 5) and *ripples* (Figs. 6 and 7). Arctangent was constructed so that the gray values range fill the range from 0 to 255; the pixel at position (p, q) has a gray value proportion to $\operatorname{atan}\left(\frac{p}{q}\right)$. This provides a ramping from the two extreme colors in the top and left edges, exposing directional artifacts of an algorithm, and a range of gray-scale gradients. Fig. 5 illustrates that LPS error diffusion appears to introduce some undesired texture artifacts into the image. The Floyd-Steinberg rendering of arctangent clearly shows artifacts that LPS overcomes; namely, tearing at the 50% gray level, worms, and a delayed introduction of black pixels in the very light gray region. LPS, unfortunately, introduces some evident "grain" and undesired stripes. (Choices of error diffusion algorithms are, unfortunately, image dependent.)

Ripples was constructed so that its pixel values filled a midrange of values, 0-127. The pixel at position (p, q) has a gray value proportional to $\cos(\alpha(p^2 + q^2))$, where α was chosen to make the maximum frequency in the image barely visible. The ripples appear in a range of orientations to expose any directional artifacts in the algorithms. Here, we are quite pleased with the performance of LPS error diffusion. The high frequency information shows clearly in this experiment.

---

[1]Image courtest of Heidelberger Druckmaschinen A.G.

As we see it, the primary deficiency in the LPS error diffusion algorithm is a small reduction in spatial resolution and the introduction of some low frequency texture in the halftoned image. This deficiency is an appropriate area for further investigation. Several *ad hoc* attempts, using alternative diffusion masks, increased the spatial resolution and lowered the texture pattern, but the cost was the appearance of false contours, which was original mask nicely avoided.

We experimented with masks such as the following,

$$D = \frac{1}{44} \begin{bmatrix} & & 0 & 1 & 0 & & \\ & 0 & 3 & 7 & 3 & 0 & \\ & 1 & 7 & P & 7 & 1 & \\ & 0 & 3 & 7 & 3 & 0 & \\ & & 0 & 1 & 0 & & \end{bmatrix}$$

but undesirable artifacts and image degradation were still evident.

Our algorithm requires more storage than Floyd-Steinberg, because the error needs to be saved for a longer period of time—until all pixels have been rendered. We used a data structure with a size equivalent to the size of the entire image to store our error. Error storage for the Floyd-Steinberg technique is merely the size of a single scan line.

Our algorithm has significant advantages over Floyd-Steinberg relative to the worm and tearing artifacts, and LPS compares favorably overall. Further studies will include a more strict quantization of the image transformation, psychophysical experiments, the texture problem, and applications in color halftoning.

REFERENCES

[1]   Anderson, Peter G. "A Fibonacci-based pseudo-random number generator. Fibonacci Numbers and Their Applications, Volume 4. Edited by G.E. Bergum, A.F. Horadam and A.N. Philippou. Kluwer Academic Publishers, Dordrecht, The Netherlands, 1990: pp. 1-8.

[2]   Anderson, Peter G. "Multidimensional golden means. Fibonacci Numbers and Their Applications, Volume 5. Edited by G.E. Bergum, A.F. Horadam and A.N. Philippou. Kluwer Academic Publishers, Dordrecht, The Netherlands, 1992: pp. 1-10.

[3]   Anderson, Peter G. "Advances in linear pixel shuffling. Fibonacci Numbers and Their Applications, Volume 6. Edited by G.E. Bergum, A.F. Horadam and A.N. Philippou. Kluwer Academic Publishers, Dordrecht, The Netherlands, 1994: pp. 1-22.

[4]   Floyd, R.W. and Steinberg, L. "An adaptive algorithm for spatial gray scale." *In SID International Symposium Digest of Technical Papers*, Society for Information Display (1974): pp. 36-37.

[5]   Knuth, Donald E. "Digital halftones by dot diffusion." *ACM Transactions on Graphics*, Vol. *6.4* (1987): pp. 245-273.

[6]   Ulichney, Robert <u>Digital</u> <u>Halftoning</u>. The MIT Press, 1987.

AMS Classification Numbers: 68U99, 11B39

Figure 1: Floyd-Steinberg error diffusion. "Worm" artifacts appear in the hat, "tearing" in the background.

Figure 2: LPS error diffusion

0	13	7	1	14	8	2	15	9	3	16	10	4	17	11	5	18	12	6
9	3	16	10	4	17	11	5	18	12	6	0	13	7	1	14	8	2	15
18	12	6	0	13	7	1	14	8	2	15	9	3	16	10	4	17	11	5
8	2	15	9	3	16	10	4	17	11	5	18	12	6	0	13	7	1	14
17	11	5	18	12	6	0	13	7	1	14	8	2	15	9	3	16	10	4
7	1	14	8	2	15	9	3	16	10	4	17	11	5	18	12	6	0	13
16	10	4	17	11	5	18	12	6	0	13	7	1	14	8	2	15	9	3
6	0	13	7	1	14	8	2	15	9	3	16	10	4	17	11	5	18	12
15	9	3	16	10	4	17	11	5	18	12	6	0	13	7	1	14	8	2
5	18	12	6	0	13	7	1	14	8	2	15	9	3	16	10	4	17	11
14	8	2	15	9	3	16	10	4	17	11	5	18	12	6	0	13	7	1
4	17	11	5	18	12	6	0	13	7	1	14	8	2	15	9	3	16	10
13	7	1	14	8	2	15	9	3	16	10	4	17	11	5	18	12	6	0
3	16	10	4	17	11	5	18	12	6	0	13	7	1	14	8	2	15	9
12	6	0	13	7	1	14	8	2	15	9	3	16	10	4	17	11	5	18
2	15	9	3	16	10	4	17	11	5	18	12	6	0	13	7	1	14	8
11	5	18	12	6	0	13	7	1	14	8	2	15	9	3	16	10	4	17
1	14	8	2	15	9	3	16	10	4	17	11	5	18	12	6	0	13	7
10	4	17	11	5	18	12	6	0	13	7	1	14	8	2	15	9	3	16

Figure 3: The numbers in this table satisfy $T_{ij} - (9i + 13j)\%19$. The parameters (9, 13, 19) are a consecutive triple in the sequence G_n. The pixels in a 19×19 image would be processed, in LPS order, as shown here; namely, all 19 positions labeled 0 are processed first, then those labeled 1, and so on. The outlined section centered at $\boxed{8}$ pictures the error diffusion neighborhood (the region wherein we disperse the quantization error) for the pixel in the center.

Figure 4: Floyd-Steinberg error diffusion of arctangent. The jarring texture on the diagonal of this image is "checkerboarding." In higher resolution images, this appears as "tearing." It is especially apparent in images produced on less expensive laser and ink jet printers which produce round black pixel; these black pixels by necessity encroach on the areas of neighboring white pixels, resulting in nonmonotonic (let alone nonlinear) printer responses.

Figure 5: LPS error diffusion of arctangent.

Figure 6: Floyd-Steinberg error diffusion of ripples.

Figure 7: LPS error diffusion of ripples.

ON VECTOR SEQUENCE RECURRENCE EQUATIONS IN FIBONACCI VECTOR GEOMETRY

J. C. Turner

1. INTRODUCTION

In [4] we introduced some ideas, methods and results in a topic which we chose to call Fibonacci vector geometry. The basic idea was to study triples of consecutive terms taken from Fibonacci sequences, regarding the triples as position vectors in Euclidean space, and seeking geometric properties of their locii and related figures.

For example, the triple $F_n \equiv (F_{n-1}, F_n, F_{n+1})$ was called the nth Fibonacci vector; and it was observed that all Fibonacci vectors lie in the plane $z = x + y$, which was designated π_0. We called the sequence $\{F_n\}$ the basic Fibonacci vector sequence. It was shown how the integer points in π_0 are arranged in a honeycomb pattern, and we described some properties of these points; we named the plane the *Fibonacci honeycomb* because of the pattern and properties. Further, we showed how vector sequences could be plotted in the plane, with their points joined by line-segments; and we called the resulting figures *Fibonacci vector polygons*. The Lucas vectors, defined similarly, also lie in the Fibonacci honeycomb. Figure 1 below shows the plane π_0, together with axes X and Y, and plots of the Fibonacci and Lucas vector polygons.

In this paper we shall develop these ideas further. In particular, we shall study vector sequences in the general plane $z = cx + dy$, defining first a vector/matrix mode of generating vector sequences which are 'inherent to the plane'. Then we shall concentrate on geometric properties of vector sequences generated by another type of recurrence relation, which is a direct vector analogue of the Fibonacci number recurrence.

This paper is in final form and no version of it will be submitted for publication elsewhere.

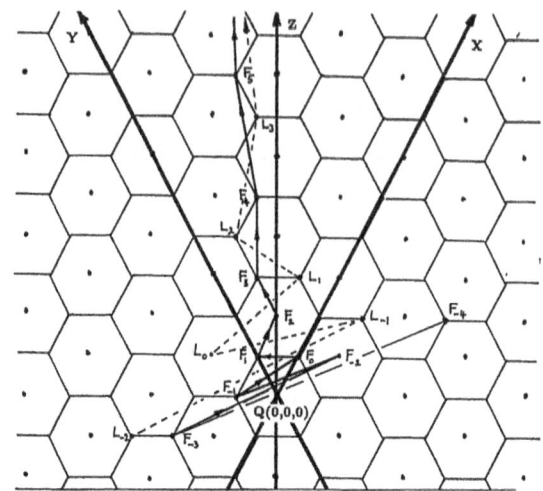

Figure 1. The plane π_0 and the Fibonacci and Lucas polygons

2. INTEGER VECTOR SEQUENCES

We shall use the following terminology: $x = (x, y, z)$ is an *integer vector* if all of its coordinates are integers; it is a *coprime vector* if also $\gcd(x, y, z) = 1$. We shall generally omit the adjective 'integer' when it is clear that integer vectors are being treated. $\{x_n\}$ will denoted an *integer vector sequence* if x_n is an integer vector for each given value of n. It will be a coprime vector sequence if all its terms are coprime vectors.

Eight examples of integer vector sequences follow, with discussion on some of their properties.

Examples of vector sequences:

(1) $(0,1,1), (1,1,2), (1,2,3), \cdots, (F_{n-1}, F_n, F_{n+1}), \cdots$ (basic Fibonacci)

(2) $(2,1,3), (1,3,4), (3,4,7), \cdots, (L_{n-1}, L_n, L_{n+1}), \cdots$ (basic Lucas)

(3) $(0,1,2), (1,2,5), (2,5,12), \cdots, (P_{n-1}, P_n, P_{n+1}), \cdots$ (basic Pell)

(4) $(0,1,1), (1,2,3), (3,5,8), \cdots, (F_{2n-2}, F_{2n-1}, F_{2n}), \cdots$

(5) $(0,1,1), (2,3,5), (8,13,21), \cdots, (F_{3n-3}, F_{3n-2}, F_{3n-1}), \cdots$

(6) $(0,1,2), (1,1,3), (1,2,5), \cdots, (F_{n-1}, F_n, F_{n+2}), \cdots$

(7) $(0,1,3), (1,3,8), (3,8,21), (8,21,55), \cdots, (F_{2n-2}, F_{2n}, F_{2n+2}), \cdots$

(8) $(1,1,2), (3,5,8), (4,6,10), (7,11,18), (11,17,28), \cdots$

We observe that vector sequences (1), (2) and (3) are, respectively, the basic Fibonacci, Lucas and Pell sequences. (4) is a subsequence of (1), and in vector form it is: F_1, F_3, F_5, $\cdots$, $\{F_{2n-1}\}$, $\cdots$. All eight are integer vector sequences: only example (8) is **not** a coprime sequence.

It is of interest ot note that in both the vector sequences (4) and (5), the union set of all coordinates used in the vectors is the set of Fibonacci numbers $\mathfrak{F} = \{0, 1, 1, 2, 3, 5, 8, \cdots\}$. In (5) the set $\mathfrak{F}$ is used exactly, each Fibonacci number appearing once only among the coordinates; whereas in (4) there is redundancy — the set $\{F_{2n}\}$ is used twice. A problem for study is to determine other vector sequences which use only Fibonacci numbers as coordinates; in which planes these sequences occur, and how much redundancy they have. Vectors in (6) use only Fibonacci numbers, but they do not lie in the Fibonacci plane. Vector sequence (7) is another example of such sequences; this time the vectors all lie in the plane $z = -x + 3y$, and there is both deficiency and redundancy in the use of the Fibonacci numbers in their coordinates. Other examples of vector sequences with these properties will be given below.

Finally, vectors in sequence (8) lie in the same plane as do those of sequence (1); but only the first two vectors have all their coordinates Fibonacci numbers. Its general term is $(F_{n-2}F_2 + F_{n-1}F_5)$.

3. VECTOR SEQUENCE PLANES

The vectors in sequences (1) to (8) above lie in planes of type $z = cx + dy$. Thus vectors of (1) and (2) are all in the plane $\pi_0 \equiv z = 1x + 1y$. Vectors of (3) lie in the plane $z = 1x + 2y$; we shall refer to it as *the Pell plane*, and designate the nth vector in the sequence by P_n.

We can use $\pi(c, d)$ to symbolize the planes $z = cx + dy$. Then the Fibonacci and Pell planes are, respectively, $\pi(1, 1)$ and $\pi(1, 2)$. The only other plane represented in the examples is $\pi(-1, 3)$, which includes sequence (7). We shall call the planes $\pi(c, d)$ *sequence planes* when c and d are both integers; we shall see below how vector sequences can be generated in them in a special way.

Before moving on to discuss vector recurrence relations, we include an example of an integer vector sequence which is **not** in a sequence plane as just defined, thus:

$$(1, 2, 2), \ (3, 2, 3), \ (4, 4, 5), \ (7, 6, 8), \ \cdots, \ (L_n, \ 2F_n, \ F_{n+2}), \ \cdots \tag{3.1}$$

The vectors in this sequence all lie in the plane $2x + 3y = 4z$, which is not a sequence plane since it cannot be expressed in the form $z = cx + dy$ with integer coefficients.

In Sections 4 and 5 we shall define two types of recurrence relation, by which the vector sequences exemplified above are generated. First we must look at certain transformation matrices, which are associated with the plane equations in interesting ways.

4. INHERENT TRANSFORMATIONS OF PLANES

Suppose that the vector $x = (x, y, z)$ belongs to the plane $\pi(c, d)$. Then we call* the 3×3 matrix H the *inherent transformation matrix* of the plane if $x' = (y, z, cy + dz)$ and if $xH = x'$; that is, if H transforms x into the given x'. We call this a *forward transformation* of x. It may quickly be checked that:

$$H = \begin{pmatrix} 0 & 0 & 0 \\ 1 & 0 & c \\ 0 & 1 & d \end{pmatrix}.$$

We note that $\det(H) = 0$, so H is a singular matrix; therefore it has no inverse. However, we can choose (see below) a pseudo-inverse H^- which will perform a *backward transformation* of the kind we want. The desired pseudo-inverse is one which satisfies $x'H^- = x$; it has the following form, as may be checked easily:

$$H^- = \begin{pmatrix} -d/c & 1 & 0 \\ 1/c & 0 & 1 \\ 0 & 0 & 0 \end{pmatrix}.$$

To see that H and H^- are pseudo-inverses of one another, it is only necessary to check that both have zero determinant and that $HH^-H = H$ and $H^-HH^- = H^-$. The matrix H^- is therefore a reflexive generalized inverse of H.

The inherent matrix $H(1,1)$ for the honeycomb plane has the following properties, in its powers H^n and H^{-n}.

Theorem 4.1:

(i) $\quad H^n = \begin{pmatrix} 0 & 0 & 0 \\ F_{n-2} & F_{n-1} & F_n \\ F_{n-1} & F_n & F_{n+1} \end{pmatrix}$; $\quad H^{-n} = \begin{pmatrix} F_{-n-1} & F_{-n} & F_{-n+1} \\ F_{-n} & F_{-n+1} & F_{-n+2} \\ 0 & 0 & 0 \end{pmatrix}$

---

*We use the symbol H for this matrix as a tribute to the Australian mathematician A.F. Horadam, who has published so many beautiful results about pairs of Fibonacci sequences.

(ii)
$$H^n + H^{-n} = \begin{pmatrix} F_{-n} \\ F_{1-n} + F_{1+n} \\ F_n \end{pmatrix}, \quad \text{for } n = 1, 2, 3, \cdots$$

[Recall that $F_n = (F_{n-1}, F_n, F_{n+1})$.]

It is easy to show that the middle row of this matrix is $L_{n-1}F_0$ if n is odd, and $F_{n-1}L_0$ if n is even.

(iii)   The characteristic polynomial of H^n is $-\lambda(\lambda - \alpha^n)(\lambda - \beta^n)$; hence its characteristic roots are 0, α^n, β^n where α is the golden ratio, and $\alpha\beta = -1$.

Proof (i): Both formulae, for H^n and H^{-n}, are proved easily using induction on n.

Proof (ii): Using (i), only trivial checking of the terms of the formulae is required.

Proof (iii): Using H^n from (i), and I12, I13 from [3, p. 57] we have:

$$|H^n - \lambda I| = \begin{vmatrix} -\lambda & 0 & 0 \\ F_{n-2} & F_{n-1} - \lambda & F_n \\ F_{n-1} & F_n & F_{n+1} - \lambda \end{vmatrix} = -\lambda(\lambda^2 - L_n\lambda + (-1)^n)$$

The roots of the quadratic factor in the above expression are $\lambda = L_n \pm \sqrt{L_n^2 - 4(-1)^n}/2 = (L_n \pm F_n\sqrt{5})/2$, (using identity I12). Hence, using $\alpha^n = F_{n-1} + F_n\alpha$, we obtain the required results: the characteristic roots are 0, α^n, β^n.

Now we are ready to define our first type of recurrence relation, for generating vector sequences.

5. VECTOR RECURRENCE RELATIONS

Type I: The matrix/vector equation, order 1:

We shall study how we can use inherent transition matrices (and others that are closely related to them) to generate the kinds of vector sequence which we gave as examples in Section 2.

Using H and H^-, for any given sequence plane, we can generate vector sequences which are entirely in that plane, if we take a starting vector which is also in the plane. The two recurrence equations which we need to do this are:

Forward: $x_{n+1} = x_n H(c, d)$ (5.1)

Backward: $x_{n-1} = x_n H^-(c,d)$.   (5.2)

These two relations, together with a starting vector x_1 in $\pi(c,d)$, will generate a doubly-infinite vector sequence, all in the same plane.

Looking back to the examples (1) to (8) in Section 2, we see the following illustrations of (5.1) and (5.2) in operation. Thus:

(1)′   Using $H(1,1)$, with $x_1 = (0,1,1)$, we get the forward sequence shown.   And, using $H^-(1,1)$, the following backward sequence is produced:

←···, $(-2,1,-1)$, $(1,-1,0)$, $(-1,0,1)$, $(0,1,1)$. The two together comprise the full, basic, Fibonacci vector sequence.

(2)′   Similarly, using $H(1,1)$ and $H^-(1,1)$, with $x_1 = (2,1,3)$, the two-way basic Lucas vector sequence is generated.

(3)′   The two-way Pell vector sequence shown is generated using $H(1,2)$ and its pseudo-inverse, together with $x_1 = (0,1,2)$, in relations (5.1) and (5.2).   The full *basic* Pell vector sequence, has the following central terms:

$$\leftarrow \cdots (-2,1,0),\ (1,0,1),\ (0,1,2),\ (1,2,5),\ (2,5,12),\ \cdots \rightarrow$$

(4)′   The vectors in the sequence (4) are the basic Fibonacci vectors taken two-apart. We could write this sequence thus:

$$F_1,\ F_3,\ F_5,\ \cdots,\ F_{2n-1},\ \cdots \rightarrow$$

The square of $H(1,1)$ will effect the transformations needed for this sequence. Similarly, $H^{-2}(1,1)$ will effect the backward transformations.   Hence this two-way vector sequence is given by the following recurrence relations, of order 1:

$$x_{n+1} = x_n H^2(1,1) \text{ and } x_{n-1} = x_n H^{-2}(1,1), \text{ again with } x_1 = (0,1,1).$$

If we were to use (1,1,2) as initial vector, we should generate the vector sequence of all even-subscripted basic Fibonacci vectors.   Hence the union of these two subsequences would use up all the Fibonacci numbers, in their coordinates, three times.

We remark, looking back at Theorem 4.1, that we can generate the Fibonacci vectors n-apart by using $H^n(1,1)$ and its inverse; it seems most appropriate that their characteristic roots should include α^n and α^{-n}, linked to forward and backward generation respectively. However, it is not so surprising, for the general solution to the vector/matrix equation is intimately linked with the characteristic roots of H. The following theorem shows how.

Theorem 5.1: (General solution of inherent Type I, Binet form)

Let $x_{(n+1)} \in \pi(1,1)$ and $H = H(1,1)$. Then:

(i)
$$x_{n+1} = x_1 \frac{1}{\sqrt{5}}[\alpha^{n-2}H(I+\alpha H) - \beta^{n-2}H(I+\beta H)].$$

(ii)
$$x_{n+1} \to x_1 \frac{1}{\sqrt{5}}\alpha^{n-2}H(I+\alpha H), \text{ as } n\to\infty.$$

Proof:

If we repeatedly apply the recurrence $x_{n+1} = x_n H$, for $n = 1, 2, 3, \cdots, n$, we obtain $x_{n+1} = x_1 H^n$.

Inserting the Binet identity $F_n = (\alpha^n - \beta^n)/\sqrt{5}$ into the expression for H^n given in Theorem 4.1, we get the following:

$$\sqrt{5}H^n = \begin{pmatrix} 0 & 0 & 0 \\ \alpha^{n-2} & \alpha^{n-1} & \alpha^n \\ \alpha^{n-1} & \alpha^n & \alpha^{n+1} \end{pmatrix} - \begin{pmatrix} 0 & 0 & 0 \\ \beta^{n-2} & \beta^{n-1} & \beta^n \\ \beta^{n-1} & \beta^n & \beta^{n+1} \end{pmatrix}$$

$$= \alpha^{n-2}H(I+\alpha H) - \beta^{n-2}H(I+\beta H).$$

Part (i) of the theorem follows immediately. And, letting $n\to\infty$ and noting that $\beta < 1$, part (ii) also follows.

General Type I vector recurrences

The inherent transmission matrix H is a specially chosen matrix, related directly to the equation of the plane of operation. We can, however, replace H by any 3×3 matrix (say T), nd choose the elements of T so that different kinds of vector sequence result from the recurrence relation. It is clear that setting rules for the choice of T's elements is tantamount to defining classes of integer vector sequences. We shall call the recurrence equation: $x_{n+1} = x_n T$ the *general Type I vector recurrence*, and give one new example. [N.B.—the use of H^2 for T is an example already discussed.]

Example:

Working in the Pell plane $\pi(1,2)$, with starting vector $x_1 = (1,2,5)$, and transmission matrix:

$$T = \begin{pmatrix} 0 & 1 & 2 \\ 1 & 1 & 3 \\ 0 & 0 & 0 \end{pmatrix} \text{ with inverse } T^- = \begin{pmatrix} 0 & 0 & 0 \\ 3 & -2 & -1 \\ -1 & 1 & 1 \end{pmatrix},$$

the following double vector sequence is generated:

$$\leftarrow\cdots (1,1,3),\ (1,2,5),\ (2,3,8),\ (3,5,13),\ \cdots\rightarrow$$

This sequence uses the Fibonacci numbers twice, for its coordinates, with all the vectors being in the Pell plane. Some pleasing comparisons of T with $H(1,1)$ can be made: for example, they have the same characteristic equation.

Type II: The Fibonacci vector recurrence equation, order 2:

The Type II vector recurrence relation (in plane π_0) is simply the analogue of the one which generates the well-known Fibonacci numbers: the numbers in that relation become vectors, and the number addition operation becomes vector addition. Thus the Type II vector recurrence relative to plane π_0 is: $x_{n+2} = x_n + x_{n+1}$. Although this is the vector recurrence equation relative to the Fibonacci plane, the vector sequence which it generates is always entirely in the plane which is determined by $Q(0,0,0)$ and the two starting vectors.

Similarly, the Type II recurrence equation relative to plane $\pi(c,d)$ is

$$x_{n+2} = cx_n + dx_{n+1}.$$

For examples, we observe that each of the vector sequences (1), (2), (3), (5), and (7) may be generated by a Type II recurrence. Notice that, for example, vector sequence (3) is in the Pell plane, whereas it is generated by the Type II recurrence which is relative to the Fibonacci plane.

This type of vector sequence has been treated in [1] and [4] in some detail; and some geometric properties of it are given below, in Section 6. For now, we will give one useful theorem which relates Type I and Type II generations of vector sequences in the general plane $\pi(c,d)$.

Theorem 5.1:

If $x = (x,y,z)$ is in the plane $\pi(c,d)$, then the sequence generated by the Type I inherent recurrence $x_{n+1} = x_n H(c,d)$ is the same as that generated by the Type II relative recurrence of that plane, viz. $x_{n+2} = cx_n + dx_{n+1}$.

Proof: The first three terms of the sequence generated by the Type I recurrence are x, xH, and xH^2. Using the general form of the inherent matrix $H(c,d)$ (see Section 4), we find that $xH^2 = cx + d(xH)$ if (and only if) $z = cx + dy$—that is, if x is in the plane $\pi(c,d)$.

6. SOME GEOMETRIC OBSERVATIONS

For lack of space, we shall show only a very few geometric properties of the kind of vector sequence which can be generated by recurrence relations of Types I and II. Indeed, we can do little more than draw diagrams and announce the sequence properties that they display.

Demonstration 1:

Figure 2 below shows the general development of a general Type II vector sequence, say $G(a,b)$. We assume that a and b are non-collinear, integer vectors directed from origin Q.

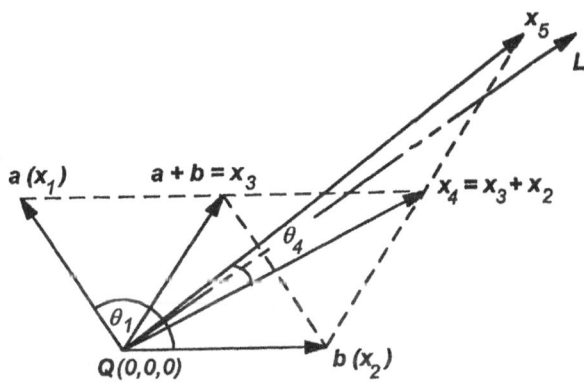

Figure 2. Type II production of vector sequence $G(a,b)$

Observations (i):

Because of the repeated operation of the vector addition law, the angle θ_n reduces each time that n increases,; it is clear that $\lim_{n\to\infty} \theta_n = 0$. Further, successive vectors in the sequence oscillate above and below a line through $Q(0,0,0)$. We call this the limit-line L. Note that all vectors with odd subscripts are above, and all with even subscripts are below, the line L. We can write, formally, that $\lim_{n\to\infty} x_n = L$.

It is easy to show, algebraically, that if a and b are any two vectors in Z^3, then L has equations $x/(a_1 + \alpha b_1) = y/(a_2 + \alpha b_2) = z/(a_3 + \alpha b_3)$, where α is the golden ratio. If the sequence is in the honeycomb plane, then the following theorem holds.

Theorem 6.1: If a and b are consecutive Fibonacci vectors in π_0, then the limit-line L is independent of them.

Proof: Let $a = (a_1, a_2, a_1 + a_2)$; then, since a and b are consecutive Fibonacci vectors, $b = (a_2, a_1 + a_2, a_1 + 2a_2)$. Now, considering the direction-ratios of L, the ratio of the first two components is:

$$\frac{a_2 + \alpha b_2}{a_1 + \alpha b_1} = \frac{a_2 + \alpha(a_1 + a_2)}{a_1 + \alpha b_1} = \frac{\alpha(a_1 + \alpha a_2)}{a_1 + \alpha a_2} = \alpha \text{ (using } \alpha^2 = 1 + \alpha).$$

Similarly, it follow easily that the ratio of the second two components is also α. Hence direction-ratios of L are $1, \alpha, \alpha^2$, and so L is $x/1 = y/\alpha = z/\alpha^2$, independent of the two starting vectors.

Comment: This result is analogous to the fact that in any Fibonacci number sequence, the ratio F_{n+1}/F_n tends to α, independently of the choice of starting numbers.

Observations (ii):

(a)   The sequence of triangles $Qx_1 x_2$, $Qx_2 x_3$, $Qx_3 x_4$, $\cdots$ have equal areas.

(b)   The sequence of quadrilaterals $Qx_1 x_3 x_2$, $Qx_2 x_4 x_3$, $Qx_3 x_5 x_4$, $\cdots$ are parallel-ograms of equal area.

(c) $x_1 x_3 x_4$, $x_2 x_4 x_5$, $x_3 x_5 x_6$, $\cdots$ are straight-line segments, with x_3, x_4, x_5 etc. being their mid-points.

(d)   All terms of the vector sequence $G(a, b)$ are in the plane determined by the three point Q, a, b.

Observations (iii):

The well-known Simson's identities (see I13, [2, p. 57]) may be deduced directly from (ii) (a) and the geometry of the plane, thus: —Suppose that $a = x_1$ and $b = x_2$ are consecutive Fibonacci vectors, in the plane $\pi_0 \equiv x + y - z = 0$. Then x_n and x_{n+1} lie in π_0, and the areas of all triangles $Qx_n x_{n+1}$ are proportional to the magnitudes of the vector cross-products $x_n \times x_{n+1}$ (recall that $\Delta = \frac{1}{2}|x_n| \cdot |x_{n+1}| \sin\theta$); but the cross-product coordinates are proportional to $(1, 1, -1)$, the direction-ratios of a normal to the plane π_0. Putting these two results together, we immediately get the Fibonacci identities:

$$F_{n-1} F_{n+1} - F_n^2 = (-1)^n = F_{n-1} F_{n+2} - F_n F_{n+1}$$

It is curious how the inductive process of arriving at proof of these identities is all taken care of by the geometry of the plane and the Type II vector sequence generated in it.

It is also of note that we can immediately generalize Simson's identities to other sequences, in other planes. For example, if a and b are both in the Pell plane $\pi(1,2)$, we know that the normal coordinates are proportion to $(1, 2, -1)$ (the plane is $x + 2y - z = 0$), and there follow from this two identities in the basic Pell sequence of numbers.

<u>Demonstration 2:</u>

In Figure 1, Section 1, we showed how the Fibonacci and Lucas vector polygons zig-zagged upwards in the honeycomb plane, tending towards the limit line $L \equiv x/1 = y/\alpha = z\alpha^2$.

We now wish the reader to form in his/her mindspace (see [3]) a picture of the complete set of Fibonacci vector polygons in the plane π_0. Imagine that every integer vector in the plane $x + y = z$ simultaneously generates (by the inherent matrix/vector type I recurrence equation) a Fibonacci vector sequence. Joining consecutive points by line-segments, we now have a 'mind-picture' of the complete set of Fibonacci vector polygons.

It should be clear that (i) each integer point lies on only one such polygon, and (ii) all the polygons will criss-cross each other infinitely often, but only at non-integer points; and (iii) we know that as they all progress 'upwards' they will tend to each other and the line L. The total result is a very complex networking of intersecting polygons in π_0 — a seemingly incomprehensible mish-mash of them.

To reduce the mental strain, instead of thinking of the formation of Fibonacci polygons, we can imagine a point map of the whole plane to itself. That is, imagine every point, such as x_n, moving to its image point x_{n+1} under the Type I inherent transformation.

To help see where all the points go, we show in (a) below how the transformation $x_{n+1} = x_n H$, when applied simultaneously to all points of π_0, causes the plane to transform. The points move as a sequence of sectors, say $\{S_i\}$. Thus section S_0 maps to section S_1, which maps to sector S_2, and so on. The boundaries of these sections are the lines (rays from $Q(0,0,0)$): $x/F_{n-1} = y/F_n = z/F_{n+1}$, with $n = 0, 1, 2, \cdots$. (These sections are marked in Figure 3(a) below.)

Another helpful thing to do, is to think of the mapping obtained by using H^2 rather than H as the transformation matrix. With this, sectors map to alternate sectors thus: $S_1 \rightarrow S_3 \rightarrow S_5, \cdots$, and $S_0 \rightarrow S_2 \rightarrow S_4 \cdots$.

If we 'join up' the consecutive pairs of dots in *this* H^2 mapping, we find that each vector polygon is now separated into two branches, *with no crossings at all.* in Figure 3(b) we show the two branches which together contain the basic Fibonacci vector sequence. They form a kind of tilted funnel, or chimney, up which the basic Fibonacci polygon zig-zags. We have also included in the diagram the *negative* of this chimney — i.e. the mappings of all points $-F_n$ under H^2.

Therefore, we propose to call the pair of branches the *upward basic Fibonacci Chimney*, and its negative the *downward basic Fibonacci chimney*. Similarly, we can imagine the chimneys formed from the vector sequence $\{2F_n\}$, and from $\{3F_n\}$, and so on, applying the H^2 transformation to all of their points. The result is a sequence of nested chimneys, which we shall designate by $\{\chi_m\}$. [The notation seems most appropriate: using 'chi-m' for 'chimney'.]

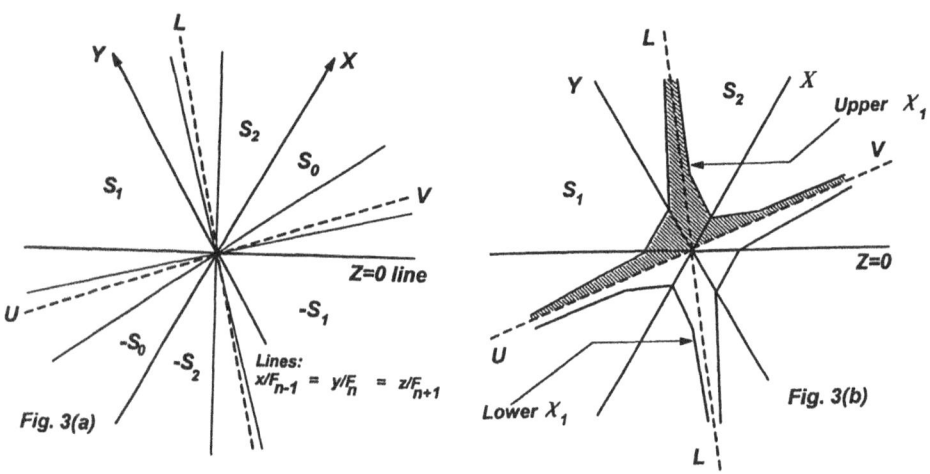

Figure 3(a). The points transformation of π_0, using H

Figure 3(b). The upper and lower chimneys $\pm \chi_1$

In Figure 4 below, we give two diagrams which show how the nests of chimneys are arranged in the plane π_0. In Figure 4(a), we show the first four upper chimneys only. On the branches, we show points from other Fibonacci vector sequences; and we claim that in all of the branches of the whole set of chimneys, there lie all the integer-vectors in the plane π_0 (above the UV-limit-line).

To prove this, we only have to look in the section S_1, and observe that there are m 'nearest points' on the edge which joins mF_{-1} to mF_1 (in the left branch of that chimney). And the transformation by H takes this edge, and all of the evenly spaced integer points in it, into the image edge in section S_2, in the right branch of χ_m. Letting $m = 1, 2, 3, \cdots$, and then generating all chimneys from S_1, we see that each integer point above UV will be sited on some Fibonacci chimney branch. It follows that all integer-vectors below UV will occur on the downward Fibonacci chimneys.

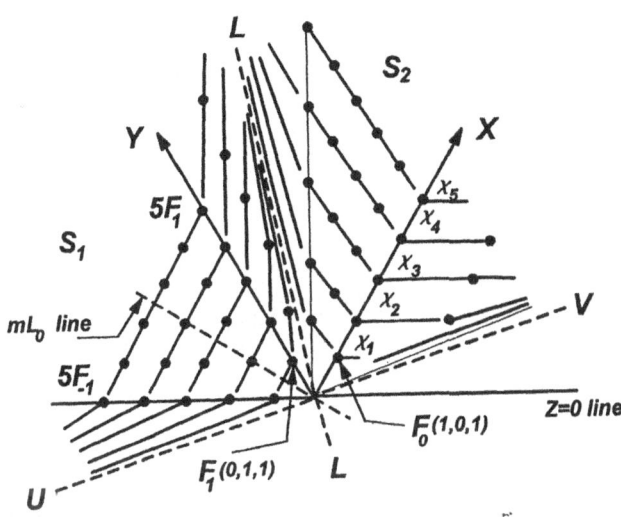

Figure 4(a). The first five upward Fibonacci chimneys

In Figure 4(b), we give a picture of the way in which the Fibonacci chimneys cover the integer-points of the plane π_0. The branches are actually piece-wise linear, but we have 'smoothed' their sides together, to produce a memorable diagram; the similarities to hyperbolic curves are striking. Note the asymmetry of the chimney directions with respect to the axes, and to limit-line UV. Note too how the branches of the chimneys are different on either side of limit-line L; but they each have symmetry axes, which are perpendicular.

This is our simplified, pictorial representation of the set of all Fibonacci vector sequences, to be stored in our mindspaces. (We have drawn in the first three central hexagons, dotted, to create a spider's web image of the system.)

A final remark is that we can think of the two branches of a chimney χ_m as being the envelope of all the m Fibonacci vector polygons which zig-zag upwards within the chimney.

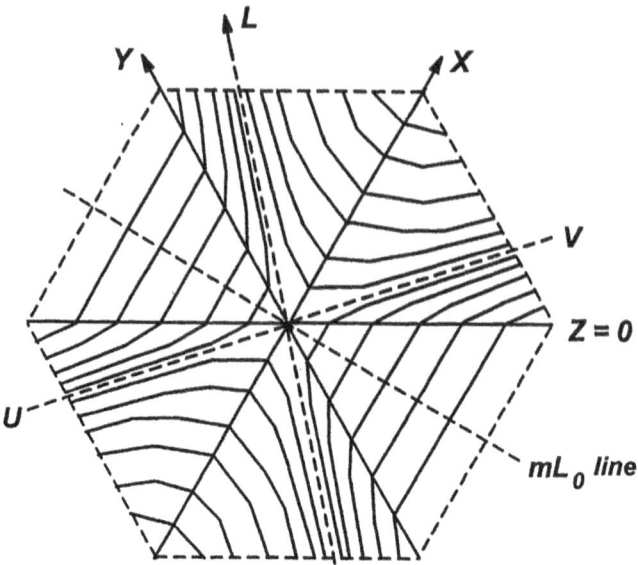

Figure 4(b). The Fibonacci chimney diagram (or web) (smoothed)

Demonstration 3:

The Lucas vector polygon zig-zags upwards in the second Fibonacci chimney. Since $L_0 = \frac{1}{2}(2F_{-1} + 2F_1)$ it is the mid-point of the first side (in sector S_1) of the left branch of the chimney. Figure 5 shows the first four sides of the polygon. We have marked the incidence and reflection angles, respectively i nd r, occurring when the polygon changes direction at point L_2. To describe such changes of direction, when the vector polygon 'bounces off the chimney walls', we define the *chimney reflection (or zig-zag) ratio* to be: $f_n = b_n \sin r / (a_n \sin i)$. It can be computed for each reflection point, and the sequence $\{f_n\}$ provides a measure of the stretching and flattening-out of the polygon as n increases. Below the diagram, we give a formula for f_n, and note that f_n tends to α.

By taking the dot-products of the edges $L_{n-1}L_n$ and L_nL_{n+1} with the chimney side $2F_{n-1}2F_{n+1}$, and forming the ratio of the results, we get the formula:

$$f_n = \frac{F_{2n+1} + F_{2n+3} + F_{2n+5} + (-1)^n}{F_{2n} + F_{2n+2} + F_{2n+4} + (-1)^{n-1}}.$$

It is easy to show † that $f_n \to \alpha$, the golden ratio, as $n \to \infty$ (which we could easily deduce by geometric observations on the rising Lucas vector polygon, but perhaps not rigorously).

---

†Divide numerator and denominator by F_{2n}, and let $n \to \infty$.

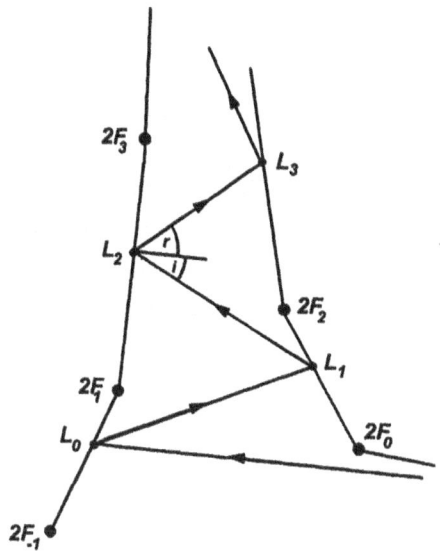

Figure 5. The chimney χ_2 and the Lucas polygon

7. SUMMARY

In this paper we have defined two types of vector recurrence equation, and given many examples of vector sequences generated by them. These examples have all been related to Fibonacci sequences of one form or another; and we have tried to show that there is much of interest and value to be gained from studies of them. In particular, we have shown how inherent transmission matrices of a family of sequence planes generate vector sequences in those planes. We have given an attractive way of viewing (imagining) the whole set of Fibonacci vector sequences in the honeycomb plane $x + y - z = 0$; nd we have demonstrated how the Lucas vector polygon 'rises' up its Fibonacci chimney.

We used the term 'transmission matrix' because we wish to encourage the notion of dynamic generation of vector sequences. Fibonacci vector sequences may be though of as locii of points moving in zig-zag paths, up and down chimneys! This view has similarities with that usually taken in Markov chain theory, where sequences of state probability vectors are generated using probability transition matrices. Perhaps fruitful parallels between Markov chain theory and Fibonacci vector geometry can be drawn.

REFERENCES

[1]   Atanassov, K. & V., Shannon, A.G. and Turner, J.C. Visual Perspectives on Number Sequences. In preparation, 1998.

[2]   Hoggatt, V.E. Jr. Fibonacci and Lucas Numbers. The Fibonacci Association, 1979: pp. 56-67.

[3]   Rucker, R. Infinity and the Mind. Penguin Books, 1997: pp. 36-38.

[4]   Turner, J.C. and Shannon, A.G. "Introduction to a Fibonacci Geometry." Applications of Fibonacci Numbers, Volume 7. Edited by G.E. Bergum, A.N. Philippou and A.F. Horadam, Kluwer Academic Publishers, Dordrecht, The Netherlands, 1998: pp. 435-448.

AMS Classification Numbers: 11B37, 11B39, 11A35

CONSTRUCTING IDENTITIES INVOLVING Kth-ORDER F-L NUMBERS BY USING THE CHARACTERISTIC POLYNOMIAL

Chizhong Zhou

1. INTRODUCTION

For convenience, we quote some notations and symbols in [20].

Let the sequence $\{w_n\}$ be defined by the recurrence relation

$$w_{n+k} = a_1 w_{n+k-1} + \cdots + a_{k-1} w_{n+1} + a_k w_n, \tag{1.1}$$

and the initial conditions

$$w_0 = c_0, w_1 = c_1, \cdots, w_{k-1} = c_{k-1}, \tag{1.2}$$

where $a_1, \cdots, a_k$, and $c_0, \cdots, c_{k-1}$ are complex constants. Then we call $\{w_n\}$ a kth-order Fibonacci-Lucas sequence or simply an F-L sequence, call every w_n an F-L number, and call

$$f(x) = x^k - a_1 x^{k-1} - \cdots - a_{k-1} x - a_k \tag{1.3}$$

the **characteristic polynomial** of $\{w_n\}$. A number a satisfying $f(a) = 0$ is called a **characteristic root** of $\{w_n\}$. If $a_k \neq 0$, we may consider $\{w_n\}$ as $\{w_n\}_{-\infty}^{+\infty}$. The set of F-L sequences satisfying (1.1) is denoted by $\Omega(a_1, \cdots, a_k)$ and also by $\Omega(f(x))$. Let $\{u_n^{(i)}\}$, $0 \leq i \leq k-1$, be a sequence in $\Omega(f(x))$ with the initial conditions $u_n^{(i)} = \delta_{ni}$ for $0 \leq n \leq k-1$, where δ is the Kronecker function. Then we call $\{u_n^{(i)}\}$ the ith **basic sequence** in $\Omega(f(x))$.

As well as of being interest for their own sake, F-L identities have applications in many areas of mathematics, including number theory and combinatorics. Because of this, the search for new F-L identities remains important. Much has been done for the case $k = 2$. See, for instance, [1], [7], [9], and [12]-[15], [19]. For $k > 2$ there has been some focus recently on special sequences [4]-[6], [11], [16]-[18], while the general case has been considered in [20] and [21], the

---

This paper is in final form and no version of it will be submitted for publication elsewhere.

latter publication being in Chinese.

In this paper, we present a method for constructing identities for arbitrary k and arbitrary F-L sequences. The method is based on the characteristic polynomial of an F-L sequence. In Section 2, we prove a key theorem—Theorem of Constructing Identity (simply, TCI), and give some examples to illustrate the use of it. By using TCI, it is very easy to construct many identities involving higher-order F-L numbers. The convenience of using TCI is similar to that of using a symbol algorithm[8]. In Section 3, by using TCI, we give a series of identities involving expressions of kth-order F-L numbers w_{m+n}, w_{m-n}, w_{-n}, w_{mn+r}, $\Sigma_{i=0}^n w_{i+r} t^i$, $\Sigma_{i=0}^n w_{mi+r} t^i$ etc., which are all useful formulas. We shall observe that many results appearing in our references are corollaries of them. By using TCI and some additional special techniques for the case $k=2$, we can generalize identities which hold only for some special F-L numbers to those which hold for general F-L numbers, and we can construct many new identities. However, due to limited space, we leave the case $k=2$ to another paper.

2. THE THEOREM OF CONSTRUCTING IDENTITIES

In this paper we always assume that $a_k \neq 0$ for $\Omega(a_1, \cdots, a_k)$, so we may consider $\{w_n\}$ as $\{w_n\}_{-\infty}^{+\infty}$. If $a_k = 0$ we point out that, with the exception of those identities which involve w_n for some $n < 0$ our identities are valid for $\{w_n\}_0^{+\infty}$.

Lemma 2.1[20]: Let $\Omega = \Omega(a_1, \cdots, a_k)$. Let $\{u_n^{(i)}\}$, $0 \leq i \leq k-1$, be the ith basic sequence in Ω and let $\{w_n\}$ be an arbitrary sequence in Ω. Then $\{w_n\}$ can be represented uniquely by $\{u_n^{(0)}\}$, $\{u_n^{(1)}\}$, $\cdots$, $\{u_n^{(k-1)}\}$, as

$$w_n = \Sigma_{i=0}^{k-1} w_i u_n^{(i)} \text{ for } n \in Z. \tag{2.1}$$

Theorem 2.2[20]: Let $\{u_n^{(i)}\}$, $0 \leq i \leq k-1$, be the ith basic sequence in $\Omega(f(x))$, where $f(x)$ is denoted by (1.3). Then

(a) $\quad x^n \equiv \Sigma_{i=0}^{k-1} u_n^{(i)} x^i \pmod{f(x)}$ for $n \in Z$. $\tag{2.2}$

(b) If, besides (2.2), we have $x^n \equiv \Sigma_{i=0}^{k-1} v_n^{(i)} x^i \pmod{f(x)}$, where each $v_n^{(i)} (i = 0, \cdots, k-1)$ is independent of x, then $u_n^{(i)} = v_n^{(i)}$, $i = 0, \cdots, k-1$.

Theorem 2.3 (TCI): Let $f(x)$ be denoted by (1.3). Suppose

$$\Sigma_{i=0}^s d_i x^{n_i} \equiv \Sigma_{j=0}^t e_j x^{p_j} \pmod{f(x)}, \tag{2.3}$$

where n_i, $p_j \in Z$, and d_i, e_j are all independent of x; $i = 0, \cdots, s$, and $j = 0, \cdots t$. Then for all $\{w_n\} \in \Omega(f(x))$ the identity

$$\Sigma^s_{i=0}d_i w_{n_i} = \Sigma^t_{j=0}e_j w_{p_j} \tag{2.4}$$

holds correspondly. Conversely, if (2.4) holds for all $\{w_n\} \in \Omega(f(x))$, then (2.3) holds.

Proof: Assume that (2.3) holds. Then (2.2) implies

$$\Sigma^s_{i=0}d_i \Sigma^{k-1}_{r=0} u^{(r)}_{n_i} x^r \equiv \Sigma^t_{j=0}e_j \Sigma^{k-1}_{r=0} u^{(r)}_{p_j} x^r \pmod{f(x)}. \tag{2.5}$$

Because the degree of the polynomial in x in each side of (2.5) is less than k, we obtain

$$\Sigma^s_{i=0}d_i u^{(r)}_{n_i} = \Sigma^t_{j=0}e_j u^{(r)}_{p_j} \quad (r = 0, \cdots, k-1), \tag{2.6}$$

whence Lemma 2.1 implies that (2.4) holds for all $\{w_n\} \in \Omega(f(x))$. Conversely, if (2.4) holds for all $\{w_n\} \in \Omega(f(x))$, then, it holds for every basic sequence $\{u^{(i)}_n\}(i = 0, \cdots, k-1)$ in $\Omega(f(x))$. That is, (2.6) holds, whence (2.5) also holds, and so (2.2) implies that (2.3) holds. $\square$

Remark: In reality, for (2.3) to hold it is sufficient that (2.4) holds for k linearly independent sequence in $\Omega(f(x))$.

Example 2.4: Let $f(x) = x^3 - 2x^2 + 1$. Let $\{w_n\}$ be an arbitrary sequence in $\Omega(f(x))$. That is,

$$w_{n+3} = 2w_{n+2} - w_n. \tag{2.7}$$

(i)   From $x^3 \equiv 2x^2 - 1 \pmod{f(x)}$ we obtain

$$x^{3n+r} \equiv \Sigma^n_{i=0}(-1)^{n-i}\binom{n}{i}2^i x^{2i+r}(n \geq 0) \pmod{f(x)}.$$

Then TCI gives

$$w_{3n+r} = \Sigma^n_{i=0}(-1)^{n-i}\binom{n}{i}2^i w_{2i+r}(n \geq 0). \tag{2.8}$$

(ii)   From $x^2 \equiv (1+x^3)/2 \pmod{f(x)}$, by the same way as in (i), we get

$$w_{2n+r} = \frac{1}{2^n}\Sigma^n_{i=0}\binom{n}{i}w_{3i+r}(n \geq 0). \tag{2.9}$$

(iii)   Similarly, from $x^{-2} \equiv 2 - x \pmod{f(x)}$ we get

$$w_{-2n+r} = \Sigma^n_{i=0}(-1)^i\binom{n}{i}2^{n-i}w_{i+r}(n \geq 0). \tag{2.10}$$

(iv)   We have $x^4 \equiv 2x^3 - x \equiv 2(2x^2 - 1) - x = 4x^2 - x - 2 \pmod{f(x)}$, whence by the polynomial theorem we obtain

$$x^{4n+r} \equiv \sum_{i+j+s=n}\binom{n}{i,j}(4x^2)^i(-x)^j(-2)^s x^r$$

$$= \sum_{i+j+s=n}(-1)^{j+s}\binom{n}{i,j}2^{2i+s}x^{2i+j+r}(n \geq 0) \pmod{f(x)}.$$

Then TCI given

$$w_{4n+r} = \sum_{i+j+s=n}(-1)^{j+s}\binom{n}{i,j}2^{2i+s}w_{2i+j+r}(n \geq 0). \tag{2.11}$$

(v)   We have $x^6 \equiv (2x^2 - 1)^2 = 1 - 4x^2 + 4x^4 \equiv 12x^2 - 4x - 7 \pmod{f(x)}$,

whence
$$x^{6n+r} \equiv \sum_{i+j+s=n} (-1)^{j+s}\binom{n}{i,j}12^i \cdot 4^j \cdot 7^s x^{2i+j+r} \pmod{f(x)}.$$

Then TCI gives

$$w_{6n+r} = \sum_{i+j+s=n} (-1)^{j+s}\binom{n}{i,j}12^i \cdot 4^j \cdot 7^s w_{2i+j+r} \quad (n \geq 0). \tag{2.12}$$

(vi) From (iv) and (v) we have $x^4 \equiv 4x^2 - x - 2$ and $x^6 \equiv 12x^2 - 4x - 7 \pmod{f(x)}$. Eliminating x^2 from the last two expressions, we can obtain $x^4 \equiv (x^6 + x + 1)/3$, $x \equiv 3x^4 - x^6 - 1$ and $x^6 \equiv 3x^4 - x - 1 \pmod{f(x)}$. By the same method as above we get

$$w_{4n+r} = \sum_{i+j+s=n} \binom{n}{i,j}3^{-n}w_{6i+j+r} \quad (n \geq 0), \tag{2.13}$$

$$w_{n+r} = \sum_{i+j+s=n} (-1)^{n-i}\binom{n}{i,j}3^i w_{4i+6j+r} \quad (n \geq 0), \tag{2.14}$$

and

$$w_{6n+r} = \sum_{i+j+s=n} (-1)^{n-i}\binom{n}{i,j}3^i w_{4i+j+r} \quad (n \geq 0). \tag{2.15}$$

Remark: We can construct more identities, but we give only the above. By the way, we point out that the results of the example have included all the results in [18]. The sequence $\{C_n\}$ dealt with in [18] satisfies $C_n = C_{n-1} + C_{n-2} + r$. From $C_{n+2} = C_{n+1} + C_n + r$ and $C_{n+3} = C_{n+2} + C_{n+1} + r$ we obtain $C_{n+3} = 2C_{n+2} - C_n$. This means that $\{C_n\}$ satisfies (2.7), and so $\{C_n\} \in \Omega(f(x))$. Hence (2.8)—(2.15) holds for $\{C_n\}$. Taking $r = 6k$ in (2.13) and (2.14) we get the Theorem 1 in [18]. Also (2.13) and (2.15) are just the results of the Theorem 2 in [18]. Finally, (2.8) gives Theorem 3 in [18].

3. SOME GENERAL IDENTITIES INVOLVING K^{th}-ORDER F-L NUMBERS

Let $f(x)$ be denoted by (1.3), and $x_1, \cdots, x_k$ be the roots of $f(x)$. Let $m > 0$ be an integer. We define

$$f^*(x) = 1 - a_1 x - a_2 x^2 - \cdots - a_k x^k = \sum_{i=0}^{k}(-a_i)x^i \quad (a_0 = -1), \tag{3.1}$$

$$f_m(x) = (x - x_1^m)\cdots(x - x_k^m) = x^k - b_1 x^{k-1} - \cdots - b_{k-1} x - b_k, \tag{3.2}$$

and

$$f_m^*(x) = 1 - b_1 x - b_2 x^2 - \cdots - b_k x^k = \sum_{i=0}^{k}(-b_i)x^i \quad (b_0 = -1). \tag{3.3}$$

Theorem 3.1: Let $\{u_n^{(i)}\}$ $(i = 0, \cdots, k-1)$ be the ith basic sequence in $\Omega = \Omega(a_1, \cdots, a_k)$, and $\{w_n\}$ be an arbitrary sequence in Ω. Then

(i) $w_{m+n} = \sum_{i=0}^{k-1} u_n^{(i)} w_{n+i} \quad (m, n \in Z);$

$$\tag{3.4}$$

(ii) $\quad w_{m-n} = a_k^{-n} \displaystyle\sum_{i_0+\cdots+i_{k-1}=n} (-1)^{n-i_0} \binom{n}{i_0,\cdots,i_{k-1}} a_1^{i_1}\cdots a_{k-1}^{i_{k-1}}$

$$w_{m+(k-1)i_0+(k-2)i_1+\cdots+i_{k-2}} \quad (m \in Z,\, n > 0); \tag{3.5}$$

(iii) $\quad w_{mn+r} = \displaystyle\sum_{i_0+\cdots+i_{k-1}=m} \binom{m}{i_0,\cdots,i_{k-1}} (u_n^{(k-1)})^{i_0}\cdots(u_n^{(0)})^{i_{k-1}}$

$$w_{(k-1)i_0+(k-2)i_1+\cdots+i_{k-2}+r} \quad (m > 0,\, n, r \in Z). \tag{3.6}$$

Proof:

(i)   From (2.2) we have $x^m \equiv \sum_{i=0}^{k-1} u_m^{(i)} x^i \pmod{f(x)}$, whence $x^{m+n} \equiv \sum_{i=0}^{k-1} u_m^{(i)} x^{n+i}$ $\pmod{f(x)}$, and so TCI gives (3.4).

(ii)   From (1.3) we have $x^{-1} \equiv a_k^{-1}(x^{k-1} - a_1 x^{k-2} - \cdots - a_{k-2}x - a_{k-1}) \pmod{f(x)}$, whence

$$x^{m-n} \equiv x^m a_k^{-n} \sum_{i_0+\cdots+i_{k-1}=n} \binom{n}{i_0,\cdots,i_{k-1}} (x^{k-1})^{i_0} (-a_1 x^{k-2})^{i_1}\cdots$$

$$(-a_{k-2}x)^{i_{k-2}}(-a_{k-1})^{i_{k-1}}$$

$$\equiv a_k^{-n} \sum_{i_0+\cdots+i_{k-1}=n} (-1)^{n-i_0} \binom{n}{i_0,\cdots,i_{k-1}} a_1^{i_1} a_2^{i_2}\cdots$$

$$a_{k-1}^{i_{k-1}} x^{m+(k-1)i_0+(k-2)i_1+\cdots+i_{k-2}} \pmod{f(x)},$$

and so (3.5) is proved by TCI.

(iii)   We have $x^{mn+r} = (x^n)^m x^r \equiv \left(\sum_{i=0}^{k-1} u_n^{(i)} x^i\right)^m x^r \pmod{f(x)}$, and the rest of the proof can be done by the same way as in (ii). $\square$

Theorem 3.2: Let $\{w_n\} \in \Omega(a_1,\cdots,a_k) = \Omega(f(x))$, and $x_1,\cdots,x_k$ be the roots of $f(x)$. If $m > 0$, then $\{w_{mn+r}\}_n \in \Omega(f_m(x))$, where $f_m(x)$ is denoted by (3.2). That is,

$$w_{m(n+k)+r} = b_1 w_{m(n+k-1)+r} + \cdots + b_{k-1} w_{m(n+1)+r} + b_k w_{mn+r}. \tag{3.7}$$

Proof: From (3.2) we have $f_m(x^m) = (x^m - x_1^m)\cdots(x^m - x_k^m) \equiv 0 \pmod{f(x)}$, that is

$$x^{mk} - b_1 x^{m(k-1)} - \cdots - b_{k-1} x^m - b_k \equiv 0 \pmod{f(x)}.$$

Multiplying the two sides of the last expression by x^{mn+r} and using TCI we get (3.7). $\square$

Theorem 3.3: Let $f(x)$ and $f^*(x)$ be denoted by (1.3) and (3.1) respectively. Let $x_1,\cdots,x_k$ be the roots of $f(x)$, and $\{w_n\}$ be an arbitrary sequence in $\Omega(f(x))$.

(i)   If $t \neq x_s^{-1}$ for all $s = 1,\cdots,k$, then

$$\sum_{i=0}^{n} w_{i+r} t^i = \sum_{i=0}^{k-1} (\Psi(0,i)t^i - \Psi(n+1,i)t^{n+1+i})/f^*(t), \tag{3.8}$$

where

$$\Psi(n+1,i) = \sum_{j=0}^{i} (-a_j) w_{n+1+i-j+r} \quad (i=0,\cdots,k-1); \tag{3.9}$$

(ii)   If, for some s, x_s is a root of $f(x)$ with multiple l, then

$$\sum_{i=0}^{n} \frac{w_{i+r}}{x_s^i} = (-1)^l \left(\sum_{i=0}^{k-1} \binom{n+k-i}{l} \Psi(0,i) x_s^{k-i} \right.$$

$$\left. - \sum_{i=0}^{k-1-l} \binom{k-1-i}{l} \Psi(n+1,i) x_s^{-n-1+k-i} \right) \bigg/ \sum_{i=l}^{k} \binom{i}{l}(-a_i) x_s^{k-i} \quad \text{(if } l=k, \text{ then}$$

$\sum_{i=0}^{k-1-l}$ is an empty summation). \hfill (3.10)

Proof: (i) We have $(1-tx)\sum_{i=0}^{n} t^i x^{i+r} = x^r - t^{n+1} x^{n+1+r}$. \hfill (3.11)

Let $g(x) = \sum_{i=0}^{k-1} t^i \sum_{j=0}^{i} (-a_j) x^{i-j}$. Then

$$(1-tx)g(x) = 1 - a_1 t - a_2 t^2 - \cdots - a_{k-1} t^{k-1} - t^k (x^k - a_1 x^{k-1} - \cdots - a_{k-1} x)$$

$$\equiv 1 - a_1 t - a_2 t^2 - \cdots - a_{k-1} t^{k-1} - a_k t^k = f^*(t) \pmod{f(x)}. \tag{3.12}$$

Multiplying (3.11) by $g(x)$ we get

$$f^*(t) \sum_{i=0}^{n} t^i x^{i+r} \equiv (x^r - t^{n+1} x^{n+1+r}) g(x)$$

$$= \sum_{i=0}^{k-1} t^i \sum_{j=0}^{i} (-a_j) x^{i-j+r}$$

$$- \sum_{i=0}^{k-1} t^{n+1+i} \sum_{j=0}^{i} (-a_j) x^{n+1+i-j+r} \pmod{f(x)}.$$

Since $t \neq x_s^{-1}$ for all $s=1,\cdots,k$, then $f^*(t) \neq 0$, and so, by TCI and the last congruence, (3.8) with (3.9) is proved.

(ii)   If, for some s, x_s is a root of $f(x)$ with multiplicity l, then x_s^{-1} is a root of $f^*(x)$ with multiplicity l, whence

$$f^*(x_s^{-1}) = f^{*\prime}(x_s^{-1}) = \cdots = f^{*(l-1)}(x_s^{-1}) = 0, \ f^{*(l)}(x_s^{-1}) \neq 0. \tag{3.13}$$

Rewrite (3.8) as

$$\sum_{i=0}^{n} w_{i+r} t^i = t^{n+1+k-l} \sum_{i=0}^{k-1} (\Psi(0,i) t^{-(n+1+k-l-i)} - \Psi(n+1,i) t^{l-k+i})/f^*(t).$$

Then

$$\sum_{i=0}^{n} \frac{w_{i+r}}{x_s^i} = \lim_{t \to x_s^{-1}} \sum_{i=0}^{n} w_{i+r} t^i$$

$$= \lim_{t \to x_s^{-1}} t^{n+1+k-l} \lim_{t \to x_s^{-1}} \left(\sum_{i=0}^{k-1} (\Psi(0,i) t^{-(n+1+k-l-i)} \right.$$

$$\left. - \Psi(n+1,i) t^{l-k+i} \right) \Big/ f^*(t)). \tag{3.14}$$

Because of (3.13), the second limit of the right side of (3.14) can be found by using the L'Hôpital rule, i.e., by differentiating its denominator and numerator up to l times respectively, and making t approach x_s^{-1}. Hence, from (3.1), (3.14) becomes

$$\sum_{i=0}^{n} \frac{w_{i+r}}{x_s^i} = x_s^{-(n+1+k-l)} \left(\sum_{i=0}^{k-1} (\Psi(0,i)(-n-1-k+l+i)_l x_s^{n+1+k-i} \right.$$

$$\left. - \sum_{i=0}^{k-1-l} \Psi(n+1,i)(l-k+i)_l x_s^{k-i} \right) \Big/ \sum_{i=l}^{k} (-a_i)(i)_l x_s^{-(i-l)},$$

whence it is easy to obtain (3.10). $\qquad\square$

Corollary 3.4: Let $f(x)$ and $f^*(x)$ be denoted by (1.3) and (3.1) respectively. Let $x_1, \cdots, x_q$ be all the distinct roots of $f(x)$. Put $d = \min_{1 \leq s \leq q} \{ |x_s^{-1}| \}$. Let $\{w_n\}$ be an arbitrary sequence in $\Omega(f(x))$. Then

$$\sum_{i=0}^{\infty} w_{i+r} t^i = \sum_{i=0}^{k-1} \sum_{j=0}^{i} (-a_j) w_{i-j+r} t^i \Big/ f^*(t) \quad \text{for } |t| < d. \tag{3.15}$$

Proof: Assume that the multiplicity of x_s is $k_s (s = 1, \cdots, q)$, $k_1 + \cdots + k_q = k$. It is known that (see [2], page 104.)

$$w_n = \sum_{s=1}^{q} p_s(n) x_s^n, \tag{3.16}$$

where $p_s(n)$ is a polynomial in n of degree $k_s - 1$. If $|t| < d$, then $|tx_s| < 1$ for all $s = 1, \cdots, q$, whence

$$|w_{n+1+i-j+r} t^n| \leq \sum_{s=1}^{q} |p_s(n+1+i-j+r)| \cdot |tx_s|^n \to 0(n \to \infty),$$

and, by (3.8) and (3.9), (3.15) follows. $\qquad\square$

By using Theorem 3.2 we can generalize Theorem 3.3 and Corollary 3.4 immediately.

Theorem 3.5: Let $x_1, \cdots, x_k$ be the roots of $f(x)$ denoted by (1.3), and $\{w_n\}$ be an arbitrary sequence in $\Omega(f(x))$. Let $m > 0$ be an integer, and $f_m(x)$ and $f_m^*(x)$ be denoted by (3.2) and (3.3), respectively.

(i)   If $t \neq x_s^{-m}$ for all $s = 1, \cdots, k$, then

$$\sum_{i=0}^{n} w_{mi+r} t^i = \sum_{i=0}^{k-1} (\Psi_m(0,i) t^i - \Psi_m(n+1,i) t^{n+1+i}) / f_m^*(t), \tag{3.17}$$

where

$$\Psi_m(n+1,i) = \sum_{j=0}^{i}(-b_j)w_{m(n+1+i-j)+r} \quad (i = 0,\cdots,k-1); \qquad (3.18)$$

(ii)   If, for some s, x_s^m is a root of $f_m(x)$ with multiple l, then

$$\sum_{i=0}^{n}\frac{w_{mi+r}}{x_s^{mi}} = (-1)^l\left(\sum_{i=0}^{k-1}\binom{n+k-i}{l}\Psi_m(0,i)x_s^{m(k-i)}\right.$$

$$\left. -\sum_{i=0}^{k-1-l}\binom{k-1-i}{l}\Psi_m(n+1,i)x_s^{m(-n-1+k-i)}\right)\bigg/\sum_{i=l}^{k}\binom{i}{l}(-b_i)x_s^{m(k-i)},$$

(if $l = k$, then $\sum_{i=0}^{k-1-i}$ is an empty summation).   (3.19)

Corollary 3.6: Let $f_m(x)$ and $f_m^*(x)$ be denoted by (3.2) and (3.3) respectively. Let $x_1^m,\cdots,x_q^m$ be all the distinct roots of $f_m(x)$. Put $d_m = \min_{1\le s\le q}\{|x_s^{-m}|\}$. Let $\{w_n\}$ be an arbitrary sequence in $\Omega(f(x))$. Then

$$\sum_{i=0}^{\infty}w_{mi+r}t^i = \sum_{j=0}^{k-1}\sum_{j=0}^{i}(-b_j)w_{m(i-j)+r}t^i/f_m^*(t) \text{ for } |t| < d_m. \qquad (3.20)$$

Remark: For arbitrary $\{w_n\} \in \Omega(f(x))$ (3.15) and (3.20) can also be considered as the expressions of the generating functions of $\{w_{n+r}\}_n$ and $\{w_{nm+r}\}_n(m > 0)$ respectively.

If w_n is a polynomial in n of degree p, then its $(p+1)^{\text{th}}$-order difference turns to zero, that is

$$\Delta^{p+1}w_n = \sum_{i=0}^{p+1}(-1)^i\binom{p+1}{i}w_{n+p+1-i} = 0. \qquad (3.21)$$

This means that the characteristic polynomial of $\{w_n\}$ is

$$f(x) = \sum_{i=0}^{p+1}(-1)^i\binom{p+1}{i}x^{p+1-i} = (x-1)^{p+1}. \qquad (3.22)$$

Clearly, $f_m(x) = f(x)$. Hence, taking $k = l = p+1$ and $x_1 = \cdots = x_k = 1$ in Theorem 3.5 and corollary 3.6, we get

Theorem 3.7: Let w_n be a polynomial in n of degree p, and $m > 0$ be an integer.

(i)   If $t \ne 1$, then

$$\sum_{i=0}^{n}w_{mi+r}t^i = \sum_{i=0}^{p}(\varphi_m(0,i)t^i - \varphi_m(n+1,i)t^{n+1+i})/(1-t)^{p+1}, \qquad (3.23)$$

where

$$\varphi_m(n+1,i) = \sum_{j=0}^{i}(-1)^j\binom{p+1}{j}w_{m(n+1+i-j)+r}(i = 0,1,\cdots p); \qquad (3.24)$$

(ii) $$\sum_{i=0}^{n}w_{mi+r} = \sum_{i=0}^{p}\binom{n+1+p-i}{p+1}\sum_{j=0}^{i}(-1)^j\binom{p+1}{j}w_{m(i-j)+r^i} \qquad (3.25)$$

(iii) $\quad \sum_{i=0}^{\infty} w_{mi+r} t^i = \sum_{i=0}^{p} \sum_{j=0}^{i} (-1)^j \binom{p+1}{j} w_{m(i-j)+r} t^i/(1-t)^{p+1}$ for $|t| < 1$. (3.26)

Example 3.8: Let $f(x) = (x-2)^2(x-1)^2 = x^4 - 6x^3 + 13x^2 - 12x + 4$, and $\{w_n\} \in \Omega(f(x))$, e.g.,

$$w_{n+4} = 6w_{n+3} - 13w_{n+2} + 12w_{n+1} - 4w_n. \tag{3.27}$$

From (3.9) and (3.27) we have

$$\Psi(n+1,0) = w_{n+1+r}, \ \Psi(n+1,1) = w_{n+2+r} - 6w_{n+1+r},$$

$$\Psi(n+1,2) = w_{n+3+r} - 6w_{n+2+r} + 13w_{n+1+r} = 12w_{n+r} - 4w_{n-1+r},$$

and

$$\Psi(n+1,3) = w_{n+4+r} - 6w_{n+3+r} + 13w_{n+2+r} - 12w_{n+1+r} = -4w_{n+r},$$

Hence Theorem 3.3 gives:

(i) If $t \neq 1/2, 1$, then

$$\sum_{i=0}^{n} w_{i+r} t^i = (w_r + (w_{1+r} - 6w_r)t + (12w_{r-1} - 4w_{r-2})t^2 - 4w_{r-1}t^3 - w_{n+1+r}t^{n+1} -$$

$$(w_{n+2+r} - 6w_{n+1+r})t^{n+2} - (12w_{n+r} - 4w_{n-1+r})t^{n+3} +$$

$$4w_{n+r}t^{n+4})/(1 - 6t + 13t^2 - 12t^3 + 4t^4); \tag{3.28}$$

(ii)

$$\sum_{i=0}^{n} w_{i+r}/2^i = \left(\binom{n+4}{2} w_r 2^4 + \binom{n+3}{2}(w_{1+r} - 6w_r)2^3 + \binom{n+2}{2}(12w_{r-1} - 4w_{r-2})2^2 + \right.$$

$$\binom{n+1}{2}(-4w_{r-1})2 - 3w_{n+1+r}2^{-n-1+4} -$$

$$\left. (w_{n+2+r} - 6w_{n+1+r})2^{-n-1+3} \right) \bigg/ 4; \tag{3.29}$$

Specifically, taking $w_n = n(1+2^n)$ and $r = 0$, we get

$$\sum_{i=0}^{n} i(1+2^i)(-3)^i = (-243 - (-3)^{n+1}(196n + 49 + (112n+16)2^{n+1}))/784, \text{ (by (3.28))}$$

and $\sum_{i=0}^{n} i(1+2^i)/2^i = \binom{n+1}{2} + 2 - (n+2)/2^n$ (by (3.29)).

Example 3.9: Let $w_n = (n \mid h)_p = n(n-h)\cdots(n-(p-1)h)$. Then by Theorem 3.7 we obtain

(i) If $t \neq 1$, then

$$\sum_{i=0}^{n} (mi+r \mid h)_p t^i = \sum_{i=0}^{p} (\varphi_m(0,i)t^i - \varphi_m(n+1,i)t^{n+1+i})/(1-t)^{p+1} \quad (m > 0) \tag{3.30}$$

where

$$\varphi_m(n+1,i) = \sum_{j=0}^{i} (-1)^j \binom{p+1}{j}(m(n+1+i-j)+r \mid h)_p \quad (i=0,1,\cdots,p); \qquad (3.31)$$

(ii)
$$\sum_{i=0}^{n} (mi+r \mid h)_p = \sum_{i=0}^{p} \binom{n+1+p-i}{p+1}\sum_{j=0}^{i}(-1)^j\binom{p+1}{j}(m(i-j)+r \mid h)_p; \qquad (3.32)$$

(iii)

$$\sum_{i=0}^{\infty}(mi+r \mid h)_p t^i = \sum_{i=0}^{p}\sum_{j=0}^{i}(-1)^j\binom{p+1}{j}(m(i-j)+r \mid h)_p t^i \Big/ (1-t)^{p+1} \text{ for } |t| < 1. \qquad (3.33)$$

Specifically, taking $h = 0$ we can obtain explicit expressions for the class of summations $\sum_{i \geq 0}(mi+r)^p t^i$, which are generalizations of the results in [4] and [5] and are simpler than the explicit expression in [4] and [5]. For example, by using (3.32) we obtain

$$\sum_{i=0}^{n}(3i+1)^3 = \sum_{i=0}^{3}\binom{n+4-i}{4}\sum_{j=0}^{i}(-1)^j\binom{4}{j}(3(i-j)+1)^3$$
$$= \tfrac{1}{4}(n+1)(3n+2)(9n^2+15n+2). \qquad (3.34)$$

Remark: In [10], the explicit expression for $\sum_{i=0}^{n}(mi+r \mid h)_p t^i$ has been given for $m = 1$, $r = 0$, and $t \neq 1$. However, it is more complicated than (3.30) with (3.31).

ACKNOWLEDGEMENT

I am very grateful to the anonymous referee whose helpful and detailed comments greatly improved the presentation of this paper.

REFERENCES

[1]   Bergum, G.E. and Hoggatt, V.E. Jr. "Infinite Series with Fibonacci and Lucas Polynomials." *The Fibonacci Quarterly*, Vol. *17.2* (1979): pp. 147-151.

[2]   Brualdi, R.A. Introductory Combinatorics. The Hague: North-Holland, 1977.

[3]   Carlitz, L. & Ferns, H.H. "Some Fibonacci and Lucas Identities." *The Fibonacci Quarterly*, Vol. *8.1* (1970): pp. 61-73.

[4]   de Bruyn, G.F.C. "Formulas for $a + a^2 2^p + a^3 3^p + \cdots + a^n n^p$." *The Fibonacci Quarterly*, Vol. *33.2* (1995): pp. 98-103.

[5]   de Bruyn, G.F.C. and de Villiers, J.M. "Formulas for $1^p + 2^p + 3^p + \cdots + n^p$." *The Fibonacci Quarterly*, Vol. *32.3* (1994): pp. 271-276.

[6]   Gauthier, N. "Derivation of a formula for $\sum r^k x^r$." *The Fibonacci Quarterly*, Vol. *27.5* (1989): pp. 402-408.

[7]   Hoggatt, V.E. Jr. Fibonacci and Lucas Numbers. Boston: Houghton Mifflin, 1969; rtp. Santa Clara, Calif.: the Fibonacci Association, 1979.

[8]   Hoggatt, V.E. Jr. and Lind, D.A. "Symbolic Substitutions into Fibonacci Polynomials." *The Fibonacci Quarterly*, Vol. *6.1* (1968): pp. 55-74.

[9]   Horadam, A.F. "Partial Sums for Second-order Recurrence Sequences." *The Fibonacci Quarterly*, Vol. *32.5* (1994): pp. 429-439.

[10]   Hsu, Leetsch C. "On a Kind of Generalized Arithmetic-geometric Progression." *The Fibonacci Quarterly*, Vol. *35.1* (1997): pp. 62-67.

[11]   Lee, Jinzai and Lee, Jiasheng. "A Complete Characterization of B-power Fractions that Can Be Represented as Series of General n-bonacci Numbers." *The Fibonacci Quarterly*, Vol. *25.1* (1987): pp. 72-75.

[12]   Long, C.T. "Some Binomial Fibonacci Identities." <u>Applications of Fibonacci Numbers</u>, Volume 3. Edited by G.E. Bergum, A.N. Philippou and A.F. Horadam, Kluwer Academic Publisher, Dordrecht, The Netherlands, 1990: pp. 241-245.

[13]   Long, C.T. "Discovering Fibonacci Identities." *The Fibonacci Quarterly*, Vol. *24.2* (1986): pp. 160-167.

[14]   Melham, R.S. and Shannon, A.G. "Some Summation identities Using generalized Q-Matrices." *The Fibonacci Quarterly*, Vol. *33.1* (1995): pp. 64-72.

[15]   Vajda, S. <u>Fibonacci & Lucas Numbers, and the golden section, theory and applications</u>. Ellis Horwood Ltd. 1989.

[16]   Waddill, M.E. "The Tetranacci Sequence and Generalizations." *The Fibonacci Quarterly*, Vol. *30.1* (1992): pp. 9-20.

[17]   Waddill, M.E. "Using Matrix Techniques to Establish Properties of a Generalized Tribonacci Sequence." <u>Applications of Fibonacci Numbers</u>, Volume 4. Edited by G.E. Bergum, A.F. Horadam and A.N. Philippou. Kluwer Academic Publisher, Dordrecht, The Netherlands, 1991: pp. 299-308.

[18]   Zhang, Zhizheng. "Some Properties of the Generalized Fibonacci Sequence $C_n = C_{n-1} + C_{n-2} + r$." *The Fibonacci Quarterly*, Vol. *35.2* (1997): pp. 169-171.

[19]   Zhang, Zhizheng. "Some Identities Involving Generalized Second-Order Integer Sequences." *The Fibonacci Quarterly*, Vol. *35.3* (1997): pp. 265-268.

[20]   Zhou, Chizhong. "A Generalization of the 'All or None' divisibility Property." *The Fibonacci Quarterly*, Vol. *35.2* (1997): pp. 129-134.

[21]   Zhou, Chizhong. <u>Fibonacci-Lucas Sequences and their Applications</u> (in Chinese). MR 95m: 11027. Hunan: Hunan Science and Technology Press, 1993.

AMS Classification Numbers: 11B37, 11B39, 05A19

SUBJECT INDEX

The manufacturer's authorised representative in the EU is Springer
Nature Customer Service Centre GmbH, Europaplatz 3, 69115 Heidelberg,
Germany. If you have any concerns regarding our products, please
contact ProductSafety@springernature.com

Printed and bound by CPI Group (UK) Ltd, Croydon, CR0 4YY
23/04/2026
02095593-0006